FUNDAMENTALS OF ROBOTICS ENGINEERING

FUNDAMENTALS OF ROBOTICS ENGINEERING

Harry H. Poole

Poole Associates

VNR VAN NOSTRAND REINHOLD
New York

Copyright © 1989 by Van Nostrand Reinhold
Library of Congress Catalog Card Number 88-20861
ISBN 0-442-27298-7

Printed in the United States of America

Designed by Beehive Production Services

Van Nostrand Reinhold
115 Fifth Avenue
New York, New York 10003

Van Nostrand Reinhold (International) Limited
11 New Fetter Lane
London EC4P 4EE, England

Van Nostrand Reinhold
480 La Trobe Street
Melbourne, Victoria 3000, Australia

Macmillan of Canada
Division of Canada Publishing Corporation
164 Commander Boulevard
Agincourt, Ontario M1S 3C7, Canada

16 15 14 13 12 11 10 9 8 7 6 5 4 3 2 1

Library of Congress Cataloging in Publication Data
Poole, Harry H.
    Fundamentals of robotics engineering.
    Includes index.
    1. Robotics.   I. Title.
TJ211.P65   1989   629.8'92   88-20861
ISBN 0-442-27298-7

CONTENTS

PREFACE

Robotics engineering has progressed from an infant industry in 1961 to one including over 500 robot and allied firms around the world in 1989. During this growth period, many robotics books have been published, some of which have served as industry standards. Until recently, the design of robotics systems has been primarily the responsibility of the mechanical engineer, and their application in factories has been the responsibility of the manufacturing engineer. Few robotics books address the many systems issues facing electronics engineers or computer programmers.

The mid-1980s witnessed a major change in the robotics field. The development of advanced sensor systems (particularly vision), improvements in the intelligence area, and the desire to integrate groups of robots working together in local work cells or in factory-wide systems have greatly increased the participation of electronics engineers and computer programmers. Further, as robots gain mobility, they are being used in completely new areas, such as construction, firefighting, and underwater exploration, and the need for computers and smart sensors has increased.

Fundamentals of Robotics Engineering is aimed at the practicing electrical engineer or computer analyst who needs to review the fundamentals of engineering as applied to robotics and to understand the impact on system design caused by constraints unique to robotics. Because there are many good texts covering mechanical engineering topics, this book is limited to an overview of those topics and the effects they have on electrical design and system programs.

The text also covers related areas such as sensor subsystems and robot communication protocols. It discusses robot standards and describes representative robotic applications and the impact they have on system design requirements. It provides much needed information in the areas of ultrasonics and mobile robots.

The book extrapolates into future robotics techniques and applications by reviewing current work in research and university laboratories around the country and, to a lesser extent, advanced applications in Europe and Japan. New techniques, new applications, and trends are also covered. By presenting information about some work still in the research stage or only being applied in other countries at the present time, *Fundamentals of Robotics Engineering* should help engineers to be up to date in this important technological area.

The book concludes with a series of appendixes including a partial list of robotics manufacturers, major university research laboratories, international robotics organizations, and a glossary of robotics terms.

Any book can only be of finite size. In the number of pages that an author has to explain a subject, he or she has a number of difficult choices to make. Should the book approach the subject from a theoretical or a practical viewpoint? Should the book treat a few topics in depth or should it treat many topics in less detail? Should actual company products be described, or should more general approaches be covered?

Fundamentals of Robotics Engineering has been written with the needs of the robotics system designer in mind. It can also be used to supplement undergraduate and graduate courses in robotics. Therefore the practical approaches available in the marketplace are well covered, and the many alternatives that a system designer can select from are included. In addition, other promising techniques, although perhaps not yet commercially available, are also discussed. As such, the approach taken in this book is primarily a more general and practical one, although theory is also well covered.

ACKNOWLEDGMENTS

Any book is the product of the efforts of many individuals and the cooperation of many companies. It is, unfortunately, impossible to list everyone who has contributed. However, the author would like to thank certain individuals without whose able assistance this book could not have been completed.

Most of the art work was supplied by Paula Bushee and James Hess. Douglas Poole helped to obtain photographs. Carolyn Turcio provided some of the text for chapter 11 and several of the drawings. Of prime importance was the assistance of my wife, Marjorie Poole, in helping me meet the many deadlines that authors are faced with. To these individuals and to the countless others at firms around the world who supplied information and pictures, I give my heartfelt thanks.

North Attleboro, Massachusetts   HARRY H. POOLE

FUNDAMENTALS OF ROBOTICS ENGINEERING

PART

I

AN OVERVIEW OF ROBOTICS

Robotics is the science or technology that deals with robots. Defining a *robot,* however, is not easy. The public image of a robot, produced through myriad motion pictures and books that give them human characteristics and apparently superhuman intelligence is partially responsible for the difficulty. A constantly changing technology, which provides robots with increasing levels of abilities, also contributes to the problem. For example, in 1987 a California doctor demonstrated a robot at Robots 11 that can assist surgeons during brain biopsies. It is, therefore, hard to define exactly what potential abilities may be found in robots and which applications are realistic and practical. It is also difficult to keep various industry standard definitions up to date, as most continue to stress one part of a robot—its manipulator.

Because it is an engineering systems book on robotics, this book attempts to cover enough technology to describe what the robot does today and what it can be built to do tomorrow. Chapter 1 begins with the history of robotics to develop a working definition of robots and to provide a sound background. It then discusses the growth of the robotics industry and introduces robot engineering. Chapter 2 examines and classifies current robots, using six different criteria, providing necessary points of reference for the balance of the book.

INTRODUCTION

As part of a 1983 seminar on robot sensing and intelligence,[1] Dr. Antal Bejczy of the Jet Propulsion Lab gave the following definition of robots: "There are three parts to the technical definition of robots. First, robots are general purpose mechanical machines. Second, they are programmable to perform a variety of work within their mechanical capabilities. Third, they operate automatically." Dr. Bejczy also stated that automation is the key for advancing the state of the art in robotics. This useful definition thus combines mechanical engineering, computer programming, and electrical engineering. Taking the lead from Dr. Bejczy, this book concentrates on the second and third parts of his definition, the programming and automation of today's robots.

1.1 BRIEF HISTORY OF ROBOTICS

Stories of mechanical men, predecessors of the robot C3PO of *Star Wars* fame, have been around since the days of Greek myths, and motion pictures have been showing fictional robots from the earliest film days (Georges Méliès, the French pioneer filmmaker, made *The Clown and the Automaton* in 1897). Over the last several centuries, skilled craftsmen developed early versions of mechanical robots. Although the word *robot* was not coined until the twentieth century, mechanical humanoid robots, called *automata,* were built as early as the 1500s, when moving figures were used to perform useful work, such as striking a bell on the hour.

By the 1700s, clever inventors had added many advanced capabilities to these devices, culminating perhaps in the automatons developed by Pierre

Figure 1-1. Early automaton. *(Courtesy of Musée d'Art et d'Histoire, Neuchâtel, Switzerland)*

Jacquet-Droz, a creator of puppets. One of these devices, named "The Scribe" (or writer), can still be seen at work in a Neuchâtel (Switzerland) museum. It is a well-constructed model of a young man seated at a desk (Fig. 1-1). When activated, the automaton dips its quill pen into an inkwell and writes previously selected text of up to 40 letters in length. What makes this device of particular interest to the robotics engineer is that it is programmable. With appropriate changes of cams and levers (similar to changing instructions), the automaton can be made to write any text.

Although numerous automatons were built over the years, they were of interest primarily as entertaining novelties, and, unfortunately, their technology was not applied to industry. Not until the mid-1950s was robotic technology adapted by industry to handle the more dangerous manufacturing tasks.

The term *robot* was first used in two plays written by the Czech author Karel Capek. *Opilec* was the first, written in 1917. It is less well known than his later masterpiece *R.U.R.*, an abbreviation of Rossum's Universal Robots. An English version of this latter work is available.[2]

R.U.R. was first produced in Prague in 1921, and then attained worldwide attention by being presented in New York in 1922. The play developed the concept of small, manlike creatures obeying their master's instructions. The term *robot,* derived from the Czech word *robota,* means forced labor or compulsory service, and was the name given to these creatures. Subsequently, films and science fiction further publicized robots.

During the Second World War, the meaning of robot was extended to include an automatic pilot, such as found in robot bombs. This was also the period of two related steps—the development of the computer and the use of the first remote manipulators, now known as *teleoperated robots* or *telechirs*. Hence, robot technology evolution dates from the Second World War. Further steps in the development cycle included servo systems, numerical control, process control, industrial robots, and, finally, advanced robots. In 1954, the first industrial robot was built, and the first two industrial robot patents were applied for. It was the dawn of the robotics age.

A radiator manufacturing company, in 1954, asked Planet Corporation of Lansing, Michigan, to design a system to handle hot castings while they were water quenched and then placed on a press die. Planet, which took three years to solve the problem, demonstrated its robot technology at the International Trade Fair on Automation in Stockholm in 1957, after an invitation from the U.S. Department of Commerce. Planet named the robot Planobot, and it was publicized as the first industrial robot. Planobot (Fig. 1-2) was powered hydraulically and controlled through an electronic sequencer, going through a maximum of 25 steps to carry out a task. With five degrees of freedom, linear and rotary joint action, and an ability to be easily modified for changes in the production run, this model was a forerunner of the loading and unloading robots later developed successfully by Unimation.

During 1954, two robotic patents were applied for. In March, British inven-

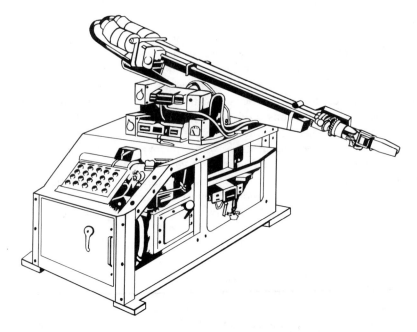

Figure 1-2. First industrial robot. *(Courtesy of Carolyn Turcio)*

tor Cyril W. Kenward applied for a British patent. The patent, issued in 1957 as number 781465, is titled *Manipulating Apparatus*. Kenward's device was a two-armed manipulator with four degrees of freedom moving in a Cartesian coordinate system. Two months later, American inventor George C. Davol applied for a patent. Davol's patent, titled *Programmed Article Transfer,* was granted in 1961 and served as the impetus behind Unimation. Founded in 1958, Unimation was the first company to specialize in robotics. The company name is a contraction of "universal automation." Another entry into the robotics field during 1958 was AMF, which began developing its Versatron robot. Versatron was subsequently made a division of Prab Robots.

The first Unimation industrial robots were built in 1961 and, like Planet's robot, were used in the unpleasant and dangerous task of loading and unloading die casting machines. Their first robot, originally delivered to a General Motors automobile plant in New Jersey, has been moved to an exhibit at the Smithsonian Institution in Washington.

After a slow start, interest began to build in 1965 as many companies from different nations entered the field. The first robots were pick-and-place robots with little intelligence, but they were very reliable, even under poor working conditions. In 1966 a Norwegian agricultural machinery firm (Trallfa), built a robot to paint wheelbarrows. In 1969, Unimation introduced the first spot-welding robot, and by 1972 Unimation was offering three series of robots. By the end of that year, *Fortune* magazine indicated that 500 robots had been installed in factories.

Universities were also becoming involved in robotics research. Progress in robot mobility, vision systems, and artificial intelligence was being made at Stanford Research Institute (now SRI International) in the late 1960s. The first mobile robot was developed at Stanford under a Department of Defense grant. Called Shakey because of the intermittent manner in which it moved and the vibration of the vision camera support structure, the robot showed an amazing degree of intelligence for its time. In one famous task, Shakey was given the job of reaching an object placed on a platform. The robot was on wheels, so it could not climb the platform. Looking around, Shakey spotted a ramp. It pushed the ramp to the platform, propelled itself up the ramp, and reached the object.

Another pioneer, MIT, also had robotics projects underway as part of its work in artificial intelligence. Particularly significant was MIT's development of a mechanical hand, suggested by a thesis in 1961, and of early robot languages, such as MHI and MINI.

Under license from Unimation, Kawasaki Heavy Industries in 1969 developed the first industrial robot in Japan. Other Japanese firms saw the potential of robots in industry and, in 1971, formed the Japanese Industrial Robot Association, four years ahead of its counterpart in the United States. Japan has pursued the robotics area so aggressively that they now lead all other countries in robotics, both as manufacturers and users. They also lead the world in several areas of robot technology.

The first adaptive robot, built in 1970 by the Electric-Technical Laboratory, a Division of the Japanese Ministry of International Trade and Development (MITI), had a vision system that could distinguish nine colors. The robot could also detect and correct errors in its actions. MITI is still a leader in robotics research.

France joined the international push into robotics during the early 1970s when it set up research and development facilities under government support. The USSR developed its first robot during 1971 and 1972. Other European countries, including West Germany, Sweden, and Italy, also became active participants in the industry during the early 1970s.

Early robots were hydraulic-based. The first all-electric model was built in 1973 by ASEA, a Swedish firm. In that year the concept of revolute joint construction was introduced by Cincinnati Milacron, and the Trallfa robot from Norway became the first continuous-path robot.

The first multiarmed, overhead robot (*Sigma*) was built by Olivetti in 1975. In 1978, Unimation introduced a smaller robot, the PUMA (programmable universal machine of assembly). A later model (the PUMA 700) is shown in Figure 1-3. This robot was designed to handle small parts and is still in

Figure 1-3. PUMA robot. *(Courtesy of Unimation, Inc.)*

widespread use. By 1980, Unimation was shipping 600 robots a year, about half of the robots built in the United States.

The next big step forward in robot technology came from the Japanese in 1979, with the selective compliance assembly robot arm (SCARA). Developed at Yamanashi University, this robot combined inflexible motion in the vertical plane with compliance, or flexible positioning in the horizontal plane, allowing parts to be inserted much more easily during assembly tasks. Now offered by many manufacturers, the SCARA concept was an important step in introducing robots to assembly tasks. Figure 1-4 shows a SCARA robot from G.E.

Although early robots freed men from dangerous tasks, they were not yet cost-effective. Three problems contributed to their slow adoption by industry. The first was cost, both initial and operating. It was more expensive to keep the robot operating than it was to perform the task manually, and that did not even include the high initial cost of the robot. Second, the technology was very new, and problems in control, gripper capability, and robot accuracy limited these devices to only a few applications. Finally, there was the general conservatism of businesses, which tend to retain successful manufacturing techniques and to reject new and untried methods. Note that, contrary to popular opin-

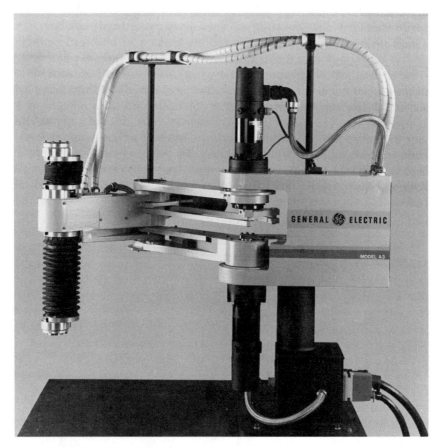

Figure 1-4. SCARA robot. (*Courtesy of General Electric Company*)

ion, unions did not initially object to robots, since they did only dangerous jobs and were not being installed to save the company money.

Hence, the sale of industrial robots got off to a slow start. Over the 10-year period since Unimation began installing robots, perhaps 500 had been installed in the United States. In 1979, however, the number of robot installations had increased to 2500 in the United States and over 10,000 worldwide. The main reason for this increase was a change in costs. Robot operating costs had increased only 20%, whereas the cost of labor had increased over 400%. Robot technology was here to stay.

Robotics had gone through three decades of development. The period from 1955 to 1965 was the concept feasibility stage, with companies still developing the base technology. From 1966 through 1975, international development accelerated as numerous foreign firms entered the arena. From 1976 through 1985, environmentally sensitive robots were developed by adding computers and limited capability external sensors. Advances in sensor technology, image processing, and artificial intelligence should make the next 10 years the decade of intelligent robots.

The earliest industrial robots were really advanced numerical control machines, also called *hard automation*. In hard automation, it is quite difficult to change the role of the robot during its lifetime, and these robots were designed for one unchanging task.

However, the latest systems are much more adaptable, leading to the concept of *flexible automation*. In the flexible manufacturing area (FMA), the range of tasks that the robot can perform has been substantially increased. Flexible automation evolved through considering the robot as part of a complete system and through advances in robotic programming. With modifications of existing programming (either offline or in teach mode), the robot can be adapted to a change in requirements without extensive equipment modifications.

Along with this progress in technology, the average person's preconceptions, fears, and hopes for robots were highly influenced by science fiction films, especially the intelligent C3PO from *Star Wars*. This public interest in turn spurred toymakers to offer an even wider selection of robot toys. New companies were then formed to produce educational or hobby robots. Small robot arms, mobile robots with and without arms, and line following–line drawing turtle-shaped robots were offered during the 1980s.

For example, one of the earliest personal robots was RB5X, offered by RB Robotics. This robot had a working arm and speech synthesis capability and was controlled by programs written in BASIC. Thousands of these robots were sold, and many individuals learned robot basics from them. Unfortunately, RB Robots, like many other small robot start-up firms of this period eventually failed.

An educational robot that has done quite well is the HERO 2000 (Fig. 1-5). This robot also has an operational arm and voice system. It comes with a patrol

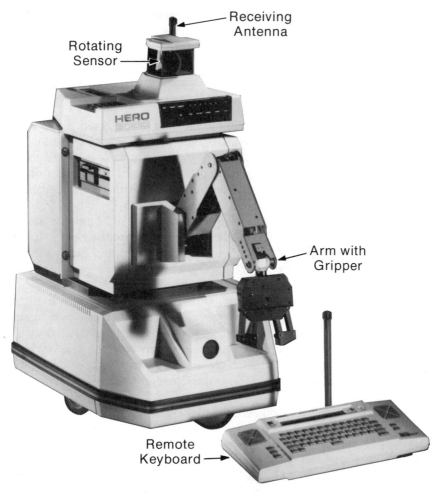

Receiving Antenna

Rotating Sensor

Arm with Gripper

Remote Keyboard

Figure 1-5. Personal robot. (*Courtesy of Heath Company*)

program incorporating both sound and light sensors in a limited collision avoidance system. Of major significance is the excellent quality of documentation available. Another strong point is its ability to download programs, a necessary feature for any serious utilization.

Except for Heath robots, personal robots have not done as well as personal computers, mainly because personal robots cannot perform many useful tasks, due to their limited strength, limited accuracy, and the lack of usable software. In addition, personal computer manufacturers have provided an open architecture, thus supporting third-party developers. To date, few personal robot manufacturers have been willing to do this.

Nevertheless, advances in sensor technology and robot programming, coupled with lower-cost computer technology, should make possible effective personal robots in the near future, although a robot that can do more than very limited household tasks is still far in the future.

1.2 WORKING DEFINITION OF ROBOT

The term *robot* means different things to different people. Even if one limits the term to industrial robots, and eliminates meanings implicit in robots found in motion pictures and toys, there are still some definition problems. Current definitions are reactive to past technology and do not address issues of the robot of tomorrow. The result is that definitions, by the time they are adapted, are out of date.

Each major robot association in the world has its own definition of industrial robot, and, although similar, there are sufficient differences to cause confusion over what should be called a robot for statistical purposes. The result has been confusion when the number of robots are compared between countries.

The Robotics Industry Association (RIA) of the United States, until recently, defined a robot as a "reprogrammable multifunctional manipulator designed to move material, parts, tools, or specialized devices through variable programmed motion for the performance of a variety of tasks." The key word in this definition is "manipulator." Since a manipulator is an arm, this definition says that all robots must have an arm and that the arm is the principle part of the robot.

In late 1987, the RIA modified this definition to a "reprogrammable multifunctional *machine* designed to *manipulate* material, parts, tools, or specialized devices through variable programmed motion for the performance of a variety of tasks" (italics added). This definition is better, but it still requires the robot to have an arm and to move something.

The British Robot Association (BRA) uses a similar definition. According to its 1985 report, "An industrial robot is a reprogrammable device designed to both manipulate and transport parts, tools or specialized manufacturing implements through variable programmed motions for the performance of specific manufacturing tasks."[3]

The British version does not require that the robot be a manipulator, only that it be able to manipulate. In both definitions, the arm is only one part, although an important one, in any robot. The British definition does add a transport capability, but it requires that the robot perform manufacturing tasks. Technically, then, robots performing nonmanufacturing industrial or commercial tasks would not qualify under this definition.

The Swedish Industrial Robot Association (SWIRA) defines a robot as "an automatically controlled, reprogrammable, multi-purpose manipulative machine with or without locomotion for use in industrial automation applications." This definition, which is being considered by the International Standards Organization (ISO), has additional improvements. It allows the robot to be mobile, and it includes all industrial applications, not just manufacturing ones.

In Japan, the Japanese Industrial Robot Association (JIRA) defines six classes of industrial robots in the JIS B0134-1986 standard. In increasing levels of capability, they include

Class 1: Manual manipulator—controlled by the operator (such as a teleoperated robot)

Class 2: Fixed sequence robot—a stand-alone robot operating in sequence and performing a predetermined and unchanging task

Class 3: Variable sequence robot—similar to class 2, but with preset data that is easily modified

Class 4: Playback robot—the robot is trained by a human operator and then repeatedly performs the required steps in sequence

Class 5: Numerical control robot—the human operator controls the robot through changing a program or entering numbers, rather than through a training mode

Class 6: Intelligent robot—the robot has the means to understand its environment and adapt to changing conditions as it completes its task

Most robots in Japan belong to classes 2 or 4. In a study performed in 1982, 42% of the robots were class 2 and 39% were class 4. The RIA does not consider the first two classes to be true robots because they are not reprogrammable. Within the other classes, the RIA does not distinguish between types of robots.

The French organization (AFRI) has four classifications. French companies are leaders in teleoperated robots, and their first robot classification (type A) is identical to JIRA class 1. Type B covers JIRA classes 2 and 3, and is subdivided into B1, equivalent to JIRA class 2, and B2, equivalent to JIRA class 3. The AFRI type C robot combines JIRA classes 4 and 5 and is often considered to define first-generation robots. Type C robots may also contain locomotion. This type is further divided into C1, for robots with less than five programmable joints, and C2, for robots with five or more programmable joints. Type D is identical to JIRA class 6 and defines second-generation robots. A third-generation robot would have three-dimensional vision and understand natural languages. No commercial robots now available meet the French definition of a third-generation robot.

Several difficulties are inherent in many of these robot definitions. First, how should remote-controlled devices, often called teleoperated robots or telechirs, be classified? The French and the Japanese call such devices robots, although of the lowest class. Most other organizations, including the RIA, have resisted classifying them as true robots, primarily because they require a human in the control loop as the brains of the robot. In other words, a telechir is not autonomous. However, the technology, ability, and appearance of some of these devices are very close to true robots, and some do have some autonomous capability and remote capability (such as NASA's moon rover). As long as a device includes *some* autonomous capability, it should not be barred from classification as a robot.

Second, how should automated guided vehicles (AGVs) be classified? These mobile carts carry parts, mail, and equipment between workstations in the factory automatically, although they usually follow an explicit or implicit track and have little ability to go around objects. Many of these units can also load and unload parts at each workstation. These mobile units are autonomous, can perform useful work in an industrial setting without operator intervention, and may contain a manipulator, yet they are not classified as an industrial robot by any current definition (except perhaps the Swedish).

Third, there must be a method of describing robots with features similar to industrial robots, such as a programmable arm, but that cannot be used in an industrial environment due to their lack of reliability in the workplace, their limited accuracies, and the small payloads that they can handle. Since their primary function may be educational, on either an individual or school level, they should not be called industrial robots, yet in every other way they qualify as robots. The RIA addresses this by including a separate classification within its membership. Originally called the National Personal Robot Association, this name was changed in 1986 to the National Service Robot Association. However, RIA has not yet modified its definition to include a robot of this type.

There is also a question of robot generations. If the French type C robot is a first-generation model, it leaves the status of the French type B (JIRA classes 2 and 3) unclear. Possibly these could be called prerobots.

Finally, no definition of robot describes the robot of the future. Overall, the definition should be less restrictive, not limited to industrial use, not centered around a manipulator, and not limited to today's technology.

In the opinion of many individuals in the field, all current definitions of robots are too restrictive. The definitions may have been satisfactory when they were adopted, but they had no provision to incorporate the changes in technology or application that developed, and thus the definition is now generally considered out of date. The ISO and the RIA, among other organizations, recognize this, and they are attempting to develop a more up-to-date definition of a robot.

Until a better definition is developed, this book adopts Dr. Bejczy's definition,[1] repeated here for emphasis. "There are three parts to the technical definition of robots. First, robots are general purpose mechanical machines. Second, they are programmable to perform a variety of work within their mechanical capabilities. Third, they operate automatically."

There is no robot currently accepted by the RIA as an industrial robot that does not fit that definition. It also picks up several current and future devices that should also be classified as robots, such as AGVs, educational robots, and mobile robots. It also includes those telerobots that are capable of some programmable action, but eliminates those telechirs that have no inherent programmable capabilities and only serve as remote manipulators. This definition also extends the applicability of robots by not restricting them to industrial environments. Finally, the definition omits those devices that should probably not be called robots, such as numerical control machines and all types of hard automation.

The key operative in Dr. Bejczy's definition is the phrase "they operate automatically," which is the starting point for the development of intelligence in robots. It is certainly possible to build a robot with no degree of intelligence and no input from its environment, and yet have it operate automatically. Indeed, robots are still being built that will continue on their assigned tasks oblivious to even the most drastic changes in their environments. Nevertheless, most robots are beginning to be supplied with some level of intelligence, in which they can sense and react to changes in their environment. These early "intelligent" robots will open up new areas of applications while emphasizing the safety issues of working with humans.

1.3 GROWTH OF THE INDUSTRY

It is difficult to provide accurate figures to track the growth of the robotics industry. For example, some sources do not distinguish between robots built, robots sold, robots shipped, and robots installed. They do not always state whether the figures are cumulative or for that year only, nor do they distinguish between imported robots and exported robots. In addition, since the definition of robots and the type of statistics reported change between countries, information gathered from national organizations cannot be easily compared. Finally, some country organizations have corrected earlier released figures, and these corrections have not always been picked up in the literature. This book, as far as was possible, includes only comparable figures. Thus, the figures from European countries and from Japan include only robots of AFRI definitions, B2, C, and D, which match fairly well the RIA definition of a robot. Nevertheless, the figures have been collected from different sources and, in a few cases, adjusted to match data from conflicting sources. Therefore the data should be considered only as trend information rather than completely accurate or even consistent with other sources.

To provide an international organization to collect and publish worldwide statistics, a number of countries banded together in 1986 to form the International Federation of Robotics (IFR). An offshoot of the informal organization that had been running the International Symposium on Industrial Robots (ISIR) for many years, IFR provides various worldwide functions and brings some stability to robot statistics. Sweden is serving as the secretariat of this organization.

Table 1-1 presents the number of installed robots in 16 countries for 1981, 1983, and 1985. For comparison with the United States and Japan, Europe is shown both in total and divided by country. More detailed data may be found in Table 1-2 for six of the largest users of robots over a seven-year period.

Robots have been used in many types of tasks, but they function in some applications better than others. In a 1985 study, the most common uses for robots were examined in the largest users: the United Kingdom, West Germany, Japan, and the United States. This analysis indicated that spot welding,

Table 1-1. Installed Robots

	1981	1983	1985
Japan	14,250	29,113	70,000
Total Europe	8,000	13,500	28,400
USA	4,700	9,400	20,800
West Germany	2,300	4,800	8,800
France	1,000	2,210	5,904
Italy	990[a]	1,510	4,000[b]
Great Britain	713	1,753	3,208
Sweden	750	1,492	2,046
Czechoslovakia	330		
Belgium		514	1,000[a]
Poland	240		
Norway	210	275	350[a]
Spain	400	433	675
Australia	62	300	
Switzerland	50	110	600[a]
Finland	116	112	261
Denmark	66	76	164[a]
Netherlands	56	120	350[a]

Note: Most values supplied by appropriate national robot organizations.
[a] Estimate by AFRI.[4]
[b] Estimate by SWIRI.

Table 1-2. Robot Growth

	Japan	USA	England	West Germany	Sweden	France
1979	4,200^a	2,500	185	850	600	480^a
1980	7,560^a	3,300	371	1,255	675^a	580
1981	14,250	4,700	713	2,300	750	1000
1982	21,000	6,300	1152	3,500	1273	1600^a
1983	29,113	9,400	1753	4,800	1452	2210
1984	45,616	14,400	2623	6,600	1745	3350
1985	70,000	20,800	3208	8,800	2046	4904^a
1986	90,000	25,000	3800	12,500	2500	5270

Note: Most values supplied by appropriate national robot organizations.
^{a} Estimates combining several sources.

arc welding, machine loading, and assembly are four of the five top current uses for robots everywhere. The fifth common application is injection molding in the United Kingdom and Japan, and spray painting–surface coating in the United States and West Germany.

Another study (in Japan) examined the principle use of various types of robots. In 1982, the largest portion (70%) of fixed-sequence robots was used in loading and unloading tasks. The largest portion (88%) of playback robots was used in welding. Finally, the largest portion of intelligent robots (71%) was used in assembly applications.

By 1986 it was estimated that the installed base of robots had grown to over 150,000 worldwide, from a humble beginning of only 500 in 1972. The number of firms producing robots has also risen. In 1970 there were only five companies actually manufacturing robots. By 1980 there were about 20 firms, two of them having gross sales of $6 million or more a year. The expansion continued through 1986. In that year there was a total of 45 U.S. firms manufacturing complete robots, with a total of 150 worldwide. In addition to manufacturers, there are many robot system integrators, and a complete peripheral industry of firms supplies major robot subsystems, such as vision systems and end-effectors. One survey has identified 350 public and private organizations, including universities, in 23 countries conducting robotics research.[5]

Although there have been recent signs of a temporary slowdown (1986 and 1987 had almost zero growth), the overall growth of robotics has been very strong for many years. From 1979 through 1985, robot installation has increased at an average rate of 41% per year. Even more significant, robotic installations in Japan have grown at a rate of 60% a year.[6] In addition, solid growth is predicted to resume again during 1988 in the United States.[7]

The average price of a robot has also climbed, but not as rapidly. In the early 1970s, a typical robot sold for perhaps $25,000. In 1986, robots with much more capability sold for an average price of $75,000. Not all types of robots have increased in price. Some prices have even come down significantly. For example, the average cost of welding robots, which had been close to $150,000 in 1982 was reduced to about $78,000 in 1986.

In any discussion of typical robot prices, the resulting figures depend on whether a complete robotics system is being discussed (including end-effectors and, perhaps, a vision system) or just the basic robot. For servo-based robots, several sources indicate an average system figure of $112,000 and an average base robot price of $71,300 for robots sold in 1985. If installation and other associated costs are included, these figures would be higher.

Unfortunately, this attempt to hold prices down to stimulate sales adversely affected profits, and many companies remained unprofitable for years. Others dropped out of robotics manufacturing. There was simply not enough profit to cover the development costs of the robots. New firms entering the market have also had an impact on both profitability and market share.

In 1980, two companies shared most of the market. By concentrating on larger accounts and staying away from the custom engineering required by many small users, these companies began competing with newer firms for what was only a limited segment of the total market. Some firms also stayed with hydraulic robots much too long, when many users were demanding the quieter and cleaner electric robots. This factor also reduced the potential customer base.

Table 1-3 provides data on changing U.S. market share for some major companies. The table demonstrates how market share has dropped significantly as new venture-funded firms were founded and some of the larger industrial giants entered the field. Note, however, that the table only presents information on U.S. market share and thus does not present a complete picture of sales by Japanese and European firms. For example, ASEA had worldwide sales of $150 million in 1985, about the same as GMF.

Robot sales figures are available from several sources, with differing assumptions. To gain a complete picture, the Census Bureau surveyed 72 robot manufacturers and system suppliers in 1984 and 1985 to ascertain both total robot sales figures and the types of robots being sold.[11] In 1984 they determined that there were approximately 2,500 servo-controlled industrial robots sold, for a gross figure of $206 million; in 1985 the figures had increased to 3,272 robots and $296.9 million. Out of the 72 companies surveyed, 26 build servo-controlled robots, 10 build non-servo-controlled robots, and 7 build education and hobby robots (others supplied various types of accessories).

The Census Bureau has introduced current industrial reports (CIRs) on industrial robots. According to the Bureau's figures, 6,534 robots, worth $225.5 million, were shipped in 1984, whereas only 5,796 robots, $317.7 million, were shipped in 1985, indicating a slowdown in the industry as well as an interest by users for more sophisticated and complex robots. In compari-

Table 1-3. Market Share (percentage of robots sold in United States)

	1980	1981	1982	1983	1984	1985	1986
Unimation/Westinghouse	44.3	44.0	33.2	15.0	11.3	7.6	6.4
Cincinnati Milacron	32.2	32.3	16.8	21.3	13.3	10.3	10.4
Prab	6.1	5.3	6.5	6.3	2.9	2.7	2.5
DeVilbiss	5.5	4.2	12.4	3.9	7.6	5.5	4.5
ASEA	2.8	5.8	5.0	6.3	7.6	6.6	7.7
Automatix	0.4	1.9	4.2	5.3	4.4	4.2	3.8
GMF Robotics			0.2	9.3	26.0	31.4	29.6
General Electric			2.9	5.7	4.9	2.6	2.8
GCA/CIMCORP				2.5	3.4	5.7	6.5
American Robot/CIMFLEX				1.2	2.0	2.9	5.1
Kuka						5.0	3.8
Others^a	8.7	6.5	18.8	23.2	18.9	15.5	16.9

Notes:
(1) Data adapted from Prudential-Bache Securities newsletters.[8,9,10]
(2) Figures cover U.S. sales only and may include vision systems.
(3) Information is presented only for those firms that captured at least 5% of the U.S. market during one or more years.

^{a} Includes Adept, Advanced Robotics, Copperweld, Cybotech, Graco Robotics, IBM, Intelledex, Seiko, and other producers.

son, RIA reported 5,136 robots, valued at $332.5 million, shipped in 1984 and 6,209 robots valued at $442.7 million, in 1985. RIA figures for 1986 showed zero growth, with a total of 6,219 units valued at $440.9 million.

Many robots purchased by U.S. companies were imported. Imports reached 1,718 units totaling $43.7 million in 1983 (with 1,412 from Japan), 3,411 units totaling $92 million in 1984 (with 2,800 from Japan), and 4,313 units totaling $126 million in 1985. Many of these robots are not servo controlled (which can be confirmed by examining the average price, $25,000). Therefore these figures must be reduced to compare them with U.S. manufacturers. This comparison shows that imports supply almost half of our domestic robot market.

Future sales are even more difficult to estimate and require a large amount of guesswork. In high technology fields actual growth sometimes outruns predictions (such as in the personal computer field), while some areas never seem to perform as well as predicted (e.g., videotext). Numerous projections have been made. A study by Hunt and Hunt in 1982 described six projections for the year 1990. These projections ranged from 50,000 to 150,000 robots installed in the United States. A 1982 study by the University of Michigan offered a conservative total of 31,000 installations. Conigliaro estimated there would be 125,000 installed by 1990.[12] The RIA predicted 90,000. A recent textbook projects a 25% growth through 1987, followed by 40% growth to 1990, with a total of 43,600 robots.[13] One projection has the total number of robots in the world by 1993 reaching 268,000. These predictions did not anticipate the much higher than expected growth in 1984 and 1985, nor the almost zero sales growth in 1986 and 1987. The net effect of these years is that they just about cancel each other, with average growth over the four years approximately 21% per year.

More recent projections vary widely, based upon how they interpret the downturn in growth during 1986. Conigliaro now sees a reduction in sales of 10% in 1986 (compared to 1985) and an even greater reduction in 1987.

Business Communications Company (Stamford, CT) predicts a growth rate of 18% through the early 1990s after a nearly flat 1987. Data Quest sees the United States pulling even with Japan in sales by 1990, at a sales growth of 27% a year. The Freedom Group is even more optimistic and expects an installed base of 109,000 robots in the United States by 1990, with an average growth of 40%.

To narrow this wide range of figures, take as an example one set of 1982 projections. Table 1-4 summarizes data from one University of Michigan study. Although this projection underestimated the number of robotics installations by 1985 in many countries, if the downturn in 1986 and 1987 is considered the figures for 1990 (except for Japan) may still be correct. A 1985 University of Michigan forecast[15] estimated that the robot population of the

Table 1-4. Example of Predicted Robot Installations

	1985	1990
Japan	16,000	29,000
USA	7,700	31,000
West Germany	5,000	12,000
Sweden	2,300	5,000
Great Britain	3,000	21,000[a]
Norway	1,000	2,000
Switzerland	600	5,000
Finland	950	3,000
Poland	200	1,400
Belgium	200	
Yugoslavia	150	300
Denmark	110	250

Note: These figures were taken from a 1982 University of Michigan study.

[a] Prediction by Policy Studies Institute, an English firm, is for about half this figure.[14]

United States would reach 128,000 by 1995. The projections made in 1982 and 1986, are surprisingly similar. In 1982 predictions ranged from a total of 31,000 to 150,000 robots installed by 1990. In 1986, predictions ranged from 38,000 to 109,000, within the range of the predictions made four years earlier.

This author believes that robot installations will grow at a slow pace until mid-1989 (average of 10% a year), then increase to 25% a year through at least 1997 (see Fig. 1-6). Thus we forecast about 44,000 robots installed at the end of 1990, which is slightly lower than many other projections. The impact on

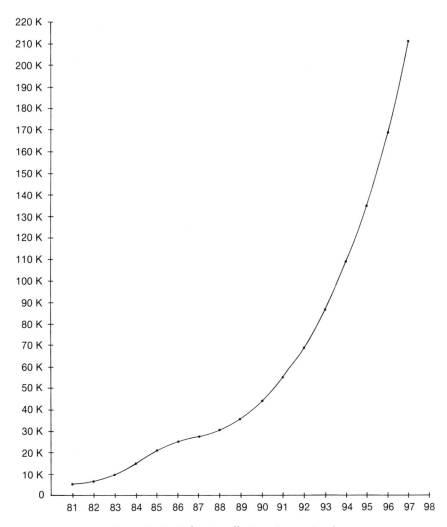

Figure 1-6. Robot installations (projections).

U.S. manufacturers will not be as favorable, because lower-cost imported robots will take an increasingly larger share of the domestic market. Worse yet, we believe that foreign firms will excel in many new application areas and new technologies, since many foreign governments support industrial robotics research at a much higher level than does the U.S. government.

One area of agreement among most experts is the projected increase in number of robots sold with vision systems. Most sources agree that within five years, 25% of all robots sold will include vision systems. This increase is due to cost reductions and improvements in available vision technology. Robots equipped with vision obviously have a wider range of applications.

The Automated Vision Association (AVA) released a study in June 1986, indicating that a typical machine vision system now costs $25,000 to $50,000, but will drop to between $15,000 and $35,000 by the end of 1990. Currently, vision systems average 25% of a typical production cell cost. This percentage should drop to 20% by 1990.

Predictions have also been made on the growth of technology. For example, it is estimated that the use of color will expand from 2% in 1985 to 15% by 1990 and that the use of three-dimensional (3D) information will increase from 5% in 1985 to 30% in 1990. Based on this, Delphi forecasts an average growth rate of 60% a year for the machine vision industry.

What types of applications will be addressed by these future systems? Chapter 17 explores this area in depth. A few examples here will suffice. Auto industry applications accounted for most early robot purchases. However, by 1990, it is expected that machine vision purchases by the electronic industry should equal those of the auto industries. Regarding complete robot systems, automotive users share of the market should drop from 51% in 1986 to 26% in 1995 while electronics industries will grow to 14%.

Robots are used mainly in large factories. Two thirds of all robot users have only one or two robots. But it is the 8% who have more than 10 robots that accounts for 75% of the total robot use hours. It is estimated that over 90% of the 65,000 industrial firms in the United States have yet to install their first robot. A major reason for this imbalance is payback. Studies have shown that payback can take 3–5 years for small firms, whereas it is typically 1.5–2 years for the largest firms.[16]

The three biggest problems found in robot installations to date have been the high cost of development, unexpected installation and integration problems, and inadequate support from manufacturers after sales. On the other hand, it is significant that 27 users have considered the installation of robots in their factory worthwhile for each user that has not.

Chapter 15 explores future technical trends in robotics. This section summarizes a few of these trends and adds three nontechnical ones. Fusaro[17] discusses several trends, among them the trend to direct-drive robots, an increase in offline programming, and the demise of the hydraulic robot. Other technical trends include an increase in robot complexity and capability, an increase in the use of mobile robots, and the use of intelligent sensors, all of which will significantly increase the applications in which robots can be used. Direct-drive robots are fast becoming a new industry standard; hydraulic robots have been replaced by all-electric ones except in a few specialized applications.

Another trend is the increased use of robot communications and networking. This trend is associated with the rising use of robots in manufacturing workcells, encouraged by newly developed communications standards. The availability of standard hardware interfaces and modular programs that can be added to existing robots is another trend. Although this modularity is in its early phases, it will also significantly increase the potential applications of robots.

A major nontechnical trend is an increase in robot utilization for other than economic reasons. Benefits such as product quality, consistency, reduced work-in-process inventory, improved customer responsiveness, and elimination of boring, unsafe jobs are becoming increasingly important.

Another trend is in pricing. As more robots are built, economies of scale will allow prices to be reduced. In addition, the maturing of related technological areas, such as vision processing, will also result in price reductions. This trend can be seen in the reduction of the average price of a robot during 1986 (about $66,000) compared to 1985 ($75,000).

A final trend is the increased uses of robots. Even with the rate of increase dropping, a $6 billion world market, and a $2 billion U.S. market are still expected by 1995.[18]

The toughest issue facing most firms wishing to install robots is sufficient help in implementation. There is already an insufficient number of robot system houses, and their number must increase to support the growth of robotics. System houses are particularly needed in the areas of advanced computer programming, vision systems, and intelligent sensors.

1.4 INTRODUCTION TO ROBOTICS ENGINEERING

Robots offer two major advantages over standard machines found in manufacturing. First, they provide an almost total automation capability over the manufacturing process, improving quality and output while reducing costs. Second, they provide the capability of adapting the manufacturing line more quickly to changes in the old product or for the addition of new products.

When the production units are very adaptable, they are known as flexible manufacturing systems (FMS). A flexible unit, or manufacturing cell, is made up of a few robots working together on the same task. When several cells are used together, they are called a *flexible workshop*.

The typical industrial robotics system can be divided into seven fundamental parts, as shown in Figure 1-7:

1. The articulated mechanical system, or the arm of the robot. It includes the links, joints, and wrist.
2. The end-effector, which performs the usable work by serving as a gripper or tool.
3. The actuators with their power supplies provide the mechanical power to move the robot arm. Actuators are primarily electric, but a few hydraulic and pneumatic actuators are still used.
4. The transmission system (not illustrated), which couples this power to the joints, through connections such as cables, belts, gears, or hoses.
5. Internal sensors, which monitor the motion of the robot.
6. External sensors, which collect information about the surrounding environment and provide this data to the robot.
7. The control system for the robot, which may include computer chips controlling each axis of motion, a robot "brain" or central computer directing overall activity, and the necessary interface with an operator for monitoring, reprogramming, or training.

Although often supplied as a part of a complete robot system, external sensors and grippers are usually considered to be accessories and not part of the industrial robot itself. They are included in this figure and discussed in this book because they are an important and necessary part of robot systems.

The typical robot arm has three joints comparable to a human arm: shoulder, elbow, and wrist. Designs may be based on the *revolute* approach (in which all joint movements are rotational), on *linear* motion (in which all joint movements are translational), or on some combination of translation and rotation.

The manipulator is the mechanical arm that moves the desired object through three-dimensional space. Connecting the manipulator with the object are various tools or grippers, usually called *end-effectors*. Motion is accomplished through mechanical joints and control over the motion.

Typically, robot motion is described by one of four types of coordinate systems. The one most familiar to engineers is the rectangular (Cartesian) coordinate system. In this type of system, all motion is along one of the three mutually perpendicular axes. Robots with this type of motion are the easiest to control, since each axis motion can be computed independently. Other robot systems are based on cylindrical or spherical coordinates.

As discussed more fully in chapter 2, robots are available with from three to seven (or more) axes of motion (degrees of freedom, DOF) not including

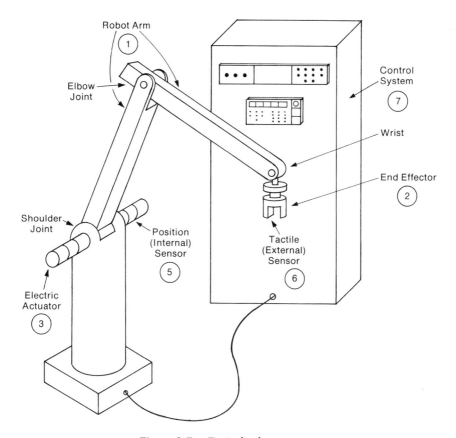

Figure 1-7. Typical robot system.

gripper motion. For example, one paint spray robot has 10 DOF, allowing it to more easily reach inside a car to paint the inside of the door. Although usually a part of the robot, the wrist joint may also be provided with some of the interchangeable end-effectors.

In order to design robot systems, a company needs a wide variety of skills. When robots were first designed, mechanical engineering skills were the primary requirement. As robots became part of industrial production, manufacturing engineering skills were needed. As servo control and advanced sensors began to be used, the need for electronics engineers to enter the field became important. Further steps required the skills of programmers, computer hardware engineers, language designers, and vision specialists. The next generation of robots will require people skilled in artificial intelligence and such specialized areas as laser technology and ultrasonics.

REFERENCES

1. "Robot Sensing and Intelligence," IEEE short course presented by satellite transmission in November 1983. Prepared by the Institute of Electrical and Electronic Engineers, Parsippany, N.J.

2. Capek, Karel. *R.U.R.,* translation by P. Selver. New York: Washington Square Press, 1933.

3. *Robot Facts 1985.* Bedford, England: British Robot Association, 1985 report.

4. *Statistiques: 1986.* Cachan, France: AFRI, 1986 (in French).

5. *The World Yearbook of Robotics Research and Development.* London: Kogan Page Ltd., 1986.

6. Aron, Paul H. Robotics in Japan: Past, present, future. *Proc. Robots 6 Conference.* Dearborn, Mich.: Society of Manufacturing Engineers, March 1982.

7. Mittelstadt, Eric, and Pat Costa. Robotics & vision: What lies ahead for automation. *Robotics World,* Vol. 5, No. 1, Jan. 1987, pp. 20–22.

8. Conigliaro, Laura, *Robotics Newsletter.* New York: Prudential-Bache Securities, Dec. 13, 1982.

9. Conigliaro, Laura, and Christine Chien. *Computer Integrated Manufacturing,* CIM Newsletter. New York: Prudential-Bache Securities, June 24, 1985.

10. Conigliaro, Laura, and Christine Chien. *Computer Integrated Manufacturing,* CIM Newsletter, New York: Prudential-Bache Securities, May 23, 1986.

11. *U.S. Industrial Outlook: 1986.* Washington, D.C.: U.S. Department of Commerce, Dec. 1985.

12. Conigliaro, Laura, Robotics presentation. Institutional Investors Conference. New York: Prudential-Bache, March, 1983.

13. Groover, M. P., et al. *Industrial Robotics: Technology, Programming and Applications.* New York: McGraw-Hill, 1986.

14. Northcott, Jim, et al. *Robots in British Industry: Expectations and Experience.* London: Policy Studies Institute, 1986.

15. Smith, D., and P. Heytler, Jr. *Industrial Robots: Forecasts and Trends,* a Delphi Study. University of Michigan, 1985.

16. Ayres, Robert U., et al. *Robotics and Flexible Manufacturing Technologies: Assessment, Impacts and Forecast.* Park Ridge, N.J.: Noyes Publications, 1985.

17. Fusaro, Dave. Outlook '86. *American Metal Market/Metalworking News,* Dec. 16, 1985.

18. Monnin, Phillip V. Worldwide robotic trends. *Robotics World Magazine,* Vol. 5, No. 1, Jan. 1987. p. 4.

TYPES OF ROBOTS

Robots are complex devices, and there are many ways to describe or classify them. Six common methods are described here. The chapter also explains some of the nomenclature currently used, the advantages of alternative approaches, introduces design details that are discussed more fully in Parts II and III, and introduces some current robotic applications.

2.1 CLASSIFICATION BY DEGREES OF FREEDOM

A good place to begin a discussion of a robot is by describing its *degrees of freedom* (DOF). There is some confusion in the literature between degrees of freedom and *degrees of mobility*. Condon and Odishaw define degrees of freedom as follows: "The number of independent coordinates needed to express the position of all its parts . . . is known as the number of degrees of freedom of the system."[1] Thus coordinate independence and number of links contribute to a robot's DOF, and determining the number of independent variables needed to define motion in any robot determines the minimum DOF required.

How many independent variables are needed in a typical robot? A single point in space can be represented by three independent coordinates. In a Cartesian coordinate system, for example, a point can be described by coordinates X_1, Y_1, and Z_1. Thus, the gripper center point—in fact, any location in space—can be specified with only three DOF, one for each Cartesian axis. However, real objects are not simply a single point, but are three-dimensional and take up space. As such, they have their own internal set of X', Y', and Z' coordinate axes, which can (for convenience) be referred to the object's center point. These two sets of coordinate systems are illustrated in Figure 2-1. Note that the object's center point may not be colocated with the gripper center point. In addition, one set of coordinates may be tilted with respect to the other. Only rarely are the two sets of coordinates the same. In general, there are angles between X and X', Y and Y', and Z and Z', and these angles can change as the object is moved in space. There will also be translational offsets between the two coordinate systems if the center points do not coincide. In many applications, the coordinate system is based on the center point of the tool and called the *tool center point* (TCP).

For a typical robot end-effector (tool or gripper), the gripper or tool center point can be placed anywhere in space (requiring three DOF) and then oriented in any direction (requiring another three DOF). Therefore six DOF are required to specifically control the motion of any end-effector *with respect to the robot base* while allowing any possible position and orientation to be reached within its workspace. The first three DOF are often called the *main DOF*, since they define the total workspace that the robot is able to reach and are controlled through actuators (motors) on the body of the robot. The second three DOF are often located at or near the robot TCP and serve as a "wrist" to control gripper or tool orientation once the robot arm has moved the tool to the required location.

Each time an independent joint, whether translational or angular, is added to a robot, a DOF is added to the *complete system.* However, there is an

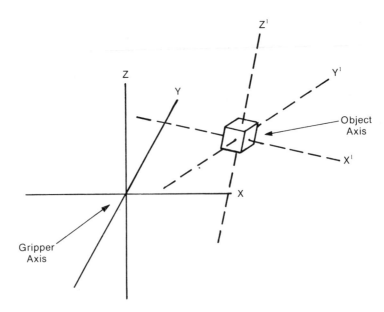

Figure 2-1. Gripper/object coordinate systems.

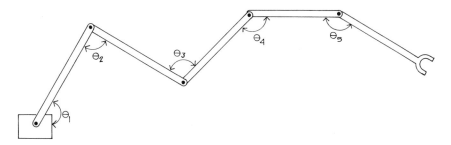

Figure 2-2. Multijointed arm.

important distinction to be made between the number of DOF of the system and the number of DOF of the end effector itself. In many cases a DOF may *not* be added to the end-effector. For example, the robot arm of Figure 2-2 has five independent joints and thus five DOF. However, the end-effector (gripper) has only three, since the robot arm can never move out of a single plane. With no Z-axis capability, it thus has only two positional DOF (X and Y) and only one orientation DOF (θ).

Some authors prefer to use the term *degree of mobility* to describe those joints that give added DOF to the system without changing the DOF of the end-effector.[2] There are valid reasons for adding redundant joints to the system, some of which are discussed in the following paragraphs. The term *degree*

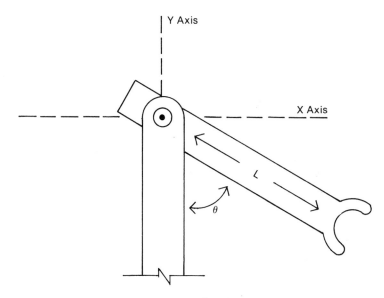

Figure 2-3. Single DOF system.

of mobility may be more descriptive of this added freedom, especially if a full six DOF is not available to the end-effector.

To explore mobility further, let us consider a typical single robot joint. As can be seen in Figure 2-3, this particular joint will act as a hinge, changing both the X and Y coordinates (or any two coordinates of the link end point) at the same time. However, these changes are not independent; that is, for any resultant X position obtained by rotating the arm, there is one and only one Y position available. With a fixed length of the link (L), the X_1 and Y_1 positions of the end point both depend on the elevation angle (θ) selected, so this figure has a single DOF. The arm end point can be considered as moving through an arc.

When a second angle is introduced *in the same plane,* a second DOF is added, since now there is more than one position of Y for each position of X, and the arm end point can be considered as moving through an area. As shown in Figure 2-2, adding a third rotary joint *in the same plane* will not add another DOF to the position of the end point or TCP because only two positional DOF are in any plane. Adding another rotary joint will add a DOF to the orientation of the TCP, although this method is not the best way to obtain the additional DOF because there would be interaction between the positional DOF and the orientation DOF. The primary reason for adding a third joint of this type is to add a degree of mobility in reaching the desired position. Figure 2-4 shows a two DOF system with only one angle α (and its mirror image) available to reach a specific (X, Y) location. An obstacle is blocking the robot's path, so the two DOF robot will not be able to reach the desired location.

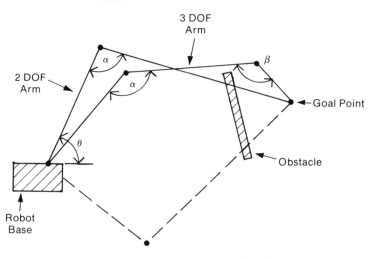

Figure 2-4. Adding a degree of mobility.

Adding a third joint, makes an almost unlimited number of angle α's available (with corresponding angles β and θ). Thus, an additional degree of mobility, while not increasing the DOF of the robot (in theory, the robot cannot move in a new direction), does give more mobility for reaching the desired end point. One reason for added mobility is due to constraints in the workplace. Robots are built to act on an object, so they must be able to reach the object. Often, however, part of the robot itself, conveyors, other production equipment in the workplace, or even another part of the object can interfere with the robot's reaching the desired point. For example, suppose a person inside a car tries to open the door by using the outside handle. With three DOF that person can reach the handle only by going through the door. With added degrees of mobility, the person can select different angles to reach the same point and route his or her hand upward through an open window and then back down to reach the outside handle. Similarly, a robot with a three DOF arm trying to paint or weld inside objects can be prevented from positioning the object properly, due to constraints on the object or in the workplace.

A second reason for added mobility is to extend the workspace limit. A robot's workspace is limited by restrictions imposed by its joints. Angular movement has limits caused by the physical construction of the joint. A human elbow, for example, is limited to a minimum of about 35° when fully closed and a maximum of 180° when fully open. Thus, with elbow motion alone, a human can move over little more than a third of a circle. To cover a wider arc, he or she must use the degree of mobility that another (somewhat redundant) joint supplies, such as the shoulder or wrist.

Although robot joints generally have larger angles available to them, they usually cannot cover a full 360°. A typical revolute robot (such as the PUMA 700 in Fig. 1-3) has a maximum base rotation movement of perhaps 320°, thus slightly restricting the workplace envelope available to it. Adding an extra rotary joint not only adds a degree of mobility but it can also remove the 320° workplace limitation and allow greater coverage.

In some applications, additional DOF at the wrist itself is desirable, to improve speed and sensitivity. Since the wrist can usually be moved much faster and more accurately (although over a smaller range) than the arm, many tasks requiring only limited motion can be performed with wrist motion alone. Arm motion would only be used for initial (gross) positioning.

Extra mobility in a robot does have disadvantages, however, including increased robot complexity and cost, increased weight of the arm, which reduces payload capacity and slows arm motion, and an increased complexity of motion computations, which also slows robot operating speed. Hence, robot manufacturers often offer similar robots with trade-offs in payload capacity and number of DOF. For example, the model T3 886 robot from Cincinnati Milacron is supplied with six axes and a 200-lb payload capability, but it is also available in a three-axes version with an increased payload capacity of 440 lb.[3] Eliminating three DOF more than doubled payload capacity.

Because of the disadvantages of speed, payload, and computational complexity, only specialized robots offer any extra positional mobility beyond what is available through the standard three DOFs. These specialized robots are used when the type of robot motion, limitations on this motion, constraints of the workplace, or function of the tool require it. In addition to three DOF for positioning, robots require from one to three orientation DOF to manipulate properly the object or tool, depending on the function it is performing. Note that the motion of a gripper (such as open and close) is not considered part of the DOF of the robot, since it is specific to the gripper and does not contribute to its position or orientation.

How many degrees of freedom are needed in typical applications? For functions as simple as limited pick-and-place robots, where object orientation is either controlled (such as soda bottles standing on a conveyor belt) or does not matter (such as moving small spheres), a simple three DOF, Cartesian coordinate robot or a limited selective compliance assembly robot arm (SCARA) robot may be sufficient. Slightly more complex assembly tasks can often be performed with four DOF robots. For general-purpose functions that use a tool that is symmetrical along one axis (such as a deburring tool), a five DOF robot with three DOF in the main body and two DOF in the wrist is sufficient. Many applications have symmetry in the object or the tool and can use five-DOF robots. A common example is simple spray painting. Another application that usually requires only five DOF is parts transfer.

For applications requiring complex control of the end-effector and where objects do not block the robot's path, six DOF (three on the main body and three on the wrist) are required. Examples of this type of application include arc welding and complex assembly functions. For specialized applications in which the robot arm must reach around and through objects that are in the way (such as applying sealer inside an automobile) or through a limited-access tunnel to approach the object (such as attaching nuts to hard-to-reach bolts), up to nine DOF may be needed.

In summary, requirements imposed on agility, coverage area, type of tool, and payload influence the number of DOF a robot must have for an application. Thus, selecting the proper robot for a task requires that the number of DOF needed by that task, as well as any future tasks that the robot is expected to perform, be determined. Robots are commercially available with as few as three DOF and as many as seven or more DOF. Two or three DOF are generally located on the arm, with the base adding one (two in the case of gantry robots). The wrist adds an additional one to three. When more than seven DOF are offered, added axes are placed in the arm.

2.2 CLASSIFICATION BY ROBOT MOTION

Joints provide DOF and affect available motion. Depending upon the design of the joints, a robot will be able to move in different ways. The types of joints selected impact the robot's coordinate system and its performance.

Motion Control

Four types of motion control are in general use: sequential, slewing, interpolated, and straight-line. All four types are illustrated in Figure 2-5. Non-servo-

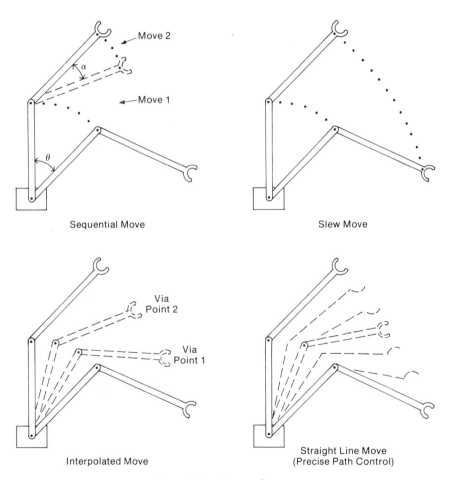

Figure 2-5. Types of moves.

controlled robots are limited to the first two types of motions. In sequential operation, each joint is moved in a predetermined sequence, one joint at a time. No coordinated (multijoint) motion is possible, so the time it takes to complete a move suffers. Only the earliest robots were limited to this method of motion control, although some teach-pendant systems still offer this type of move as an option for controlling robot motion.

With systems employing slewing, joints start their moves simultaneously. There is no control over move velocity in this approach, so there is no way to ensure that all joint movement ends at the same time. The path taken by the robot in reaching the end point is also undefined. This type of robot move is the fastest possible since each joint is moving at maximum speed. Slewing motion control is primarily used for intermediate moves (where accuracy is not important) in uncluttered environments (with no possibility of collisions along the resulting undefined path).

With fine control of intermediate motion, such as is available with servo-controlled robots, other motion is possible. Interpolated motion ensures that the joints start and stop together, thus controlling the path to a much greater extent, although precise path control is still not possible. If the controlled points are close enough, good path control is possible. When intermediate points are designated that do not require the robot to stop, motion at almost full slew speeds is possible. These intermediate points are called *via* points, and the robot is made to pass through these points accurately while continuing in motion. If via points are located close together, the effect is similar to straight-line control.

Straight-line control, somewhat slower than interpolated control, constantly checks robot motion to ensure that the robot follows the prescribed path accurately. Straight-line control requires position feedback sensors and continuous servomotor control, and is primarily used to perform a task rather than move to a designated end point. With the ability for very accurate motion control, it can be used in precision tasks such as seam welding.

Types of Joints

Moving the robot requires joints. Figure 2-6 illustrates four types of joints used on commercial robots. The first two types are adapted from a SCARA-type robot. The second set illustrates a revolute-type robot. The joint on the left is a *sliding* joint, which moves along a single Cartesian axis. It is also known as a linear joint, a translational joint, a Cartesian joint, or a prismatic joint.

The remaining joints are rotational joints. Since they all result in angular movement, their chief distinction is the plane of motion that they support. In the first type, the two links are in different planes, and angular motion is in the

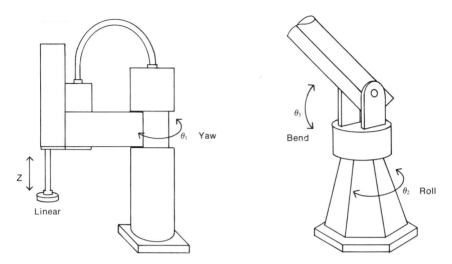

Figure 2-6. Basic robot joints.

plane of the second link (θ_1). The second type provides rotation in a plane 90° from the line of either arm link (θ_2). In this case the two links always lie along a straight line. The third joint is a hinge type (θ_3) that restricts angular movement to the plane of both arm links. In this case, the links always stay in the same plane.

The combination of joints determines the type of robots. Most robots have three or four major joints that define their *work envelope* (i.e., the area that their end-effector can reach). Actions that control wrist motion usually provide only tool orientation and do not affect the work envelope.

Robot Coordinate Systems

The function of a robot's major joints is to reach a designated point in space. To do this, robots can be designed to move with linear motion (Cartesian coordinates), angular motion (all revolute coordinates), or a combination of the two (cylindrical or spherical coordinates), depending on the combinations of joints used. Figure 2-7 illustrates the coordinate systems available with different joints. For example, the spherical-coordinate robot shown uses one revolving joint (in the base), one rotational joint (as an elbow), and one sliding joint (along the arm).

Cartesian robots have three linear joints, ensuring that the motion of the end-effector is in Cartesian coordinates. Robots of this type are good for extensive horizontal travel, such as a gantry robot, or for stacking parts in bins. Figure 2-8 shows a typical example of a small Cartesian robot. This type

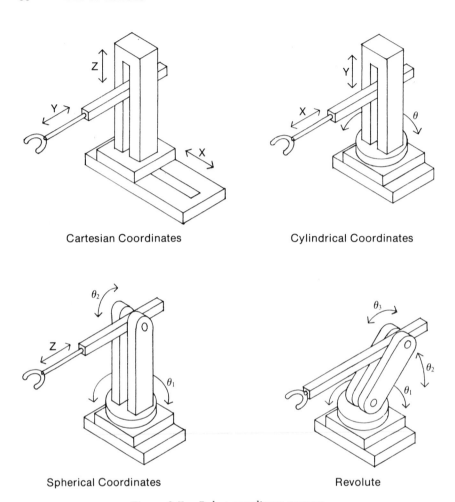

Cartesian Coordinates

Cylindrical Coordinates

Spherical Coordinates

Revolute

Figure 2-7. Robot coordinate systems.

is easy to program, has the highest repeatability, and, with larger robots, has good payload (lift) capability. However, they require larger work envelopes than do other robot types.

Cylindrical robots have two linear joints and one rotational joint, resulting in a robot work envelope that is cylindrically shaped. These robots move linearly in the horizontal and vertical axes (X and Z) and rotate about the base (θ). Their motion describes a cylinder. Many pick-and-place robots are of this type. They are especially good for transferring materials from conveyor belts or for handling small parts that require only limited horizontal motion. They provide a good capability when working against gravity. This type robot is easy

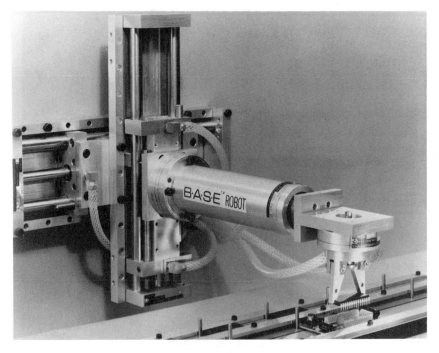

Figure 2-8. Small Cartesian robot. *(Courtesy of Mack Corp.)*

to program, has good lifting capacity, and can reach into tight spaces. Generally, it is unable to reach the floor.

Spherical robots have two rotational joints and one linear joint, and their motion is specified with θ_1, θ_2, and Z coordinates. An example of this type of robot is the Maker 110, supplied by United States Robots and shown in Figure 2-9. Most spherical robots have better lifting capability and are easier to design than are all-revolute robots (discussed next). They are also well suited for loading and unloading machine tools (punch presses), have good reach, and can fit into tight spaces, but their vertical motion is limited.

SCARA robots are also examples of this class, although their joint arrangement is different from the arrangement shown on the spherical robot in Figure 2-7 (see Fig. 1-4). A second type of SCARA is a hybrid, in which the Z-axis motion is performed in Cartesian coordinates. SCARA robots are used in simple assembly operations, such as placing components in printed circuit boards.

A revolute robot has all rotational joints and is the most flexible. Hence, this type is quite popular. A Hitachi revolute robot is shown in Figure 2-10.

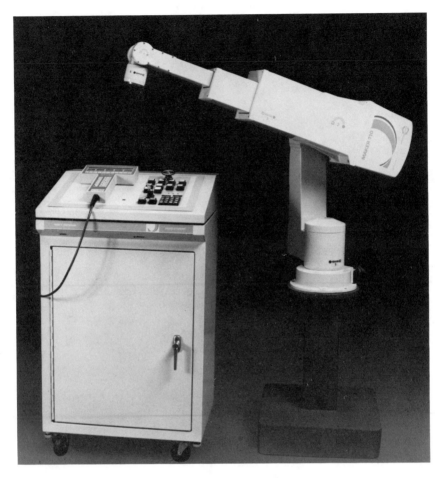

Figure 2-9. Spherical robot. *(Courtesy of U.S. Robots)*

This robot also illustrates a parallel linkage between the base and the upper arm. Parallel linkage allows both shoulder and elbow joint motion to be controlled directly by motors located in the base, therefore removing motor weight from the elbow joint. This linkage type also allows a compact mechanism, so the robot requires less space in the work area. Because of the robot's stiffness and lower forearm weight, the end-effector may be moved more quickly and more accurately. Most revolute robots have articulated motion, which means that joint movement is linked together, limiting orientation changes in the end-effectors during robot motion.

Two disadvantages of a revolute robot are a reduction in load-carrying ability when its arm is at the end of its reach capability, and the advanced

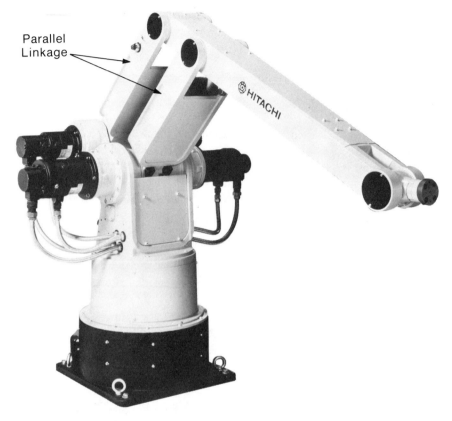

Parallel Linkage

Figure 2-10. Revolute robot. (*Courtesy of Hitachi America, Ltd*)

computer control capability required, due both to the coordinate system employed and the need for articulated motion.

Each robot type has its own strong points. The Cartesian robot has the best repeatability and is the easiest to program. Cylindrical robots are also somewhat easy to program and offer good lift capability. Both spherical and revolute robots excel in arm extension and flexibility of reach. The revolute robot can also mimic human arm motion fairly well and can thus be taught certain tasks (such as paint spraying) more easily.

2.3 CLASSIFICATION BY PLATFORM

Platform classification refers to the robot's type of base. Most robots have a fixed base and are limited to tasks involving objects that they can reach. If a robot must move to perform its tasks, then the robot must be located on a movable base or vehicle. Three types of movable bases are currently in use: gantry, AGVs, and mobile robots.

Gantry

A *gantry* robot is mounted overhead, usually on the ceiling, runs along tracks or rails, and can move over a cluttered environment. The robot arm is lowered from the mobile base to pick up and move items, especially heavy items. It may, for example, use an electromagnet to move heavy bars of steel. Gantry robots have fairly good mobility. For example, it is common for them to travel over 40 ft of X travel, 10 ft of Y travel, and 5 ft or more of Z (vertical) travel. Figure 2-11 shows an example of a gantry robot.

Figure 2-11. Gantry robot with magnetic gripper. (*Courtesy of Sterling-Detroil Co.*)

Figure 2-12. AGV with conveyor top. (*Courtesy of Control Engineering Company, a Jervis B. Webb Company affiliate*)

AGVs

An *AGV* (*automated guided vehicle*) is a small driverless truck that moves material around a manufacturing plant or warehouse by following a path in or on the floor. Figure 2-12 shows an example of an AGV. Most AGVs are designed to stop if something gets in their way, and they must wait for an operator to clear the path. Some AGVs, primarily in the experimental stage, can recognize an object and go around it.

Mobile Robots

Mobile robots are just beginning to be found in factories. Pioneered by companies like Cybermation, a mobile robot can be programmed to travel anywhere in a plant, rather than just following a path. See chapter 8 for a fuller discussion.

2.4 CLASSIFICATION BY POWER SOURCE

Historically, three types of motive power have been used to drive the manipulator arm: hydraulic, pneumatic, and electric. Gripper motive power is usually pneumatic or electric.

Hydraulic

Hydraulic actuators are smaller than electric motors with the same power output. Thus, hydraulic actuators add less mass to the robot arm, which allows it to carry more weight for the same power, a distinct advantage in the largest robots. With hydraulic robots, only a single, low-cost motor is needed (as a pump) to provide power to all of the joints. Hydraulic systems store energy and, thus, can handle transient load conditions better.

A robot driven by hydraulic systems is shown in Figure 2-13. A pump draws oil from a reservoir (sump) and pumps it into a pressure reservoir, which provides some residual power for shutdown purposes if electric power is lost. Typical pressures used are about 120 psi (pounds per square inch), with built-in safety valves if the pressure becomes too high. Solenoid-operated valves are used to control the flow of oil into the cylinders that control various robot motions.

One problem with most hydraulic systems is their tendency to leak oil, thus making the workplace messy and dangerous. Another problem is noise. Since hydraulics are generally only used on large robots, they can be quite noisy. For both reasons, hydraulic robots are being supplanted by electric models.

Pneumatic

Pneumatic robots are based on the pressure available from compressed air. Pneumatics have two principle uses: to control the joints in a small, low-power robot, and to supply motive power to operate the end-effector. Air pressure can move joints, but the effective pressure varies with load, reservoir volume, and temperature. Air is compressible, so precise control of joints is not possible through direct control over the air pressure. Therefore, pneumatic robots that require greater accuracies must use additional components, such as a digital air valve for controlling position and an air brake for holding position.

Pneumatic systems also require clean air, often cleaner than is available in a factory. Abrasive dust, moisture, and oil drops in the air can cause excessive downtime to pneumatic-robots, so extra filtering of the air is a wise precaution. Pneumatic robots are limited to smaller loads than are hydraulic systems

Figure 2-13. Hydraulic robot. (*Courtesy of AKR Robotics*)

because there is less pressure available in a pneumatic system. Continuous positional control is much easier with electric and hydraulic types than with pneumatic robots.

Pneumatic robots do not offer many of the advantages found in the all-electric robot, nor can they lift heavy weights as the larger hydraulic robots can; thus they are not supplied as often. Their major application is handling small parts, particularly in clean rooms.

An example of a pneumatic robot is the M50 from International Robomation/Intelligence. This robot has five DOF and handles payloads of 50 lb, satisfactory for many applications. It has a repeatability of 0.04 in.

Electric

Electric motors offer the convenience of obtaining their power from a wall source, and do not need bulky hydraulic or pneumatic power sources. They provide more power than pneumatic sources do, less workplace contamination than hydraulic sources do, and are usually easier to control by computer. Therefore it is no surprise that the all-electric robot has just about replaced the other two types of robots except in specialized applications.

There are many types of electrical motors, and most types have been tried in some type of robot. The two most popular types are the dc (brush type) motor and the stepper motor, both of which are described in chapter 4.

Direct Drive

Most motors have been designed for high-speed, low-torque applications; when low-speed, high-torque is needed, a speed-reducing transmission system, such as gears, is employed. This system has proven quite acceptable for tools and for many robots, but it requires power and space and reduces accuracy. To get around these limitations, direct-drive systems are being explored. In direct drive, the motor has been specially designed to produce good torque at low-speed operation. An example is the Megatorque motor supplied by Motornetics, which can produce a torque of 250 ft-lb at 60 rpm.

Self-Contained

The electrically actuated robots used in most applications derive their power from the power mains of the manufacturer's facilities. Thus, devices such as stepper motors, which have a large power drain even when not moving, can easily be used. There is a difference with mobile robots, which must carry their power source with them. Mobile power is usually supplied by batteries, although gasoline-driven robot vehicles are also under development, especially for outdoor models.

Carrying a power source with the robot adds weight and reduces the load the robot can carry. There is a trade-off in any mobile robot between operating time and load-carrying ability. One way around this is to recharge the mobile robot's batteries at every opportunity (e.g., while the robot is waiting for its next assignment).

2.5 CLASSIFICATION BY INTELLIGENCE

Current robots have a wide spectrum of intelligence, from the simplest pick-and-place robots, which may try to insert a part that is not there, to a complex piano-playing robot that can recognize a song being sung and then play the piece in time with the singer. In general, seven levels of intelligence classifications can be defined, but only the first six listed are commercially available.

1. Robots under complete manual control (underwater search of shipwrecks)
2. Unprogrammable robots (hard automation systems)
3. Robots that learn through a teaching mode (typical paint-spraying robot)
4. Robots that sense errors and shut down (most existing industrial robots)
5. Robots that accept new directions while online (workcell-type applications)
6. Robots that sense a change in their environment and change their response (some of the latest model robots with vision)
7. Robots that learn from mistakes (still in the experimental stage)

Perhaps the most intelligent robot currently commercially available is a mobile robot from Toshiba. This model is used to automatically test bank teller machines in factories. The robot travels to each electronic teller to be tested, inserts a bank card in it, enters appropriate password and transaction information with its fingers, and checks the automated teller's performance.

In most applications, the operator serves as trainer, observer, supervisor, or controller. Rarely does he serve as partner, although there are applications where human and robot workers stand side by side on an assembly line, each doing what he can do best. Dr. Larry Leifer has suggested that the robot and the human work even closer on a particular task. In a seminar presented in 1983, Dr. Leifer compared the two extremes of telerobots, where the operator almost completely controls the robot, and standard robotics, where the operator does not interfere as the robot performs its job.[4] He believes there should be a middle ground—that is, the robot should be capable of accepting human help in areas in which the human excels, such as intelligence, vision, and certain other sensor abilities. Putting a human directly into the inner control loop of a robot is an intriguing idea, but one which requires the breaking of a lot of new ground in robot-human interfaces and robot intelligence.

2.6 CLASSIFICATION BY APPLICATION AREA

Robots have been used in many applications, too many to list. Even if such a list were prepared, new applications occurring almost daily would quickly put it out of date. Thus, we have chosen 11 areas of application for discussion. Figure 2-14 shows six categories of manufacturing applications and five specialized applications areas. Later chapters look at applications in more detail. Application-oriented issues are covered in chapter 13, requirements peculiar to a number of applications are presented in chapter 14, and future applications of robots are covered in chapter 17.

Robot use is continually expanding. When new industries learn of the benefits of robots, sales in that area increase. As new technologies are perfected (such as vision systems), areas previously not open to robotics suddenly become users. On the other hand, some older industries have numerous robots installed already, and thus new sales are dropping off. For example, the automotive industry in Japan has reduced its percentage of robot purchases from 30% of total sales in 1981 to about 15% in 1985, and a similar drop has started in the United States from a high level of 51% of purchases to an expected level of 20% within a few years.

Parts Handling—Palletizing, Packaging, Bin Picking

One of the earliest uses for robots was in parts handling (loading and unloading die casting machines). In parts handling, the robot simply moves an item from point A to point B. The robot can be used for unloading incoming supplies or loading finished goods (i.e., palletizing), they provide support at the workplace but do not actually work on the parts. Parts-handling robots are usually pick-and-place types, so they need less accuracy and less intelligence than most other applications. Even in this area, however, vision systems have been integrated into some of the more complex bin-picking tasks.

Part Modification—Drilling, Cutting, Welding

Part modification initially came from successful numerical control applications. Each robot of this type is characterized by the use of special tools (no simple grippers here), very high accuracy requirements, and, in many cases, the addition of specialized sensors to help the robot locate the position to be worked on. This is particularly true for seam following or through the arc welding applications.

The cutting area is becoming more sophisticated. Water jet cutting and laser cutting methods are now being used. Even in such a "simple" application as drilling, there may be multiple tools on the same manipulator. Vision systems may be employed to find a fiducial mark on the material and to line up the robot with the part.

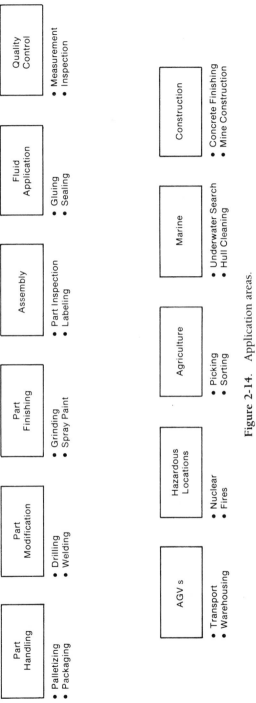

Figure 2-14. Application areas.

Part Handling
- Palletizing
- Packaging

Part Modification
- Drilling
- Welding

Part Finishing
- Grinding
- Spray Paint

Assembly
- Part Inspection
- Labeling

Fluid Application
- Gluing
- Sealing

Quality Control
- Measurement
- Inspection

AGV s
- Transport
- Warehousing

Hazardous Locations
- Nuclear
- Fires

Agriculture
- Picking
- Sorting

Marine
- Underwater Search
- Hull Cleaning

Construction
- Concrete Finishing
- Mine Construction

47

Parts Finishing—Grinding, Deburring

Parts finishing was also adapted from the numerical control machine tool era. However, there has been not as much use of advanced tools or advanced vision systems. Many grinding and deburring applications fall into a type of hard automation, most of which are characterized by limited DOF robots and the requirement for expensive fixturing to position the part.

Fluid Application—Gluing, Caulking, Sealing

For fluid applications, especially in the automotive industry, the robot arm must have extremely flexible movement. No simple four or five DOF arm can handle the maneuvering and convolutions that typical sealing robots go through to apply sealants in out-of-the-way places. Such a robot often has eight or nine DOF as well as small links near its wrist to allow the wrist to fit into tight places.

Spray Painting

Spray painting could possibly be included in either part finishing or fluid application areas, but it does have some characteristics of its own. Spray painting is an excellent task to automate, because the paint spray is hazardous and the job is boring yet requires reasonable repeatability to ensure a quality finish.

A unique characteristic of a spray-painting robot is its need for manual teach-control programming. Contrary to most other robot applications, spray painting works best if the robot is *shown* what to do (via teach mode) rather than being *told* how to do it (via offline programming). When robots gain more intelligence and speech recognition ability, there will likely be an instructor one day who will explain to the robot that "it's all in the wrist."

An example of a modern spray-painting robot is the TRACS robot from DeVilbiss, shown in Figure 2-15. With nine different axes of motion, the arm is very flexible over the areas it can reach. Mounting this robot arm on a track enables it to integrate easily with moving production lines.

Assembly—Parts Insertion, Parts Fastening, Labeling

Assembly has been a natural for robots, and devices that simply pick up parts and insert them into a printed circuit board are among the least expensive robots and are at the bottom of the robot scale. These robots may have some

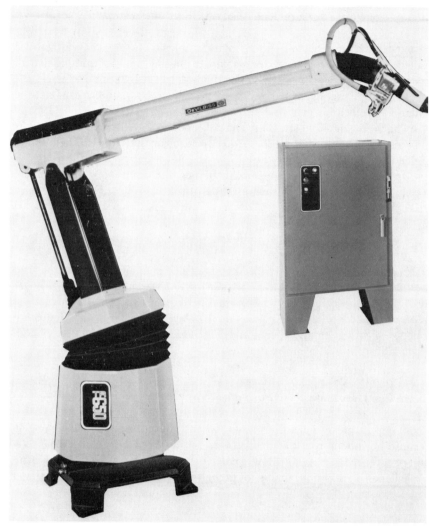

Figure 2-15. Spray-painting robot. (*Courtesy of The DeVilbiss Company*)

reasonable accuracies to meet, but have little need for high accuracy, intelligence, or such peripheral units as robot vision systems. In addition to parts insertion, other simple tasks, such as labeling and part fastening, fall into this category. The only area in which extra care may be needed is in assembling printed circuit (PC) wafers in a clean room. Although standard assembly-type robots can be used here, special care must be given to the external design of the robot to avoid contaminating the air.

Quality Control—Measurement, Visual Inspection, Testing

A robot used in quality control may have a simple task, such as checking to see if a label is attached to a bottle, or a complex one, such as determining that all solder joints have been properly soldered on a PC board. In this latter case, the robot must find missing solder joints, poor-quality joints, and places where the solder overflowed.

A characteristic of most quality control robots is intelligence. These robots show intelligence in their sensors (most have vision systems), in their ability to make decisions (this unit passed, this did not), and even in the ability to spot trends (the hole diameters, while still within allowable tolerances, are getting larger, so it must be time for a new drill bit).

AGVs—Material Handling, Warehousing, Mobile Systems

AGV applications are characterized by moving large amounts of materials (up to 4000 lb) over long distances in factories and warehouses. Their primary function can be considered just-in-time material transport. Although most AGVs are not currently classified as robots, they work closely with industrial robots, are gaining in intelligence and capabilities, and should join the rank of other industrial robots in the near future.

Hazardous Locations—Nuclear, Police, Fire

Robots operating in hazardous environments are currently all teleoperated, because no mobile robot comes close to the required intelligence level. Remote-controlled robots are now available for police use and nuclear plant surveillance. In addition, robots for firefighting and repairs in space are under development.

Agriculture—Sorting, Picking, Planting

A few research agriculture robots operate in the open field, picking fruit, harvesting grain, and planting seedlings, and these robots are discussed in chapter 17. The only robot that has been used extensively outside of factories is the sheep-shearing robot developed in Australia. To date, most practical agriculture application robots are used in vision inspection tasks where size, color blemishes, and frost injury must be looked for. In addition, there are many food product robots, including chocolate inspection robots, shiri cake-making robots, and even fish-sorting robots.

Marine Uses—Underwater Search, Boat Scraping

Robots placed in a salty atmosphere or, even worse, directly in seawater require extra corrosion protection and sealing. Just a small amount of sea moisture in the electronic circuitry can cause the need for extensive repair. Yet there are many robots in use in these difficult environments. Robots pilot ships, assist (through teleoperation) in underwater search operations, and, in a most unusual application and difficult design, scrape barnacles off boat hulls. This last application requires a mobile robot that is always on an angle on the hull, that attaches itself to the metal hull via vacuum gripper feet, that is able to determine which areas of the boat hull have not been examined, and that must decide where next to remove barnacles.

Construction

Japan has taken the lead in applying robots to construction. These robots are all mobile, work outdoors or in partially finished areas, and are very specialized. Robots have been used in only a few applications in the construction field, but it is one of the key areas expected to expand greatly over the next few years (see chap. 17). Concrete finishing robots are a current example. These robots are placed on wet cement to trowl and finish the poured concrete floor. Specially designed wheels ensure that they evenly spread their weight and do not sink into the concrete.

REFERENCES

1. Condon, E. U. and H. Odishaw, eds. *Handbook of Physics*. New York: McGraw-Hill, 1958, p. 2-15.
2. Coiffet, Phillipe, and M. Chirouze. *An Introduction to Robot Technology,* translation by Meg Tombs. New York: McGraw-Hill, 1983, p. 35.
3. Hohn, R. E. Tomorrow's technology today: Cincinnati Milacron at Robots 10. *Proc. Robots 10 Conference.* Dearborn, Mich.: Robotics International of SME, April 1986, pp. 10–26.
4. "Robot Sensing and Intelligence," IEEE short course presented by satellite transmission in November, 1983. Prepared by the Institute of Electrical and Electronic Engineers, Parsippany, NJ.

PART

II

ROBOTIC TECHNOLOGY

Part II describes the technologies associated with current robotics electronic design. It starts with an introduction to robot mechanics and continues into robot electronic design, sensor design, vision systems, and ultrasonic systems. It concludes with a discussion of the unique design problems of mobile robots. Computer-related topics are covered in Part III. Elementary or intermediate engineering areas (e.g., logic or circuit design) are not discussed. Rather, areas of technology vital to robotics system design that are not normally covered in textbooks are emphasized. Many related fields, such as television, are also not covered, even though television systems are often used with teleoperated devices, and many robot vision systems have as an intermediary product a standard analog television signal.

INTRODUCTION TO ROBOT MECHANICS

Robot mechanics covers the mechanical engineering aspects of robot design and includes factors such as inertia, stress, load-carrying ability, and dynamic response.* Mechanical engineering is also involved in the design of grippers, transmission systems, hydraulic and pneumatic couplings, and related design areas. This chapter presents four major, and a few minor, topics of interest to electronics engineers or systems designers.

3.1 ROBOT ARM KINEMATICS

In industrial robotics, robot arms and end-effectors hold and manipulate tools or objects. The mechanical arm that moves the object through three-dimensional space is called the *manipulator.* Connecting the manipulator to the object is a *wrist,* to which tools or grippers, called *end-effectors,* are attached. Control of the motion of the arm and orientation of the end-effector is part of robot *kinematics,* the branch of mechanics dealing with motion.

The arm moves by electrical or mechanical actuators that control the position of the joints. This motion is complicated by many factors, both kinematic and dynamic; hence there must be coordinate conversion in order to predict where the robot's arm will move. Since many (and in some robots all) of the joints are rotary, coordinate conversion is more complicated than with the more familiar Cartesian coordinate transformations. An additional complication is the need to solve an inverse kinematic problem for most practical applications. (This inverse problem is discussed in the next section, after we cover the more familiar direct conversions.)

Robot motion has two steps: a move to the desired end point, and the appropriate orientation of the end-effector. In practical applications these two motions are often performed at the same time, but for clarity we will cover the basic move first. End-effector orientation is controlled via wrist motion, and is discussed later.

---

* Robot mechanics is quite complex. For a rigorous treatment the reader should see a complete book on the subject.[1,2,3]

To perform useful work, the robot must move its arm and, through arm motion, its end-effector. Therefore, control over appropriate joint motion is necessary. As discussed in chapter 2, most robots use some combination of four basic types of joints. A sliding joint provides linear motion. Robots whose major joints are all sliding joints are called *Cartesian robots,* since all motion is in a Cartesian coordinate framework (X, Y, and Z). Cartesian robots have two major advantages:

1. Motion control is easier, especially when solving inverse kinematic motion.
2. The same amount of weight can be handled at any position of the arm, whereas arm position must be considered in revolute robots when calculating load.

Other types of joints provide angular movement. Robots whose major joints all provide rotation are called *revolute robots.* They have three major advantages:

1. Greater minimum and maximum reach
2. Easier arm movement around obstacles
3. No sliding joints to maintain

In addition, other robots have both sliding and rotational joints, such as a selective compliance assembly robot arm (SCARA) robot, thereby combining advantages from both basic types.

The actual *work envelope*—the area over which the robot can move—of any robot depends on the type of joints used and the distances over which they can travel. Conversely, the amount of work area needed depends on the type of task being performed and the arrangement of the workplace. Figure 3-1 illustrates typical work envelopes for two types of robots. Note that the revolute robot generally provides the largest work envelope.

Robot arms come in all sizes: the smallest can manipulate miniaturized circuit components, and the largest can lift several hundred pounds. Arms are also highly specialized, and the type of arm selected for the robot depends strongly on the robot's particular application. Direct-drive arms and compliant arms are also beginning to be used.

Geometric or Positional Control

To perform useful work, the robot must move its arm and, thereby, its end-effector. The type of joints used in a robot affect the kinematic calculations. Coordinate conversion is a necessary part of determining the final position of the end-effector after movement of some or all of the robot joints. In an all-

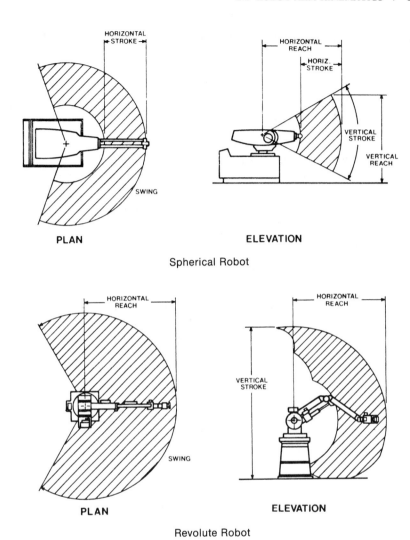

Spherical Robot

Revolute Robot

Figure 3-1. Typical work envelopes. (*From R. D. Potter, Robotics, in* The Handbook of Computers and Computing, *A. H. Seidman and I. Flores, eds. New York: Van Nostrand Reinhold, 1984, p. 317; copyright © 1984 by Van Nostrand Reinhold*)

Cartesian robot, calculating the changes in end position with changes in each joint position is straightforward, since there is no interaction between the joints. The more rotational joints there are in a robot, the more involved the coordinate conversion will be.

Most robots use rotational joints for some, or all, of their motion, which in turn changes the link angles. Each angle change produces corresponding changes in two Cartesian coordinates, such as X and Y or X and Z. The resultant motion of the tool center point is a complicated function of the length of each limb, the type of joints provided, and the motion of the joint. For appropriate robot motion control, we must be able to predict motion as a function of these variables.

If the value of each joint angle and the length of each link are known, we can calculate the position and orientation of the end-effector. Two examples will illustrate the principles involved. These examples ignore the effects of bending, backlash, overshoot, and inertia.

In simple cases, such as forward calculation of a two- or three-degree-of-freedom (DOF) arm, the calculations are easy. The following discussion illustrates the steps involved in a single-plane, two DOF case, illustrated in Figure 3-2. The location of X_1 and Y_1 can be described in terms of X_0 and Y_0, link length, and the appropriate angles:

$$X_1 = X_0 + L_1 \cos \theta_1 \tag{3-1}$$

$$Y_1 = Y_0 + L_1 \sin \theta_1 \tag{3-2}$$

Similarly, the location of X_2 and Y_2 can be described in terms of X_1 and Y_1:

$$X_2 = X_1 + L_2 \cos \theta_2 \tag{3-3}$$

$$Y_2 = Y_1 + L_2 \sin \theta_2 \tag{3-4}$$

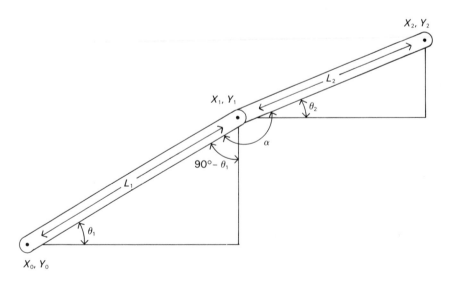

Figure 3-2. Forward coordinate conversion.

Substituting equations (3-3) and (3-4) in (3-1) and (3-2) gives

$$X_2 = X_0 + L_1 \cos \theta_1 + L_2 \cos \theta_2 \qquad (3\text{-}5)$$

$$Y_2 = Y_0 + L_1 \sin \theta_1 + L_2 \sin \theta_2 \qquad (3\text{-}6)$$

Points X_0 and Y_0 are fixed in space and thus are constants. If we define (X_0, Y_0) as the origin of our coordinate system, X_0 and Y_0 are equal to zero and can be dropped from the equations.

The second joint movement is actually over angle α, so a further conversion must be made from θ_2 to α. In the right triangle that includes L_1, the three angles are $90°$, θ_1, and $90° - \theta_1$. Hence,

$$\alpha = (90° - \theta_1) + 90° + \theta_2 \qquad (3\text{-}7)$$

or

$$\theta_2 = \alpha + \theta_1 - 180° \qquad (3\text{-}8)$$

Thus,

$$X_2 = L_1 \cos \theta_1 + L_2 \cos(\alpha + \theta_1 - 180°) \qquad (3\text{-}9)$$

$$Y_2 = L_1 \sin \theta_1 + L_2 \sin(\alpha + \theta_1 - 180°) \qquad (3\text{-}10)$$

Using the formulas for the sine and cosine of the difference between two angles, we get

$$X_2 = L_1 \cos \theta_1 - L_2 \cos(\alpha + \theta_1) \qquad (3\text{-}11)$$

$$Y_2 = L_1 \sin \theta_1 - L_2 \sin(\alpha + \theta_1) \qquad (3\text{-}12)$$

Three-dimensional space and multiple DOF arms complicate matters. For example, for a three DOF arm moving in three dimensions (see Fig. 3-3),

$$X_3 = (L_2 \cos \theta_2 + L_3 \cos \theta_3)(\cos \theta_1) \qquad (3\text{-}13)$$

$$Y_3 = (L_2 \cos \theta_2 + L_3 \cos \theta_3)(\sin \theta_1) \qquad (3\text{-}14)$$

$$Z_3 = L_1 + L_2 \sin \theta_2 + L_3 \sin \theta_3 \qquad (3\text{-}15)$$

This method of coordinate conversion may be continued for any number of DOF, but a simpler method uses homogeneous transformations. First developed by Denavit and Hartenberg[4] in 1955 to provide a mathematical approach to the transformation of points in space from one coordinate system to another, homogeneous transformations use four by four matrices to represent robot joint coordinate conversion. Conversion is required to determine the different coordinate point locations that the various joints go through as the robot arm moves. A three by three matrix is used to cover the general rotation coordinate conversion, and a fourth row and column handles linear transla-

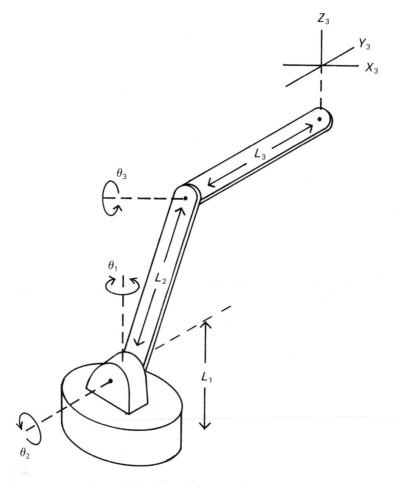

Figure 3-3. Three-dimensional conversion.

tions. The coordinates are assumed to start at $(0, 0, 0)$ at the robot base, and go through translation along any link and rotation at any joint. Since this matrix has nonzero determinants, the resulting matrices are invertable, which is desirable in handling the reverse kinematic problem.

One of the great advantages of homogeneous transformations is that a single composite matrix (the product of the individual link matrices) can be used for multiple translations and rotations. Once this matrix has been developed for a particular robot, it directly provides the desired end point. Because matrix multiplications are noncommutative, the order of operations must be observed.

Readers wanting a more detailed discussion of this important concept are referred to the original paper by Denavit and Hartenberg, or to textbooks such as the one by Critchlow,[5] for application discussions.

To make a revolute joint robot easier to control, most manufacturers provide articulated joint control. In other words, moving any joint results in compensatory motion of another joint so that, as far as possible, orientation of the end-effector does not change.

Inverse Kinematics

Forward kinematics concerns where in space the end-effector would be positioned if known angles are provided to the robot actuators. Most of the time, however, we know where the end-effector should be positioned and oriented, and we must determine the required angular changes. It is much more difficult to take end position information and calculate the resultant joint angles needed to provide this positioning. This calculation is called the *inverse kinematic problem,* and the equations, when they can be solved, are called *resolvable.*

The inverse kinematic problem must be in closed form if it is to be solved. In teach mode, where the arm is simply following joint commands, closed-form solutions are not necessary; however, if they are not available, the robot motion cannot be modified by subsequent off-line programming. To allow modifications when desired, even teach-mode robots should have closed-form solutions.

Robot Resolvability

Transformation matrices may be used to analyze multiple DOF mechanisms. Similarly, rotation matrices are often used to solve the inverse kinematics problem. Both techniques are covered in any mechanical engineering robotics text. It is important to be aware of how long it takes to solve these equations at the accuracies required. If the robot merely repeats the same motion indefinitely, these equations need be solved only once; the values can then be placed in a lookup table for future use. However, if the robot must be able to respond in real time to new factors in its environment, equation complexity and accuracy become extremely important, since they both slow up robot response.

Most robot control systems need interpolaters to control the motion of the robot between end points (or even via points) and to determine the accuracy of this movement. Most robots can perform both straight-line and smooth-curve interpolations.

One complication with robots having multiple degrees of mobility is that multiple solutions are available—many sets of joint angles could put the end-effector in the desired location (just as we may use different wrist positions to pick up an object). Not all positions are equally desirable, nor do they all offer the right flexibility, load-carrying ability, speed, accuracy, or other factors. Computation time therefore can increase sharply with an increase in the number of joints.

Wrists

An orientable robot is one whose end-effector can be put at any arbitrary angle. A Cartesian coordinate robot wrist cannot orient the end-effector, so wrists need angular joints to control tool orientation.

As discussed in section 2.1, it takes three DOF to reach any point in space and at least two DOF to permit orientation of symmetrical end-effectors. Thus, only robots with at least five DOF are orientable. Full three-axis orientation can only be accomplished if at least one axis has a revolute joint. In a typical application, the first three DOF are used to reach the desired location. Since the robot arm is heavier, slower, and more difficult to control precisely, it is more desirable to handle small-task motion and orientation by the wrist DOF.

To specify the position of the end-effector, we define a reference point, called the *tool center point* (TCP), which refers to the point at which the gripper grasps a tool or object. Since both grippers and tools may be changed for different applications, this point will also change, so its new location must be entered into the robot before it can be used in move commands. Unless otherwise specified, a move is assumed to refer to the TCP, not to the base of the robot. If possible, the axes of the three DOF providing TCP orientation should have a common intersection point (i.e., they should be concurrent). The end-effector can then be oriented without regard to coordinate translation.

Wrists typically fall into three categories: BBR (bend, bend, roll, shown in Fig. 3-4); RBR (roll, bend, roll), used on the PUMA robot; and RRR (roll, roll, roll), used on the T3 robot. The system designer must choose among different types of robots, with different work envelopes, different DOF, different types of wrist structures, and different types of grippers. It is no surprise, then, that a program written for one type robot does not work for a different type, even if they understand the same robot language.

Other problems also complicate design. Feedback sensors, which report distance traveled, rate of travel, and other factors, often cannot be collocated on the joint axis, thus making their mounting and the required data conversion more difficult. The weight of the load and joint backlash or slippage also contributes to end-effector location uncertainty. Furthermore, greater flexibil-

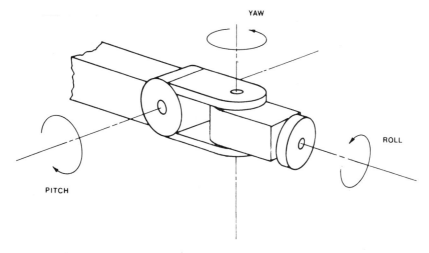

Figure 3-4. Example of BBR wrist. (*From R. D. Potter, Robotics, in* The Handbook of Computers and Computing, *A. H. Seidman and I. Flores, eds. New York: Van Nostrand Reinhold, 1984, p. 318; copyright © 1984 by Van Nostrand Reinhold*)

ity in the wrist and attached gripper requires additional actuators and cables to control it, which increases the weight of the wrist, reduces available payload, and increases arm movement time.

Actuators

Actuators are motors or other mechanisms that convert supplied power into mechanical motion. Electric devices are used on most late model robots, whereas electrohydraulic actuators were used in most older robots; they are also used in new applications needing large lifting capacities. But robots can also be pneumatically operated, and many of the early small robots were. Since air is compressible, precision control of position in a pneumatic actuator is quite difficult. Therefore pneumatic actuators are often limited to pick-and-place tasks, where on/off control is possible. Pneumatics are also used in gripper control, thereby eliminating the added motor weight.

Electric actuators (motors) are fast, accurate, easy to use, and can be driven with advanced control systems. However, they can arc or give off sparks—definitely a problem in hazardous environments—but sealed motors significantly reduce this hazard. Electric actuators are not as efficient as hydraulic actuators, particularly in larger sizes. There is also a practical upper limit to the power available.

Hydraulic-based robots have the largest lift capability and can hold their final position without expending much additional power. (Figure 3-5 illus-

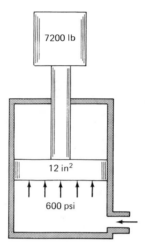

Figure 3-5. Hydraulic system principles. (*From W. G. Holzbock,* Robotic Technology: Principles and Practice. *New York: Van Nostrand Reinhold, 1986, p. 115; copyright © 1986 by Van Nostrand Reinhold*)

trates a hydraulic system where fluid with a pressure of 600 psi is holding a weight of 7,200 lb.) Hydraulic robots are also reasonably fast and can be controlled precisely, but they do leak fluid, cost more, and need more maintenance.

Pneumatic systems, on the other hand, are the least expensive, are nonpolluting (thus good for laboratory environments), and provide high speeds; however, they cannot be precisely controlled, leak air, are noisy, and need a lot of maintenance.

3.2 END-EFFECTORS

End-effectors are the grippers, tools, or other devices attached to the robot's wrist that allow it to perform various tasks. In many systems, end-effectors are simple grippers that close about an object, allowing it to be picked up and handled. However, end-effectors can be quite advanced and have multijointed fingers or tactile sensors built in. Still others serve as specialized tools.

Grippers can be subdivided into mechanical, magnetic, vacuum, adhesive, and inflatable, depending on how they grasp an object. Although grippers may be supplied with the robot, they are usually custom designed for the task (thus increasing robot system cost) or purchased from a firm specializing in grippers.

Mechanical Grippers

Most grippers are mechanical. Many of the less complex ones have two movable gripping surfaces, and some, such as the model in Figure 3-6, are specialized for a particular application. The model shown has three gripping surfaces and is designed to pick up round objects. A more conventional two-point gripper was shown in Figure 2-8.

The ideal gripper, according to Holzbock,[6] would be capable of six different types of grasp:

1. Spherical—to pick up balls
2. Cylindrical—to hold tubes or pipes
3. Tip—to grasp very small objects
4. Hook—to pick up objects by their handle
5. Planer—to lift a flat object from the bottom
6. Lateral—to pick up flat objects from the side

To this list Palm adds a palm-type grasp, such as the type of grasp used while writing with a pen.[7] However, few practical grippers offer more than one type of grasp.

Grippers are often given shaped sides to fit specific parts. Even so, there is always the possibility that the part could slip out of the robot's grip, especially during fast motion. The amount of force that the gripper must exert on the object to make sure it does not slip depends on several factors. Groover[8] developed the following formula, which neglects object acceleration:

$$F_g = \frac{W}{\mu N_f} \tag{3-16}$$

where W is the weight of the object, μ is the coefficient of friction, and N_f is the number of gripping surfaces (usually two). Engelberger suggests that the value obtained be multiplied by from 1 to 3 to handle the acceleration (depending on the direction of acceleration compared with the direction of gravity). In addition, a safety factor of perhaps 50% should be added to cover conditions when the weight has a rotational component (due to a center of mass not at the point of grippage), and to handle contingencies.

The linkage between the gripper actuator and the gripper fingers must also be taken into account. The force we have been discussing is at the gripper fingers. To compute the required gripper actuator force, we must consider the effect of gears and linkages in the gripper. Thus, to provide 5 lbf with a 10/1 gear reduction, only 0.5 lbf would be required at the actuator.

Figure 3-6. Three-finger gripper. (*Courtesy of Mack Corp.*)

General-purpose grippers handle parts up to 10 lb. If a gripper uses more force than necessary, the object might be damaged. Also, grippers that can provide excess force are heavier and more expensive.

Hands and grippers may have sensors attached to detect the force being applied to an object, the position of the object within the gripper, or even the temperature of the object. These sensors are discussed in chapter 5.

Other Types of Grippers

Vacuum grippers require a vacuum pump or a venturi connection to the plant air supply. They are ideally suited for gripping certain materials, such as plates of glass. The lifting ability of a vacuum system is

$$F = PA \tag{3-17}$$

where F is the lifting force in pounds, P is the vacuum negative pressure in pounds per square inch, and A is the effective vacuum area in square inches. Any loss in vacuum causes a loss of lift capability and, hence, potential damage to the material.

Large electromagnets have also been used as grippers. They are used when large ferrite materials, which are isolated from other magnetic materials, must be moved. A wide range of holding forces is available, with values of 450, 600, and 1100 lb typical. Magnetic holding force requirements must be considered carefully, because the values given by the electromagnet manufacturers assume a thick piece of flat, clean, smooth steel as the material to be moved. In many cases, the material is dirty or rusty, or not flat, or thin. In these cases, the holding power can be much less, and what was thought to be a good safety margin may disappear. Power must be continuously available to the electromagnetic gripper during operation. A power failure would cause the heavy object to be dropped with potentially dangerous results.

General Gripper Considerations

The end-effector must be interfaced to the robot, not only by a secure physical connection but by power to operate the gripper, signals to control it, and sensor signals from the gripper, if it is equipped with them. Physical connection is usually through a faceplate or flange on the robot's wrist, to which the gripper is bolted, or by a wrist socket. No firm standards in flange size or mounting provisions exist, but progress is being made (see chap. 12).

Physical support is very important, especially when force is applied to the end-effector as, for example, on a deburring tool. However, many things can go wrong, and to prevent damage to the robot, its end-effector, the part, or its environment, some type of overload protection is needed. This protection can be as simple as a shear pin that breaks under strain or as complex as a spring-loaded mechanism that can be set by the robot.

Groover[9] gives a very complete checklist with 40 factors to be considered in the selection and design of grippers. Some of the more significant items include the shape, size, and weight of the object to be handled, the accuracy and repeatability of the robot, the flexibility in handling required, and the gripper's operating environment.

Hands

There is no perfect end-effector, but some of the robot hands being developed have a variety of capabilities. Although any end-effector that can grasp and manipulate an object may be called a *hand,* we will use the term only for grippers that have jointed fingers.

Precise position control of the part often requires a dexterous hand. It is usually faster to use the hand rather than the heavier and bulkier arm for final positioning. Salisbury[10] has defined the eight positions by which fingers can grasp an object. Most hands capable of manipulating objects have two to four articulated fingers. Adding fingers and joints increases the hand's capabilities, but it also requires more involved control algorithms. Unfortunately, many control problems need to be solved before some of the hands being developed can be used in practical applications. These problems are primarily due to the total number of sensor signals and joint control actions that must be monitored and controlled in real time. Most implementations use a separate multiprocessor for each joint. The only successful implementation to date was reported by Asada and Ro.[11]

Examples of advanced hands include the Utah four-finger hand, being developed at the University of Utah, and the MIT three-finger hand, being developed at the Massachusetts Institute of Technology. The Utah hand has a thumb and three opposing fingers, with four joints on each finger, providing three bend DOF and one side-to-side DOF on each finger. The hand is pneumatically controlled through eight "tendons" (two per joint) and uses a Motorola 68000 for real time control. Available sensor inputs include joint angles, tendon forces, and tactile array data.

The MIT hand has only three opposing fingers, with some grasping limitations due to a lack of a thumb or fourth finger. The stiffness of each finger can be adjusted, and a sense of touch is added to detect object slippage. These two provisions allow the hand to pick up objects more readily than conventional grippers can.

A different type of approach has been taken at the University of Rhode Island, which as developed a hand using translational rather than rotational joints.[12]

Operating Power

The end-effector needs operating power. With electric units, such provision is straightforward, unless backup power, in the event of power failure, is also required. Pneumatic power requires added consideration.

Pneumatic pressure can be used to supply power to an end-effector (it is the most logical way for a spray-painting gun, for example). Either two air lines, one to open the gripper and one to close it, or a single air line, with a valve controlling the function to be performed, can be used. The pneumatic force exerted is a function of the air pressure and the surface area it is applied against. For a circular piston,

$$F = \frac{P\pi D^2}{4} = PA \qquad (3\text{-}18)$$

where F is the force applied in pounds, P is the air pressure in pounds per square inch, D is the diameter of the circular piston in inches, and A is the area in square inches.

Pneumatics can also be used to provide a vacuum-type gripper by using a venturi device connected to an air supply.

Figure 3-7. Pneumatic tool changer. (*Courtesy of Lord Corporation, Industrial Automation Division*)

Tools

For the robot to do different types of tasks, the end-effectors must be easily changed. Some tool changers even allow the robot to change its own tool. Figure 3-7 shows one type of pneumatic tool changer supplied by Lord. Two of the most common types of tools are welding electrodes and spray-painting guns. Drills, routers, and polishers are also frequently used.

Multiple grippers, most of which are double grippers, allow the robot to independently control two items, such as raw material and a finished part.

3.3 DYNAMIC CONSIDERATIONS

The biggest problem in robot system design is probably ensuring that the robot's performance remains steady during actual motion (dynamic factors). Arm bend due to loading, inertia, friction, parallax, and related problems can ruin system dynamic performance accuracy or, even worse, pose a hazard to operators or other equipment.

Another consideration is the ratio of robot weight to load weight. With robots, it might be 5/1 or 2/1. (For humans, it is closer to 1/1.) One reason for this difference is lack of robot balance. Another is the ratio of load-carrying ability to the strength of the material used in the arm. If the arm bends slightly under load, accuracy is reduced. Therefore, arms may be overweight to reduce potential bending. A counterbalance can improve weight-handling ability in the vertical direction, but it will also increase system inertia, slow robot movement, and reduce accuracy.

Arm-Stiffening Algorithms

Fortunately, many practical robot applications that require precise positioning of end-effectors (such as welding) do not require contact forces with the object. Since any appreciable amount of contact force will slightly deflect or bend the robot arm, it will thus hurt positional accuracy. The design of many control algorithms does not consider such factors as bending, deformation, elasticity, or backlash.

Kuntze and Jacubasch[13] have developed advanced control algorithms that address this problem. Their algorithm, although developed for a specific robot (a Kuka model 250 with five revolute joints) and for a specific task (the precision cutting of metal surfaces), can be used in other situations. The selected approach is the injection of a high-frequency bias signal based upon feedforward techniques. The major disturbances that result from motion are caused by inertia and gravity and can be predicted with this approach.

Compliance

Unless very precise sensing and position control are both available, it is difficult for a robot to insert a precision part into a hole. The primary difficulties are due to errors in locating the hole, locating the end-effector, and orienting the part. Of course, the part may be outside of size tolerance limits and thus can never be made to fit.

Even the most precise drilling method gives a slight error in hole size and location. This tolerance could be increased when the part is put into position, due to errors in fixture positioning. In addition, positioning of the part by the robot arm and wrist is itself prone to servo position errors and changes caused by temperature, wear and tear on drive mechanisms, and the part's position within the gripper. All of these factors contribute to lateral and orientation errors, resulting in positional and angular displacements between the part and the hole.

Angular tolerances can be quite tight, depending on the diameter of the hole, the available clearance, and the coefficient of friction between the materials. From Figure 3-8, maximum angular error (in radians) is

$$\alpha_{max} = \frac{\Delta D}{D\mu} = \frac{|d - D|}{D\mu} \tag{3-19}$$

where ΔD is the difference between part diameter and hole diameter (clearance), D is the part diameter, and μ is the coefficient of friction. Take, for example, a 0.75-in-diameter shaft with a 0.001-in clearance and a coefficient of friction of 0.2. The angular tolerance then is only 0.43°. Note that μ varies with the material and its finish.

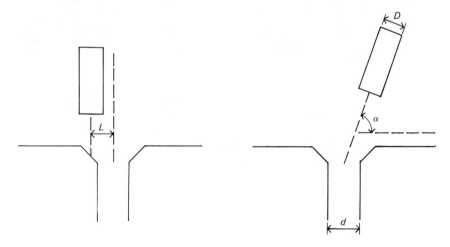

Figure 3-8. Insertion clearance parameters.

A human performing this task would sense the direction of the back pressure as well as using visual feedback to align the object. Several approaches have been adopted to get around these problems in robots. First, a slight chamfer (rounding of the edge) is put on either the part or the hole (preferably both), making it easier for the robot to find and engage the hole. Second, the wrist is given lateral compliance, that is, the ability to move slightly sideways as a result of sideways pressure. Finally, in more critical applications, either visual or force feedback will be used to slightly reposition the robot wrist. Since the part will be held near or at one end while the other is inserted, slight angular inaccuracies can also be handled this way, since the part will tend to rotate due to sideways pressure.

Two developments, both in 1979, led the way in providing robots with the necessary compliance. The Charles Stark Draper Laboratory released a paper describing their passive remote center compliance (RCC) wrist,[14] and Yamanashi University in Japan developed the principle of the SCARA robot, which added lateral compliance to arm motion.

Other systems have been built that added active compliance and touch sensors to help solve this type of problem. Compliance can also be built into the wrist joint itself. Figure 3-9 illustrates available compliant wrists.

Figure 3-9. Compliant wrist flanges. (*Courtesy of Lord Corporation, Industrial Automation Division*)

The effects of inertia and acceleration on path motion are shown in Figure 3-10, which is typical of path differences that may be found in any robot. The robot is to travel from point A to point C via point B. How close it gets to point B before turning toward point C is a function of the type of termination employed in the move. Is point B to be approached by using fine motor control? If so, the robot will reach point B with a high degree of accuracy. If course control is used, the robot will still approach point B, but with poorer accuracy.

Other path termination modes include "no settle," which removes the final settling in on the location, "no deceleration," which starts toward point B as soon as the arm begins to decelerate, and "no wait–no deceleration," which starts the turn even earlier by removing some internal delays in calculating the new move.

3.4 OBSTACLE AVOIDANCE

Three classes of obstacle avoidance are applicable to robotics. The first involves collisions between the standard industrial robot (a single robot arm on a stationary base) and obstacles, including, for example, the fixture holding

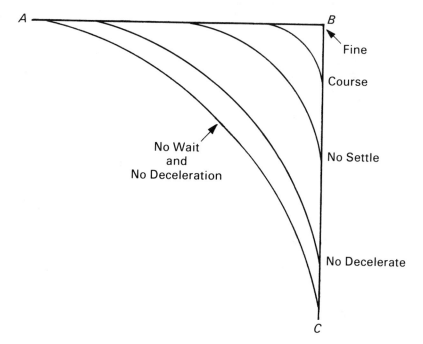

Figure 3-10. Path dynamics. (*After Karel System Reference Manual. Troy, Mich.: GMF Robotics Corp., 1986, p. 4–37; used with permission*)

the parts, a conveyor belt, various machines, the part itself (especially if it is large and complex), and even portions of the robot, which can impede the end-effector during complex motion. The second class is the potential collision of two robot arms in a work cell working on the same part. Some type of work synchronism or planned movement is always required between these two otherwise independent arms.

The third class refers to a robot base moving through its environment. The robot must not collide with material or people in its path. When movement is confined to a small area (such as for a gantry robot), this type of collision avoidance approaches the first class.

Limit switches and stops or programming the robot not to enter certain areas may be sufficient to avoid obstacles whose location is fixed. But if the obstacle moves, more involved techniques are required, and they must affect the entire robot, with attached tools, cables, or hoses, not merely the motion of the end-effector. For example, examine Figure 3-11. Although no obstacle is blocking movement of the end-effector from point A to point B, the obstacle in the way of the second link would cause a collision along the direct path. Therefore the end-effector must first be moved to point C.

This illustration demonstrates a two-dimensional movement, but in the real world, three-dimensional movement must be considered. Unfortunately, there is no efficient, general-purpose collision-avoidance algorithm for robots with rotary joints.

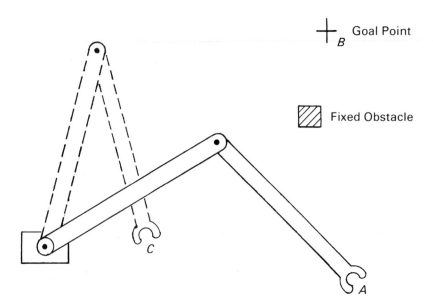

Figure 3-11. Path planning problem.

Five methods have been developed to avoid collision: robot training, "hypothesize and test," penalty functions, explicit free space, and trial and error. Training the robot means carefully controlling robot position to make sure it does not infringe on any possible location that may cause a collision.

Hypothesize and test, penalty functions, and explicit free space are all based on algorithms suggested by Lozano-Perez.[15] These approaches are carried out in real time to ensure that the robot's path is clear. In hypothesize and test, the computer first estimates a collision-free path for the arm and then tests the path for possible collisions. This approach requires geometric modeling of the environment. Penalty functions assigns a penalty when the robot approaches an object, with the penalty reaching infinity if contact is made. The computer then must determine a path with the lowest penalty. Explicit free space involves dividing the entire area into small cubes of various sizes and marking each cube as free or occupied. The robot may only take a path through known free cubes.

Trial and error can be used if the robot has a collision-avoidance sensor, such as a sonar sensor (see chap. 7), which can warn the computer if a collision is about to happen. The computer can ignore possible collisions until one becomes imminent, and it can then find a way around the obstacle through trial and error.

REFERENCES

1. Coiffet, Phillipe, and M. Chirouze. *An Introduction to Robot Technology,* transl. Meg Tombs. New York: McGraw-Hill, 1983.
2. Critchlow, Arthur J. *Introduction to Robotics.* New York: Macmillan, 1985.
3. Holzbock, W. G. *Robotic Technology: Principles and Practice.* New York: Van Nostrand Reinhold, 1986.
4. Denavit, J., and R. S. Hartenberg. A kinematic notation for lower-paired mechanisms based on matrices. *ASME Journal of Applied Mechanics,* June 1955, pp. 215–221.
5. Critchlow. *Introduction to Robotics,* pp. 189–195.
6. Holzbock. *Robotic Technology.*
7. Palm, William J. *Dexterous Hands for Robots: A State of the Art Review,* URI Report 86-0:1. Kingston, R. I.: University of Rhode Island, 1986.
8. Groover, Mikell P., et al. *Industrial Robotics: Technology, Programming and Applications.* New York: McGraw-Hill, 1986, p. 120.
9. Groover. *Industrial Robotics,* p. 139.
10. Salisbury as cited in Palm, *Dexterous Hands for Robots.*
11. Asada, H., and I. H. Ro. A linkage design for direct drive robot arms. *ASME Transactions,* Dec. 1985, pp. 536–540.
12. Palm. *Dexterous Hands for Robots.*

13. Kuntze, Helge-Bjorn, and Andres H. K. Jacubasch. Control algorithms for stiffening an elastic robot. *IEEE Journal of Robotics and Automation,* Vol. RA-1, No. 2, June 1985, pp. 71–78.
14. Whiting, D. E., and J. L. Nevins. What is the remote center compliance (RCC) and what can it do. *Proc. 9th International Symposium on Industrial Robots.* Washington, D.C.: March, 1979.
15. Lozano-Pérez, T. Task planning. *Robot Motion: Planning and Control.* M. Brady et al., eds. Cambridge, Mass.: MIT Press, 1982.

ROBOT ELECTRONIC DESIGN

A typical robot system can be subdivided into seven principal parts: the robot arm, the end-effector, and the transmission system, which are mechanical; internal sensors, external sensors, and a control system, which are electronic; and an actuator, which can be mechanical or electrical. We have already discussed the mechanical parts of the system. We now begin a discussion of the electronic components, which are the main emphasis of this book. Other related topics are covered in the next seven chapters.

4.1 ROBOT ELECTRONIC SUBSYSTEMS

Increasingly sophisticated robot systems need high-precision electronics systems to control and monitor robot performance. These systems must be easy to design, understand, and use; highly reliable; and inexpensive.

Figure 4-1 shows a simplified robot system from an electronics point of view. Each joint can be considered a separate load, along with a motor and appropriate electronics to position the joint. A central computer controls the entire system and coordinates all joint motion. The central computer, in turn, gets information from external sensors, such as a vision system, and controls external devices, such as a gripper.

When designing a robotic system, one must select the appropriate motor, determine required servo system accuracy and frequency response, select appropriate internal feedback sensors, determine the types of external sensor information required, and select an appropriate robot computer system. Note that successful robot system design requires the consideration of the interaction of all these areas. Before we examine these areas in detail, we give an overview of each subsystem function as well as some of the approaches that can be used.

Motor Considerations

Many types of motors have been used in robotic systems. Except for a few specialized applications, motors that move the robot's links or operate its end-effectors must be servo-controlled. A servo loop ensures that the motor follows commands to a sufficiently high level of accuracy. Thus, the type of motor

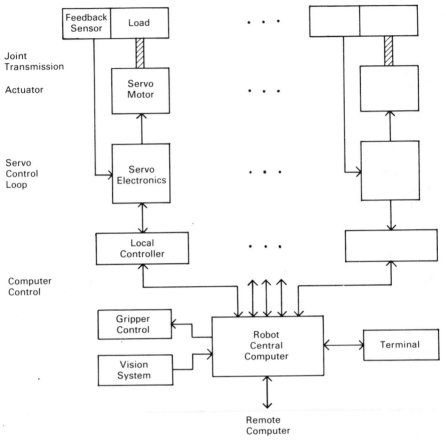

Figure 4-1. Robot electronics.

selected must be capable of working closely with the servo control system. For that reason, motors that have a linear relationship between speed and current and between torque and speed (or current) are highly desirable. In addition, since the weight of the motor system must usually be added to the weight that the arm must carry, low-weight, high-efficiency motors are needed. Because permanent-magnet motors can always be made smaller and lighter than an equivalent electromagnet-based motor, they are being used almost exclusively.

Although other types of motors are occasionally used, the major choices are between synchronous alternating current (AC) motors, standard (brush-type) direct current (DC) motors, DC brushless motors, and stepper motors, with pancake motors used less often. AC induction motors, which produce higher torques than needed for most robot requirements, are not generally used.

New developments in circuit design have revived interest in the reluctance motor, one of the earliest types, and one of the most rugged and most reliable.[1] Reluctance motors provide the same horsepower as conventional motors of the same size, but require only half of the electronics. They do need control over timing to reverse direction and therefore must have special circuitry for this function.

Synchronous AC Motors

The momentary speed of a synchronous motor is proportional to the frequency of the applied AC voltage. Because of the accuracy of this speed control, synchronous motors are used as clock motors. In a synchronous motor, the rotor contains a permanent magnet whose field is attracted to the rotating magnetic field produced by the AC field and turns with it in complete synchronization.

Synchronous motors can be a problem to start because they tend to oscillate rather than accelerate to the proper speed. This problem can be eliminated through an auxiliary squirrel cage mechanism as part of the rotor or through special electronic starting circuits.

Since motor speed is proportional to stator field frequency, the servo system must be designed to produce the proper current waveforms. The Toshiba RA 10 series motors have a rated speed of 3000 rpm and can produce up to 10 ft-lb of torque. Motornetics has produced a low-speed (60 rpm), high-torque (250 ft-lb) motor, which is ideal for direct-drive applications.

DC Motors

The standard DC motor has a commutation ring controlling the current flow to its rotor coils. As the rotor turns, the action of the commutator changes the energization of the rotor coils, keeping the magnetic field slightly offset and producing a constant torque to try to line up the magnetic fields. Figure 4-2 shows a typical DC motor and rotor.

DC motors are inexpensive, available in many sizes, and are quite reliable. With their high torque-to-volume ratio, they are smaller than other motor types for the same torque. With high precision and the ability to provide high accelerations, DC motors can reach their final position very quickly. For these reasons, most electric robots use standard DC motors. At high speeds, however, arcing and flashover can take place in the commutators, thereby damaging the brushes and reducing their life. At low speeds cogging can occur; that is, the motor does not turn smoothly, but jumps as the brushes move from one commutation contact to another.

Figure 4-2. DC motor and rotor. (*Courtesy of PMI Motion Technologies*)

DC Brushless Motors

DC brushless motors were developed to eliminate some commutator-based problems. These motors are DC motors "turned inside out." The rotor contains the permanent magnet, and the stator field is supplied by a DC voltage. Because it is still a true DC motor, commutation of the field is necessary for the motor to turn. Since the rotor cannot provide the commutator, appropriate circuitry must handle this function externally. Feedback from sensors on the motor shaft (such as a resolver or Hall-effect sensor) offer the necessary signals for electronic commutation.

Since the stator coils are closer to the external case, heat dissipation is better, so these motors can provide a higher continuous torque and more horsepower than standard DC motors. The rotor weighs less, resulting in lower inertia and faster acceleration. In addition, smoother operation at lower speeds is possible, if the external commutation system is designed for finer transitions.

Stepper Motors

Stepper motors offer precise digital position control without encoder or other feedback devices, thus simplifying the driver circuits. Except for some higher-torque motors, they are inexpensive and rugged, and do not sustain damage under stall conditions, an important consideration for robot applications.

Steppers do, however, have a few disadvantages compared with other servomotors. They do not run smoothly at low speeds, especially under 70 rpm. They producer lower torques, so slightly larger motors are required for the same load-carrying capability; and they are noisier than DC motors, especially at resonant speeds.

Conventional stepper motors are also limited in positional accuracy (typically 200 steps per revolution, or 1.8°); hence, they are used primarily for smaller robots or for open-loop servo systems. Their accuracy can be improved in two ways. (1) Gear boxes or other step-down drive arrangements can be used, which also increases output torque and reduces output speed. (2) Microstepper motors, which allow position control accuracies within a few minutes of arc, are another alternative.

Figure 4-3, adapted from technical notes provided by Compumotor Corporation, shows some typical motor curves for steppers. Figure 4-3*a* demonstrates the effect of speed on torque and power for the Compumotor C83-93 motor at a supply voltage of 90 VDC. Figure 4-3*b* illustrates the effect that reducing supply voltage has on torque produced, and Figure 4-3*c* shows the effects of lower motor inductance.

Encoders are not generally necessary with standard steppers for two reasons. First, the stepper motor was designed to position itself without requiring feedback from an encoder (note that if it does lose its position, a stepper cannot recover on its own). In addition, the accuracy of the stepper itself (1.8°) is so limited compared with the accuracy of encoders (0.1° to 0.2°) that little gain in accuracy is realized if encoders are used. However, encoders can be used to detect a stalled motor condition or slippage. Although these conditions do not damage a stepper motor, they will make the stepper reach a wrong position.

Direct Drive

The primary difference between a direct-drive motor and more conventional motors that use gears or other transmission systems is that the direct-drive motor must operate at much lower speeds and higher torques. If direct drive is attempted with a standard motor, minimum speeds are usually too high and

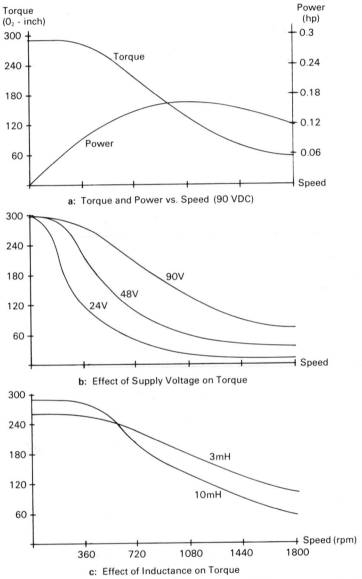

a: Torque and Power vs. Speed (90 VDC)

b: Effect of Supply Voltage on Torque

c: Effect of Inductance on Torque

Figure 4-3. Typical stepper motor curves. (*Courtesy of Compumotor; adapted from technical notes*)

available torques are too low. Special direct-drive motors, such as the synchronous AC motor from Motornetics, have been developed for robot use.

Since higher torque generally requires higher current, motor design, especially the commutation system, is affected. Brushes for high-speed motors can contain little metal and generally use only graphite. Low-speed motors, however, require lower resistance in their brushes and therefore include silver or copper, sometimes as much as 80% silver. The resulting low resistance contact ensures linear operation at low speeds.

Magnetic Materials

Motors can be fabricated from various magnetic materials. Four types are regularly used: ferrite (powdered iron and strontium oxide), ceramic, Alnico (aluminum, nickel, and cobalt), and rare-earth materials. Within the rare-earth class, samarium cobalt is perhaps the most popular, although neodymium-iron-boron magnets (pioneered by the Japanese firm Sumitomao) are now a major competitor, principally because they do not require cobalt, a scarce material, and are thus less expensive. They also offer higher magnetic flux.

Ferrite magnets suffer from flux variations with temperature and do not produce as high a value of flux density. Their main advantage is low cost. Alnico magnets can support a high flux density, but they require a small gap. The major problem with Alnico magnets is that, when used in a high-torque motor, the torque produced will often demagnetize the magnet. For this reason, they can only be used with low-torque motors.

Motors built from rare-earth magnetic materials weigh less and are smaller than other motors, but they cost much more. For a given volume, the rare-earth magnets produce four to eight times the magnetic energy as do Alnico or ceramic, but at five to ten times the cost. For example, samarium cobalt can provide 10% higher stall torque with 33% less weight and 25% less volume than a non-rare-earth magnet.

Power Amplifiers

Electric motors require an electrical power source to drive them. In most robotic systems, these sources are power amplifiers, where the input signal (usually only a few volts) receives sufficient voltage and current amplification to provide the necessary motive power. The motor driver must match the motor type and size to the available control signals. Both linear amplifiers and pulse-width-modulated amplifiers are available, the latter being more efficient since they minimize power dissipation in the amplifier. Figure 4-4 shows a typical pulse-width power amplifier.

Robot motors need 12 to 120 V and 10 to 100 A, depending on their size. Switching times and transients must be handled within a few milliseconds, thus requiring bandwidths of at least 1 kHz. These requirements are a major design burden, so power amplifiers are relatively expensive.

Servo Systems

Most early robots used open-loop control in which the input was either a position or a speed, and control electronics were trusted to bring the robot arm to the desired position. When stepper motors are used to control motion, this approach works fairly well because stepper motors provide accurate (although coarse) control over movement. But greater accuracy requires a closed-loop servo system. A closed-loop system employs feedback information to determine whether the arm has reached the desired position. If not, the servo system recognizes this failure as an error and amplifies the resulting error signal to drive the arm to the correct position.

Figure 4-4. Pulse-width power amplifier board. (*Courtesy of PMI Motor Technologies*)

Earlier motion control systems also used analog signals and amplification. It was often necessary to put a tachometer in the loop to monitor motor velocity so that the speed could be adjusted, if necessary. Today, with techniques such as proportional-integral-differential (PID) algorithms (discussed later), motor velocity can often be accurately controlled directly, so tachometer feedback may not be necessary.

The servo system receives absolute or relative position commands from a central computer, in addition to associated commands such as speed, torque, and timing controls. The servo system then compares these commands to the current motor position and adjusts its output voltage and current as necessary. Compensation filters, usually digital, must often be placed in the servo loop to ensure that the system is stable and that the output is appropriately damped.

Servo Controllers

The servo system must be supplied with position and/or velocity commands from a robot controller, which may control either single or multiple robot motors. The controller takes the position command received from a central computer and converts it to the necessary motor control signals. These signals precisely control motor position over time, thereby also controlling motor acceleration, run velocity, and deceleration.

Other signals typically provided to controllers are instructions for initialization, constants for PID filter parameters, and external signals such as wait, limit switch reached, and gripper closed. The controller must often synchronize the motions from two or more axes, which may mean delaying the execution of a motor command in one channel until an interim location has been reached by a different motor in another channel.

In some systems, a combination of analog and digital signals are used. Feedback information is usually taken from an incremental encoder, which produces a series of individual pulses (digital signals) during each rotation. In most encoder systems two signals are provided in quadrature ($90°$ phase shift between them), thus allowing direction sensing as well as relative position to be determined.

In either case, the resultant train of encoder pulses is converted to an absolute position signal. This motor position is then compared to the commanded position, and an error signal that is proportional to the amount of error is generated. This error signal can then be sent through a digital filter for necessary phase compensation, signal modifications (such as differentiation or integration), and gain adjustment. Most steps are performed digitally, with the output converted to an analog signal for input to the power amplifier–motor driver circuitry.

Sensors

Robots use two classes of sensors.[2] The first class covers *internal* state sensing and provides the robot information on the position, orientation, and velocity of each of its joints and the position of its end-effector. This type of sensing is very important for controlling robot motion. These sensors use position, velocity, and force transducers mounted directly on the robot. If this were the only class of sensor provided, the robot would be blind to its outside environment. Then, if the object the robot was trying to grasp were not in its proper position, the robot would close its gripper around empty air and attempt to manipulate this "nonobject." Because of this blind response, early robot designers recognized that robots need at least minimal information from their environment.

A second class of sensors covers *external* state sensing and provides the necessary environmental information to the robot. Depending upon the amount and type of external sensors, the robot can either shut down or take some limited action in response to an external problem. In the simplest case, the sensor could be a limit switch. In a more advanced case, a vision system might be expected to recognize the location of a part on a conveyor belt and relay information about its position to the robot.

Most sensors can only be used for a single type of function, either internal or external. Only a few can be used for both. Internal sensors are usually individual devices, which means they are interfaced to the system at the component level. External sensors are usually subsystem oriented and contain some level of intelligence and a more standardized interface. Internal sensors generally establish the robot's position around a local set of coordinate axes, whereas external sensors allow the robot to orient itself with respect to its environment (global coordinates).

Most internal sensors are feedback or position sensors. They are discussed briefly here and in more detail in section 5.1. External sensors can be proximity sensors (including contact) or distance sensors. Both types are discussed in section 4.2, and more detailed information is provided in the next three chapters.

Internal sensors measure position, velocity, force, and, occasionally, acceleration, with the most important measurement being position. Position sensing is needed for servo control and is the primary feedback source to the robot. Translational and angular positions can be obtained from potentiometers, which have been used on even the earliest robots. Magnetic means can also be used to measure translational motion. A movable magnetic coil inside fixed coils is linked to the moving part. These sensors can only be used for short ranges, less than one inch.

Rotational position information may be obtained from *optical encoders*, of which the incremental optical encoder is the most common. The *resolver* converts phase information from a reference signal to determine angular position. *Hall-effect* sensors are a third type, and a fourth, especially applicable to high-precision positioning, is the *laser interferometer*, which offers submicron resolution.

Velocity measurements can be derived from changes in position over a known time, but they are usually sensed directly through tachometers. Force sensing is often done via strain gauges, of which several types are available.

4.2 ROBOT EXTERNAL SENSING SYSTEMS

External sensing tells what is happening in the robot's working environment. These sensing systems are based on principles related to human touch, vision, and hearing, with some other techniques, such as magnetics, also used.

Tactile sensing is used to detect object presence, shape, orientation, and weight through direct contact with the object. Other sensors are used with objects at some distance from the robot. They can detect object presence, count objects, determine object coordinates, and recognize objects. Detection and counting can often be done with a simple photocell system.

To locate an object, the system often needs three-dimensional data, that is, range data as well as X and Y coordinates. Precise location of objects requires sophisticated techniques, including vision systems, infrared (IR) systems, laser ranging, and ultrasonics. If object recognition is also required, at present only machine vision can be used, although IR and acoustical techniques potentially offer recognition ability.

Tactile Sensors

Many types of tactile sensors are available, from simple microswitch-activated contact sensors to arrays of force-detecting sensors mounted on a robot "hand." This latter type of sensor is often covered by a layer of compliant, elastic material. Contact sensing can also be done optically, and contact force may be detected through piezoelectric effects or through changes in resistance of conductive elastomers.

Light-Based Systems

Object Detection

Only simple photoelectric systems are needed to detect an object or to count objects as long as the objects are clearly separated from others (i.e., there are no overlaps). Three primary methods of photoelectric detection are illustrated

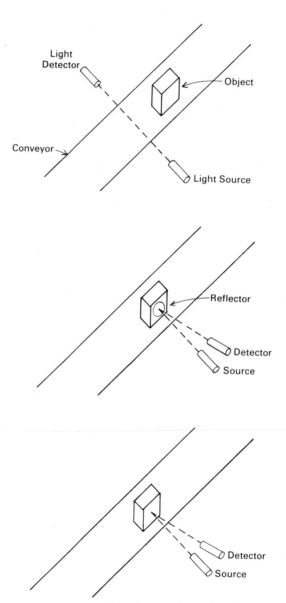

Figure 4-5. Photoelectric object detection.

in Figure 4-5. The first detects a break in a beam of light caused by the presence of an object. This method is the oldest and most commonly used. A second technique detects the absence of an object. In this approach, the object carries a light reflector and returns a beam of light only when the reflector (and therefore the object) is present to reflect the light. The third approach uses the object itself to return diffuse light to the detector. Since less energy is returned, this approach can only be used over short distances.

Simple light sensors can also be used for color or mark detection. For example, in one application color detectors are used to separate out stones that had become mixed in with nuts. Mark detection is based on the sharp reduction in light received when a black mark passes the photocell. If the mark is a different color (not black), a complementary color is used as the light source to ensure that the mark will appear black.

Even light refraction has been used in detection systems. In this approach, transparent bottles can be inspected for the proper level of fluid by checking for the highest level of refraction present. In a related application, the presence of labels was detected when the labels blocked light that would have otherwise passed through the bottle.

Some photoelectric applications, such as mark or color detection, require visible light and other systems often work in the IR spectrum. Some applications use laser light because of its narrow spectrum or its narrow beam width. Although lasers can improve performance for small objects or for objects at great distances, they are larger and more expensive than conventional photocell systems.

Ranging Systems

Range-finding techniques use noncontacting sensors and provide information about three-dimensional objects. Range sensing can be based on time of flight or triangulation. It can project a single beam or a plane of energy. Electronic or mechanical scanning can be used to cover a wider arc for area mapping, and measurements can be based on intensity or phase delays. There are three types of light-based systems, as well as a sound-based system (discussed later). Lasers or IR light can be used in point ranging, while vision systems can provide area coverage.

Laser systems use coherent light sources, usually in the visible spectrum. Laser-ranging systems are popular because they have high angular resolution due to their narrow beam width. However, it is difficult to perform time-of-flight calculations with laser systems; they could produce interference patterns, and they are generally expensive. In addition, special safety precautions must be used if lasers providing energies greater than 15 mW are used.

Laser time-of-flight techniques have been used in special applications, but they are rare since measurements accurate to the nearest nanosecond must be made to provide an accuracy of about $\frac{1}{2}$ in. A standard technique is to amplitude modulate the laser and then measure the resulting phase differences. This approach often requires processing the range data at a slow rate. It is also possible to combine lasers with vision systems. If stereoscopic imaging is done on an object while using a laser beam to provide a fiduciary mark, the required computation time is reduced significantly.

Infrared systems have been used for proximity detection and as part of a beacon ranging/tracking system. IR systems employ wavelengths from just under 1 micron (μm) to as many as 10 μm. These wavelengths are longer than can be seen by the eye, although they are perceived as heat. When used for proximity detection, IR systems operate in a passive, or receive-only, mode. When used in beacon systems, the beacon provides an active source of IR energy, and direction to the beacon can be determined. If the robot is in motion, a series of beacon angles can be used to determine range.

Vision systems provide information in two dimensions (horizontal and vertical). It is possible to obtain range information from a vision system, thus completing a three-dimensional picture, but an indirect method must be used, which adds to system complexity and yet does not provide the highest accuracy range information available. Ranging using vision systems is covered in chapter 7.

Infrared Positioning

IR systems can be used to determine the current position of a mobile robot by using several beacons. In the beacon approach, several IR transmitters are placed in known locations in a room, and the robot has a detector that can determine the angle to each beacon. With a series of known angles from beacons in known locations, the robot's position can be accurately determined. IR beacons usually operate in the near-IR range (880–950 nm) because semiconductor emitters/receivers are more efficient in this range and the beacon is less affected by smoke and dust in the atmosphere.

There is a 120° limit to most IR systems, so if the beacon is outside of this range (as it might be for a mobile robot), multiple sensors or a wide-angle (fish-eye) lens is needed to obtain 180° coverage. However, a wide-angle lens introduces severe light loss and angular distortion. Long-range beacons are used for updating angles, near-range beacons for updating position. Infrared systems can recognize beacons from as far away as 30 ft. The ratio of bright light to dim light that can be handled (the *dynamic range*) is 60 dB (or 1,000 to

1). Currently available systems allow accuracies to 3 in, but 3/4-in accuracy has been obtained experimentally.

Infrared sensors can also be used for motion detection in various security applications. Examples of IR sensors include lead sulfide (PbS) in the 1- to 3-μm range and lead selenide (PbSe) in the 3- to 5-μm range.

Experimental work in recognizing objects with IR systems is being carried out. One system produces a two-dimension display that maps the heat response of an object. In one demonstration, a normal coffeepot was imaged behind a sheet of opaque silicon. Since the wavelengths of IR are different from wavelengths of conventional light, it is possible to have a type of three-dimensional vision by imaging through an opaque covering material. For example, silicon is transparent in the 3–5-μm IR range, but opaque to conventional light and to light in the 9–14-μm IR range.

One interesting combination of IR and acoustics is thermal wave imaging, a technique still in the research stage. A hot spot focused on the object radiates a thermal wave through the material, which can be detected with an acoustical sensor due to the acoustic noise generated.

Vision Systems

Vision systems are the most common type of external sensor associated with robots, and are used in many types of applications. Vision systems are generally used to provide complex functions such as image acquisition, location, and recognition. These applications require capturing an image, similar to the way television systems do, and extensive processing to recognize and locate the object. Chapter 7 covers these systems in detail.

Sound-Based Systems

Ultrasonic Systems

Ultrasonic systems give accurate range information, better than 0.1 in at 25 ft, if temperature and humidity are compensated. They usually operate from 40 kHz to 60 kHz, above the limits of human hearing.

A problem with ultrasonic energy is surface reflectivity. Rough surfaces reflect waves of energy in all directions. Very smooth surfaces reflect the energy in a single direction, seldom the desired one. Reflection is more significant to ultrasonics than to light because an object's smoothness depends on the wavelength of the energy impinging on it. Because light has very short wavelengths, only highly polished surfaces (such as a mirror) reflect most of the energy, but ultrasonics uses much longer wavelengths, so most surfaces serve as reflectors. Reflection problems are less severe with IR and laser systems, which also have shorter wavelengths. The net result is that in ultrasonic ranging, there is a much greater chance that the object will reflect energy in such a manner that it will not be "seen" by the system. However, if an edge of the object is within the beam, this edge will often return enough energy to allow the object to be detected.

Sonar ranging systems have difficulty detecting large objects at an angle to them as well as objects with surfaces that are too porous, because they may absorb too much of the incident energy. Their angular resolution is also much poorer than that of video systems or laser range finders. Their major advantages, discussed in chapter 7, are low cost and high accuracy.

Speech Systems

One function of an external sensor that doesn't fit any of the classes already discussed is speech recognition. Voice has always been our most important medium of communication, and it is natural to consider giving instructions or data to computers and robots through speech.

In order to communicate with a robot system via speech, the robot must be able to detect speech from the surrounding noise in a factory environment, recognize the words that are spoken, and determine what action should be taken based upon them—definitely major tasks, and they have yet to be technically perfected, although some systems have been demonstrated. In one, a robot was programmed to respond to the spoken input of a person's initials by etching the initials onto a glass.

Speech systems are covered in more detail in chapter 16 as one of several evolving technologies.

4.3 MOTOR SYSTEM DESIGN

This section derives several equations to help us select a motor with the proper characteristics. These equations are accurate enough for most applications, but do not take into consideration many second-order effects (such as changes in motor inductance, flexibility of the shaft and bearings, backlash, or viscous damping). If motor selection is done conservatively (e.g., a motor with twice the required torque is selected), second-order effects can usually be ignored. For high-accuracy applications, however, they cannot be ignored when specifying the equations necessary to drive the motors.

Principles

Electric motors are based upon three principles:

1. Electric current passing through a wire produces magnetic fields.
2. Like poles of a magnet repel each other, unlike poles attract.
3. The current direction determines magnetic polarity.

The major parts of a conventional DC motor are the stationary (field) magnet (usually a permanent magnet), a set of coils located on the armature, and a commutator that supplies current to the armature coils and changes the direction of the current as the coil is rotated.

The basic principle of the DC motor is that current flowing through an armature winding sets up a magnetic field that reacts with the permanent magnet serving as the field magnet. This interaction causes the motor to rotate so that the appropriate north and south poles of the armature and field magnets can line up. As the motor rotates, commutation rings on the armature switch the exciting current to a different winding, which once again provides a force to the motor and requires it to continue turning.

When the armature revolves, it cuts lines of magnetic force between the poles of the field magnet. This action generates an electromotive force (emf) in a direction opposite to the current flow and reduces the effect of the supply voltage, thus limiting the current drawn and the motor speed. When a load is added to the motor, the inertia of the load reduces the motor acceleration and also slightly reduces its top speed.

Torque Produced

The amount of rotational force (torque) provided by a motor depends on the magnetic flux produced, which in turn depends on the armature current, number of turns in the armature coil, size of the motor, magnetic material used, and other factors. All except current are fixed once the motor design has been established, and they are usually lumped together in one constant (K_t). Under very light loads, torque is proportional to the square of the current. However, under typical operating conditions, where the motor is well loaded, the magnetic field becomes saturated and torque is directly proportional to current. Under these conditions torque is a linear function of current (i):

$$T = K_t i \qquad (4\text{-}1)$$

When a motor is first turned on, only the resistance of the armature winding (R_a) is available to limit current flow, so the motor draws the maximum amount of current at this time:

$$i_{max} = \frac{V}{R_a} \qquad (4\text{-}2)$$

where V is applied voltage. Maximum torque is also available when the current is maximum (i.e., when angular velocity is zero):

$$T_{max} = K_t \frac{V}{R_a} \qquad (4\text{-}3)$$

Maximum torque is desirable at start-up, when both friction and load inertia must be overcome. If the motor is not turning and there is a maximum current flow, then the motor is stalled. If the stall lasts more than a few seconds, the motor will overheat and possibly burn out.

As the motor gains speed, an internal generator effect produces a back emf (voltage) that opposes the externally applied voltage, and the motor current is reduced:

$$i = \frac{V - e_b}{R_a} \qquad (4\text{-}4)$$

Note that the back emf e_b increases with armature speed, and the motor reaches a limiting velocity where the back emf is great enough to prevent further increases in speed. Back emf can never exceed (or even reach) the applied voltage. One effect from equation (4-4) is that the minimum motor current occurs at maximum unloaded speed. The back emf is proportional to angular velocity (ω):

$$e_b = K_t \omega \qquad (4\text{-}5)$$

Total voltage applied provides current to the motor and counters the back emf. The necessary voltage is thus the sum of back emf and armature current times the armature resistance:

$$V = iR_a + e_b \qquad (4\text{-}6)$$

Substituting for e_b and rearranging gives

$$V = iR_a + K_t \omega \qquad (4\text{-}7)$$

$$\omega = \frac{V}{K_t} - \frac{iR_a}{K_t} \qquad (4\text{-}8)$$

For a given motor, with K_t constant, angular velocity increases as a function of the supply voltage and decreases as a function of the current. Using the value of i from equation (4-1), we can show the relationship between angular velocity and torque:

$$\omega = \frac{V}{K_t} - \frac{R_a T}{K_t^2} \tag{4-9}$$

$$T = \frac{K_t}{R_a}(V - K_t \omega) \tag{4-10}$$

where ω is angular velocity (in radians per second), V is applied voltage, R_a is armature resistance, i is armature current, T is output torque, and K_t is a torque constant. Equation (4-10) defines the approximately straight-line relationship (with negative slope) between speed and torque. The relationship of torque to current and speed is the basis for all DC motor servo control.

Some robotics tasks require speed control (such as spray painting), and others require torque control (press fittings and other assembly tasks). We can control speed (ω) by controlling voltage applied; we can control the torque (T) by controlling current applied. It is generally impossible to control both at once, although some systems automatically switch between the two types of control as speed increases.

From equation (4-8) maximum theoretical velocity occurs with minimum current and approaches

$$\omega_{\text{max}} \approx \frac{V}{K_t} \tag{4-11}$$

Angular velocity can also be expressed in rpm by

$$N = \frac{30}{\pi}\omega \tag{4-12}$$

where N is the number of revolutions per minute.

Power

Power (P) delivered by the motor is equal to total power applied minus the amount lost in the armature resistance:

$$P = Vi - i^2 R_a \qquad \text{(in watts)} \tag{4-13}$$

This power is used to turn the motor and is thus equal to torque times the change in angle per unit time (angular velocity):

$$P = T\frac{d\theta}{dt} = T\omega \tag{4-14}$$

Horsepower (hp) is another measure of power, or rate at which work is performed. Horsepower differs from torque, which is a measure of work and does not consider rate. With rate, torque can be converted to horsepower:

$$\text{hp} = \frac{TN}{K_h} \tag{4-15}$$

where the conversion constant K_h is 63,025 (often rounded to 63,000) if torque is in inch-pounds and N (motor speed) is in rpm.

One horsepower is also equal to 745 W of mechanical power. Since no motor is 100% efficient, more than 745 W of electrical power is required to produce 1 hp. In fact, the ratio of mechanical power to electrical power is often used as the efficiency of the motor. Efficiencies of an unloaded motor may be better than 90%, but when losses in transmission systems and typical loads are considered, the resultant system efficiency might be only 40%.

Torque is expended against the moment of inertia J and the total viscous friction B of the system:

$$T = J\frac{d^2\theta}{dt^2} + B\frac{d\theta}{dt} \tag{4-16}$$

where $d\theta/dt = \omega$. The robot load inertia J_r depends on the weight, size, and geometry of the load. For a typical homogeneous load,

$$J_r = \frac{WR^2}{2} \tag{4-17}$$

where W is the weight of the load in ounces and R is the radius of the load in inches. Then J_r is in ounce-inches. This load inertia must be added to the motor inertia and the gear inertia to obtain the system moment of inertia.

In many applications, lower speeds and higher torques are needed. A transmission system (such as gears) can be used to better match motor torque and speed to a load. If the transmission system has a gear ratio of $K_g = n_e/n_m$, where n_e is the number of teeth on the load shaft gear, and n_m is the number of teeth on the motor shaft gear, this gear ratio will reduce the output speed

$$\omega_{\text{out}} = \frac{\omega}{K_g} \tag{4-18}$$

while increasing output torque

$$T_{\text{out}} = TK_g \tag{4-19}$$

A transmission system will increase the moment of inertia slightly and add friction. These losses must be considered when a motor and transmission system are selected.

Acceleration

Another important factor is acceleration time—that is, how long will it take the motor to reach top speed, and how far will the motor travel during this time?

The torque required to accelerate is

$$T = \frac{J\omega}{t} \tag{4-20}$$

For stepper motors, maximum efficiency can be obtained by first accelerating the load to high speed, holding it there, and then decelerating it. The approximate time (in seconds) required to accelerate to full speed (from rest), assuming T is in ounce-inches, equals

$$t = \frac{NJ}{9.55T} \tag{4-21}$$

This formula holds when the motor is working on a constant-torque slope. It can stay on a constant-torque slope only up to a certain velocity, at which point a constant power slope mode is required. The result of this equation is that rapid acceleration is possible only if the arm and load inertia are small compared with available power.

The value of the time constant (the time to reach 70.7% of the final velocity) is also helpful in servo design. In this case, the mechanical time constant of the motor is

$$t_m = \frac{R_a J}{K_t^2} \quad \text{(in seconds)} \tag{4-22}$$

A related equation shows the interaction between servo loop gain K and the inertial time constant t:

$$t = \frac{1}{\omega_0} = \sqrt{\frac{J_r}{K}} \tag{4-23}$$

where ω_0 is the undamped oscillation frequency and J_r is the load moment of inertia. As expected, the more gain there is in the servo loop, the less time that is required to bring the motor up to speed. Time required increases with the square root of inertia.

Some servomotors are given commands in number of pulses rather than in degrees, where the number of pulses refers to data from an incremental optical encoder. For example, a relative position could be specified as follows: turn the motor for 10,000 encoder pulses at a maximum rate of 100,000 pulses/s with an acceleration of 2×10^6 pulses/s^2. Now, for a 2,000-pulse/rev encoder, the motor is being directed to make five complete turns at a maximum speed of 50 rps (3,000 rpm) and an acceleration of 1,000 rps^2.

Other Considerations

The weight and location of the motor are important. Motors can be mounted near the joint that they move or away from it, with a transmission system (such as belts or cables) connecting the motor to the joint. Each approach has disadvantages. For example, if the motor is mounted at the elbow joint, the shoulder motor has to move added weight whenever it lifts the arm. If the motor is mounted at a distance (in the base, for example), the transmission system reduces the accuracy of robot positioning. The drive motor produces a torque that must overcome the inertial effect of acceleration as well as motor inertia and viscous damping forces.

The moment of inertia (J) can vary as the robot arm moves, especially if other joints are also moving. It will always vary when changing load or using revolute motion. The damping coefficient C should also vary in proportion to $\sqrt{J}$.

Ambient temperatures also influence motor selection. Standard motor specifications assume an ambient of 40° and a maximum temperature rise under operation of 22°. Higher-temperature motors (class A and class B) are also available.

4.4 SERVO SYSTEM DESIGN

A servo system controls motor position. In closed-loop servo systems the actual motor position is compared with the desired position and corrected if necessary. Figure 4-6 illustrates the major portions of a typical servo system, called a *proportional* servo system.

Closed-loop servo systems work on a negative feedback principle. The difference between the actual output and the desired output (as represented by an input signal) is used to compute an error signal ε. An algorithm is applied to this error signal, providing necessary proportional, integral, and derivative compensation to the signal. Necessary velocity adjustments are also made after

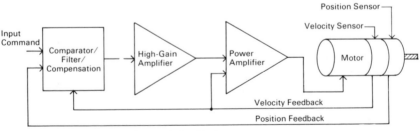

Figure 4-6. Basic servo system.

comparing the desired velocity with the actual velocity. This error signal is amplified and used to move the joint closer to the desired position. As the motor turns the shaft, a position sensor and a velocity sensor monitor these parameters and send the information to the servo system.

Most servo systems have a linear relationship between the input voltage V_{in} and the output speed. Thus, the larger the input signal, the faster the motor turns. If a feedback sensor is mounted on the output shaft and that signal is compared with the input signal, we have a basic servo feedback system. Feedback can also be derived from the speed of the motor (from a tachometer, for example) and used to control motor speed accurately, or it can be derived from the angular position of the shaft (from a resolver, for example) and used to control the position that the servo motor turns to.[3] In either case, the feedback voltage is compared with the input voltage, and the *difference* is amplified and used as a driving voltage. Thus, for position feedback, sufficient signal is available to drive the motor until it reaches the desired end position. If the motor overshoots this position, a new error signal will be generated to bring the motor back to the desired position.

The response to the error signal is delayed by inertia in the motor system and phase delays in the electronics. These phase delays cause a system to oscillate (i.e., go past the desired end point and have to return). If the delays and system amplification are too large, the system can become unstable and oscillate back and forth around the desired end point and never come to rest. Therefore, practical servo systems often require frequency compensation networks and control over the gain of the amplifier, both of which can improve the high-frequency stability.

A significant factor affecting servo system accuracy is the amount of system gain. Since there is a minimum output voltage that must be provided to have the motor move at all (due to friction and inertia), there is a direct relationship between system gain and resulting system accuracy. The greater the gain, the lower the error voltage can be while still driving the motor. Therefore, for maximum accuracy, it is desirable to have maximum gain.

On the other hand, as gain increases, stability and overshoot problems worsen. These two competing goals—minimum gain for stability and maximum gain for accuracy—can often be handled by providing high gain at low frequencies (when the servo is in a steady-state run condition) and low gain at high frequencies (when oscillation may occur). However, this type of approach, though applicable for many situations, does nothing toward smoothing the resultant output. Thus, several problems can arise.

Suppose the robot arm was to be brought very close to a large object (such as a car body) without crashing into it. If the robot used a straight proportional servo system, it would immediately see a large error (the robot's actual position compared with its desired final position) and the robot would attempt to move at top speed to reach the desired end point. Thus, the robot would start with a jerk, travel at maximum speed, and, due to inertia, might be unable to stop before it had overshot the location. One way to get around this problem is to have the robot move quickly to a safe offset point and then slowly approach the object. Another, usually better, technique is to use a PID-based servo control system.

Adding digital computers to servo systems allows them to carry out signal integration and differentiation through suitable filter algorithms.[4,5] This capability lets us add error correction terms that are functions of the differentiation and/or integration of the output signal.

In a PID servo system the feedback voltage is a combination of the actual output (proportional term) and change of speed (integral and differential terms). This combination is often effected through control over a filter circuit in the feedback loop. Under these conditions, a system at rest that is given a large error signal could integrate that error signal to produce a slower, smoother, and better-controlled start of the motion.

Basic servo systems use proportional control. An output signal is proportional to the error. With a servo amplifier gain of K, we have

$$V_{out}(t) = K\varepsilon(t) \qquad (4\text{-}24)$$

With integral control, the output is changed at a rate proportional to the error. Thus with a large error, the output voltage starts slowly and increases rapidly:

$$V_{out}(t) = K \int \varepsilon(t)\, dt \qquad (4\text{-}25)$$

Integral control helps to eliminate arm droop, which occurs when the arm moves to different positions, and changes caused by the load: When the error is zero, the output remains constant, thus counteracting any load present. Counterweights (one alternative) can balance arm position changes but do not help load changes.

A proportional system requires an error to supply an output against a load, whereas an integral system does not provide as quick a response to small errors; therefore, the two are often combined to form a PI servo system:

$$V_{out}(t) = K_p\varepsilon + \frac{K_i}{T_i} \int \varepsilon(t)\, dt \qquad (4\text{-}26)$$

where T_i is used to adjust integrator gain, K_p is the gain of the proportional term, and K_i is the gain of the integration term. In practice, this system starts smoothly but accelerates faster for larger error signals.

A differential term may also be used. Since the differential signal magnifies the response to any slight error, it is generally used when the robot reaches the vicinity of the desired end point, to increase accuracy, and to increase system sensitivity due to slight errors caused by arm bending or loading. Adding derivative control anticipates changes in the error and provides the fastest possible response to this change.

A complete PID system would provide

$$V_{out}(t) = K_p\varepsilon + \frac{K_i}{T_i}\int \varepsilon(t)\, dt + K_d T_d \frac{d\varepsilon}{dt} \qquad (4\text{-}27)$$

A PID system typically has four gain constants that must be adjusted, or *tuned*, to match the servo system to the load characteristics (inertia, friction, etc.). These adjustments are required for maximum system accuracy. They allow control over velocity loop gain, position error gain (K_p), gain at zero velocity (the integral term K_i), and gain at maximum velocity (the differential term K_d).

PI and PID controllers also affect system phase errors. A PI system increases low-frequency gain (becoming infinite at $\omega = 0$) and thus reduces or eliminates steady-state errors. By reducing high-frequency gain and providing phase lag, a PI system also increases stability margins. In a PID system, however, the differential term tends to make the system unstable, since system gain increases at high frequencies and normal phase shifts through the circuitry can provide an unstable point (causing oscillation). To get around this, a single pole can be designed into the PID filter to provide the necessary phase margin.

4.5 HALL-EFFECT TECHNOLOGY

Most types of sensor technology can be classified as internal or external sensing. An important technique used with both classes is the Hall effect. This technology was first used in the computer industry as the sensor in an all-solid-state keyboard.[6]

A Hall element is a flat piece of conductive material through which a constant reference current is passed (see Fig. 4-7). Current passes uniformly through the material, and electrical contacts at each side of the element, perpendicular to the current flow, will have no voltage across them. However, a magnetic field shifts the current flow across the Hall element, so there is now a difference voltage across the two side contacts. The stronger the field is, the greater this voltage is.

With proper placement of magnets, the Hall effect can be used in many applications, such as position, tilt, pressure, speed, and temperature, and is thus ideal for robotics. Hall-effect sensors have other advantages. (1) Since they are solid state, they have extremely long life (over 20 billion operations in one life cycle test). (2) They respond at higher speeds (100 kHz) than do many other sensors. (3) They have high repeatability and operate under wide temperature ranges. (4) The transducers can be made very small (less than 0.1 in by 0.1 in).

Hall-effect devices are often used in limit switch conditions. With a sealed bellows, they can measure temperature or pressure. Using ring magnets, they

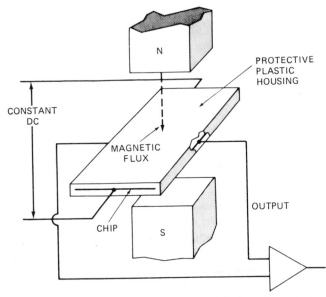

Figure 4-7. Hall-effect principles. (*From W. G. Holzbock,* Robotic Technology: Principles and Practice. *New York: Van Nostrand Reinhold, 1986, p. 255; copyright © 1986 by Van Nostrand Reinhold*)

provide a good tachometer sensor (rotary speed). One possible limitation is in rotary position measurement. Limitations in the change of magnetic field and the size of the magnet reduce the resultant accuracy below most other measurement approaches, such as optical encoders.

A basic formula for all Hall-effect sensors is

$$V_{out} = KIB \sin \Theta \qquad (4\text{-}28)$$

where B is the magnetic field in gauss and the variable K is a function of the transducer material, its thickness, length-to-width ratio, ambient temperature, and pressure. If these conditions and the input reference current I are held constant, then the output voltage is directly proportional to the magnetic field component perpendicular to the element.

Although the basic device is quite sensitive to temperature (a change of 25°C causes a change in output as much as 20%), this sensitivity can be compensated for in the sensor, and units are available that hold output to a gain shift of less than 0.1% from −25° to + 85°C.

Hall-effect sensors can either be arranged as a simple on/off device giving a digital output or to provide a varying voltage as a function of the magnetic field (analog output). In the on/off arrangement a hysteresis effect within the device has to be taken into account.

REFERENCES

1. Bass, J. T., et al. Development of a unipolar converter for a variable reluctance motor driver. *Proc. IEEE Industrial Applications Society.* 1985.
2. "Robot Sensing and Intelligence," IEEE short course presented by satellite transmission in November, 1983. Prepared by the Institute of Electrical and Electronics Engineers, Parsippany, N.J.
3. Tal, Jacob. *Motion Control by Microprocessor.* Mountain View, Calif.: Gail Motion Control, 1984.
4. Franklin, Gene F., and J. David Powell. *Digital Control of Dynamic Systems.* Reading, Mass.: Addison-Wesley, 1980.
5. Phillips, Charles L., and H. Troy Nagel, Jr. *Digital Control System Analysis and Design.* Englewood Cliffs, N.J.: Prentice-Hall, 1984.
6. "Hall Effect Transducers: How to Apply Them as Sensors," Micro Switch Division, Honeywell, Inc. 1982 (report).

ROBOTIC SENSORS

Robots need a wide range of sensor types to obtain information. Sensors provide feedback to the robot containing information about the robot's action (internal sensors) or about its environment (external sensors). Although a few types of sensors may be used for both internal and external sensing, it is more convenient to discuss these two classes separately.

5.1 INTERNAL SENSORS

Internal sensors inform the robot of the position, velocity, and acceleration of its joints. This type of sensing has also been called *internal state sensing* and *proprioceptive* sensing, a biological term referring to our ability to sense where our arms and legs are positioned. Perhaps a better term is *kinesthesis*—the sensing of the body's position and motion. Reaching for an object with our eyes closed, we rely entirely on kinesthesis, and we obtain position information from muscles, tendons, and joints. The most remarkable feature of kinesthesis is that it provides both position and movement information. It is one of our most important senses and, therefore, should be quite important to a robot also. Yet little work has been done in this area, and few robots provide true kinesthesis sensors.

Robotic systems use four classes of internal sensors to report position, velocity, force, and acceleration. Position sensors are the most common and are often used as a part of the feedback loop in a servo system.

Position Measurement

Translational and rotational movements must both be measured. A frequently used translational movement sensor is the linear potentiometer, whose resistance varies as a function of arm movement along its track. The sensor is accurate up to 0.2% of full scale, and it can be as small as 1 in (25 mm) or as long as 39 in (1 m), although the large devices are usually heavy and bulky. Linear potentiometers have high linearity ($\pm 0.05\%$), high resolution (0.001 in), and high repeatability ($\pm 0.002\%$ full scale).

Most potentiometers have a linear taper; that is, with a constant voltage placed across the unit the output voltage (connected to the sliding tap) is proportional to position. With an accurate potentiometer, the resultant motor position and an appropriate error signal can be accurately determined.

Rotational potentiometers may be single turn or multiple turn (typically ten turns) and are usually quite small. For improved accuracy, especially near the end of the resistance element, some units have two contact points 180° apart on the potentiometer arm.

The resistance element may be nickel chromium wire, cermet (*ceramic metal*), or plastic (conductive films). For wire potentiometers linearity is about 0.03%; for plastic models it is about eight times poorer (0.25%). In a single-turn wirewound potentiometer, the resultant accuracy is about 6 min of arc. The life of a wirewound or a cermet element is typically 1 million cycles, whereas conductive plastic lasts for 10 million cycles.

Magnetic means (linear variable differential transformers) also can be used to measure translational motion. A movable magnetic coil inside fixed coils is linked to the moving part. An AC reference voltage is applied to one set of coils, and the signal is retrieved from the other set. The amplitude of the output voltage and its polarity are linear functions of displacement from a center null position.

For rotational measurements many devices are available. One such device is the *optical encoder,* which uses a photoelectric method to detect the movement of an internal disk. An optical encoder can provide *incremental* or *absolute* information.

An absolute encoder gives an angular reading that corresponds to its specific position. Since encoders are sensitive to noise, absolute encoders often use a special output code called the *Gray code* (see Table 5-1). The Gray code has been designed to ensure that only a single bit changes value for each

Table 5-1. Gray Code Example

Decimal Value	Gray Code
0	0000
1	0001
2	0011
3	0010
4	0110
5	0111
6	0101
7	0100
8	1100
9	1101
10	1111
11	1110
12	1010

increment of the absolute encoder. Thus, absolute encoders are less suscepti-ble to errors in determining position. The accuracy of these encoders is a function of their speed. During periods of high velocity (5,000 rpm), the accuracy obtainable may be only 3.5°. With slow speeds (5 rpm), accuracies of 0.6° are possible.

Two types of incremental encoders are available. Single-output incremental encoders produce an output pulse each time an angular increment occurs, generally every 1.8°. Through gearing, this pulse can occur at much smaller rotational increments. However, there is no way to determine the direction of encoder rotation with only a single output. If an external load or other force moves the arm enough to produce an output pulse, the computer could not ascertain the direction in which it moved.

Therefore, most robots use a dual-output encoder, with the spacing of the photocells arranged to produce alternate outputs, as illustrated in Figure 5-1. The direction of rotation may be determined by comparing the order of black-to-white transitions from the two outputs. For example, a black-to-white tran-sition from photocell 1 followed by a black-to-white transition from photocell 2 indicates clockwise rotation. A black-to-white transition followed by a white-to-black transition indicates a counterclockwise rotation. Dual-photo-cell encoders offer another advantage. Since pulses occur twice as often, the effective resolution is doubled, and it may be doubled again through a phase detection technique.

Incremental encoders are much less expensive than absolute encoders, but they have two disadvantages. First, since absolute position is not supplied, it is up to the robot controller or other computer chip to count the pulses and convert them into the actual angle. Second, after electrical noise pulses or

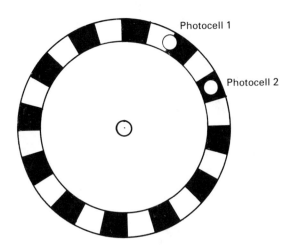

Figure 5-1. Dual output incremental encoder.

other disturbances produce an error in the computed angle, it is usually necessary to reset the robot by moving the affected joint to one end of its travel and resetting the incremental counter. Under conditions of high electrical noise (e.g., with stepper motors), even absolute encoders might produce many momentary errors in their output, thus requiring filtering of the data.

Another device used for angular position measurement is the *resolver*. A resolver has one movable magnet coil that rotates inside several fixed coils. As the device rotates, the movement changes the amount and phase of current induced in the external coils. When two base coils are used, the device is called a *resolver;* with three base coils, it is called a *synchro.*

Figure 5-2 illustrates the principle. An oscillator or other AC signal source provides the reference signal, applied in this case to the rotor, generally between 400 Hz and 10 kHz. The two output coils are 90° out of phase, so a sine wave signal and a cosine wave signal are both available for decoding. The resultant signals are analog, so they must be combined to determine absolute position and then converted into a digital signal by a special integrated circuit (IC) called a *resolver-to-digital converter.* These converters are available with 10- to 16-bit accuracy (depending upon resolver speed), corresponding to better than 1/3 arc-minute. Accuracy is affected during acceleration and over a range of temperatures, resulting in a typical resolver system error of 2 to 4 arc-minutes.

Several types of resolvers are available. Brushless resolvers are very reliable, with mean time between failures (MTBFs) rated at over a century. Their principal limitation is speed; about 10 rps (600 rpm) gives the best accuracies.

In addition to potentiometers, resolvers, and encoders, Hall-effect sensors

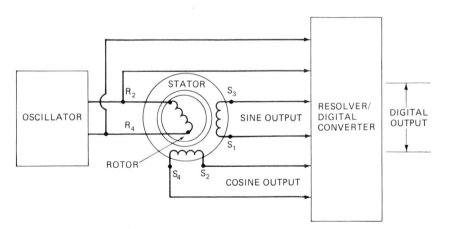

Figure 5-2. Resolver principles. (*From W. G. Holzbock,* Robotic Technology: Principles and Practice. *New York: Van Nostrand Reinhold, 1986, p. 263; copyright © 1986 by Van Nostrand Reinhold*)

can be used to determine rotary shaft position, provide overtravel sensing, and detect direction of travel. Hall-effect sensors, described in chapter 4, sense the approach of a magnetic field through the change in current flow. Regardless of the sensor used, they should be colocated directly on the axis to be measured, to increase their accuracy and reduce the effects of backlash, inertia, and linkage deformity.

Velocity Measurement

In most robots, speed of arm movement must be monitored. This measurement is almost always of rotational velocity; if linear speed is desired, then a method of converting translational motion to rotational motion, such as rack-and-pinion gearing, is required.

By using a stable time base, we can derive velocity from any position-measuring sensor (such as counting pulses per unit time from an optical encoder). In one approach light-emitting diodes (LEDs) are mounted on one side of a disk. With a hole in the disk and a photocell as a detector, the light pulses may be used for speed measurement. However, sensors that measure speed directly provide the data without imposing any computational load and usually with greater accuracy.

Tachometers are the most common speed-measuring device. DC tachometers are essentially small generators that produce an output voltage proportional to their speed. AC tachometers are not used as often, but their output can be combined with signals from a resolver, if the same excitation signal is used for both devices. Minor deviations in tachometer linearity occur, but these errors are usually no more than 0.2%, accurate enough for any speed-monitoring requirement.

Hall-effect generators have also been used to sense the speed of a rotating shaft by using the effect a rotating magnetic field has on the Hall transducer.

Force Measurement

Force detection of the load on a robot's arm is another type of internal sensing. Force measurements determine the weight of an object and can alert the robot to a potential overload condition. Without such a measurement the robot could pull itself over or damage its arm by attempting to pick up an object that is too heavy or too firmly fixed in its position.

For internal force sensing, strain gauges are the most common method employed. Other methods include Hall-effect sensors, piezoresistive devices, and monitoring the current drawn by the arm motor.

Strain-gauge technology was originally based upon the change in resistivity of a wire as it is deformed under pressure. The range of forces in which this type of strain gauge can be used depends on the coefficient of deformation, which varies from 0.5 for manganin to 3.5 for elinvar. Typical available sensors are accurate to 3% of full scale, with ranges of 0–2 lb and 0–25 lb, depending on material selected. Recently, silicon strain gauges have been developed, which are smaller, but they are still too fragile and brittle for industrial use.

The application of strain gauges requires special attention to temperature compensation, since they change resistance as a function of temperature as well as pressure. One approach is to use two strain gauges in a bridge arrangement, with only one of them subject to the pressure being measured. Temperature variations will be balanced out by the bridge, but pressure will not.

Force detection is also used in external sensing, particularly as part of a gripper. However, different types of devices are used, and these are discussed in section 5.2.

Accelerometers

Acceleration can be calculated by monitoring speed changes over time, but this approach is rarely used because of accuracy limitations. A more practical approach is using strain gauges to measure the force created by the acceleration of a known mass. The equation is

$$F = \frac{\Delta RSE}{RC} \tag{5-1}$$

where S is surface area, E is Young's modulus, R and ΔR are the resistance and change in resistance of the material, respectively, and C is the deformation constant for the strain gauge material used. By Newton's law, $F = ma$, so we can substitute the value of force from equation (5-1) and solve for a:

$$a = \frac{\Delta RSE}{mRC} \tag{5-2}$$

The most accurate method for determining acceleration is to measure the amount of opposing force necessary to maintain a mass in an equilibrium position. Accelerations ranging from a few tenths of a g to thousands of g's can be measured this way.

Angular acceleration can be calculated by measuring tangential acceleration at a known distance from the rotation center and by measuring the rotation angle during a 1-s period. The angular acceleration is then proportional to the product of the radius and the second derivative of the angle.

5.2 EXTERNAL SENSORS

External sensors are primarily used to learn more about the robot's environment, especially the object being manipulated. External sensors run the gamut of simple limit switches to those measuring force, temperature, or object slip.[1]

Contact Sensors

To sense objects about it, the robot can use both contact and noncontact sensors. Contact sensors have the disadvantage that they require physical contact. Thus they cannot give information about a part of an object that the robot is unable to reach. On the other hand, contact sensors give important information on object shape, weight, and slippage, which are very useful in handling objects. These sensors are also generally inexpensive, and the information obtained requires less processing than visual data.

The simplest contact sensor is an electrical switch. It is easy to use and inexpensive, but it cannot be used to obtain three-dimensional data or to measure pressure. Pneumatic switches can also be used, but they are more expensive. By placing contact sensors in arrays, we can obtain two-dimensional and, under some conditions, limited three-dimensional data.

Sensors that must actually touch something in order to operate are limited to circumstances that will not damage the object or the sensor. One obvious application of an array type of touch sensor is locating the exact position of a hole so that the robot can insert parts into it.

Tactile sensors are placed on the gripper or other specially designed robot hands. These sensors can be contact sensing (touch), pressure sensing (force), or slip sensing (slide), and they include all devices that mimic the operation of the human hand.[2]

In humans, tactile sensation is a property of the skin, although the skin is not uniformly sensitive. Fingertips, for example, are much more sensitive than the wrists. Touch may be sensed, without direct contact, through subtle air motion simply by moving a hair, for example, which helps explain how a person can sense someone else in a room even though that person is out of sight.

Touch sensitivity is also a function of the rate at which the skin is deformed. The skin responds to separate touches up to a frequency of about 20 Hz, at which point it changes to a feeling of vibration. Maximum vibration sensitivity is about 250 Hz. Differences between vibration frequencies can be determined. The skin soon adapts to lightweight contacts, and after a short while they are not felt at all. For example, you do not notice that you are wearing a shirt, yet a new contact, such as a leaf falling on your arm, will be detected immediately.

In robots, touch sensing can help to reduce the required amount of structure in the environment. For example, if the robot cannot see or feel a part, then the part must be located where the robot was told to expect it. If the robot can sense object location through touch, exact part placement is not quite as critical. (Vision systems also reduce the amount of structure needed, but even here, without touch, last minute minor corrections to the robot's movement are not practical because the gripper will block the camera's view.) With touch sensing, the robot's hand is telling the arm processor how to move, thereby providing a type of end-effector-driven manipulation. Naturally there are delays in this type feedback, and total delays may be as long as a few seconds.

Noncontact Sensors

To circumvent possible damage, robots can use *proximity sensors*, which sense when an object is close to, but not touching, the robot. Proximity sensors are often used for limit switches. With no actual mechanical contact required, sensor reliability is increased and safety is improved. Proximity sensors can be magnetic or optical. In the magnetic mode, they can sense at a 3/8–5/8-in distance by using ferrous actuators. Magnetic techniques include the use of reed relays, eddy current sensors, and Hall-effect devices.

Another type of noncontact sensor is the *optical sensor*, which uses the presence of an object to block out light, and usually operates in an on/off mode. More complex optical systems can determine the amount of motion through the amount of light detected. These sensors can be used in conjunction with other techniques to provide force or pressure information. Optical systems employ visible or infrared (IR) light, and a technique based upon either light reflection or light interruption is used.

Two major types of noncontact sensors—vision systems and ultrasonic systems—are covered in chapters 6 and 7.

Pressure Transducers

Pressure transducers measure external pressure or force and can detect an object's presence, shape, and orientation. Object presence can be sensed with a single measurement. Determining object shape requires an array of sensors, such as 8 × 15, to provide the necessary image resolution. Object orientation involves matching the object's stored shape template to the feedback from a similar array of sensors.

By monitoring the applied pressure, and knowing the characteristics of the object, the robot's end-effector can close on an object with sufficient force to prevent slipping and damage to the object. For example, knowing how much force to apply to an egg will allow a robot with a pressure sensor to pick up the egg. Due to variations in egg sizes, accurate feedback from the egg must occur or the egg will be damaged.

Table 5-2 lists five types of sensors and compares them to the human hand in several categories. Current technology compares favorably. For example, most sensors are faster than, and have a comparable resolution to, a human finger, but they are not as sensitive and cannot handle as much weight. An approach not included in this table uses a piezoresistive sensor. Sensor chips incorporate an integral silicon diaphragm with ion-implanted piezoresistors. One example is the Micro Switch 16PC. These resistors change value with applied stress and produce a voltage output proportional to pressure.

Three types of sensors are available commercially to determine the pressure exerted by a robot hand or gripper on an object: conductive elastomer sensors, optical sensors, and silicon strain gauges. In addition, research laboratories are investigating other types, including carbon fibers, capacitance strips, ultrasonics, and ferroelectric polymers. There are even piezodiodes with special p-n junctions that change their reverse bias characteristics as a function of pressure. The following paragraphs discuss techniques both currently available and still in the research stage.

Figure 5-3 illustrates three devices based upon piezoelectric effects. The first approach determines the change in resistance between two sets of metal strips (or wire) as pressure is applied. The strips are arranged in an array with a conductive rubber sheet placed between them. The second approach is based on the creation of a small electric charge under either temperature or pressure. This very interesting approach is described in detail later in this section. In the third approach, the applied force produces an electric voltage in the silicon, with the strength of the signal proportional to the pressure.

Table 5-2. Tactile Sensor Comparison

	Resolution (in)	Sensitivity (g)	Range (kg)	Speed (ms)	Elements
Human hand	0.03	20	100?	50	10×20
Deformable elastics	0.07	40	0.7	3	16×16
	0.3	3.2	0.7	3	8×8
Optical fiber	0.1	4	1.0	80	16×16
Conductive elastomers	0.05	230	50	33	8×16
	0.1	230	50	33	16×16
Silicon S/G	0.08	10	1.0	NA	3×3
Piezoelectric	0.11	20	80	10	8×16

Source: After P. Dario and D. de Rossi. Tactile sensors and the gripping challenge. Spectrum, August, 1985.

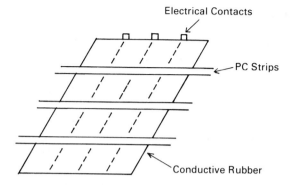

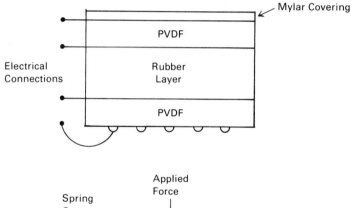

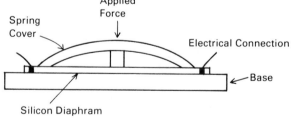

Figure 5-3. Piezoelectric approaches.

Conductive polymers change resistance under pressure, and this is the principle of most piezoelectric devices. The change in resistance then changes the current flowing through the device. By an external precision resistor, the change in current is converted to a change in voltage. This change in resistivity allows piezoelectric devices to be used as a substitute for strain gauges, and they have three key advantages: higher sensitivity (100 times), lower hysteresis, and stronger band strength (10 times).

Other types of pressure-sensing devices include carbon fibers and various types of conductive rubber and elastomer sheets. The elastomer sheet forms a pressure-sensitive variable resistor between two sets of metal contacts that provide the output. It is inexpensive but wears out after only a few hundred operations.

Another conductive silicon approach uses one round component, such as a rod, and one flat component. As they press together, the silicon rubber material deforms and the contact area increases. With this device as a leg in a voltage divider, the output voltage drops rapidly at first with small changes in pressure, and then less rapidly as the pressure increases, decreasing resistance. This device is hard to overload or damage. This voltage change is entered into the computer through an analog-to-digital (A/D) converter.

Optical tactile sensors are based on changes in light energy transmitted through the sensor due to changes in pressure. Different approaches are being investigated, including the two shown in Figure 5-4. The sensor at the top uses physical interference of a movable rod on a light beam to provide a signal. A LED/photodetector pair measures the light transmission. With the elastic material covering the rod providing a back resistance to the force applied, the amount of downward motion is proportional to the pressure. This type of sensor is available commercially in an array of 10×16 elements in approximately a 1-in^2 area.

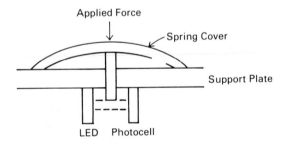

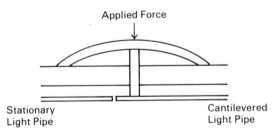

Figure 5-4. Optical tactile techniques.

At the bottom of this figure two light pipes (optical fibers) are aligned, allowing a maximum amount of light energy to be transmitted. Force on the outer elastic material (molded rubber, for example) presses one light pipe slightly out of alignment, thus reducing the amount of light transmitted and providing a signal proportional to the applied force.

Carbon fiber sensors are made of graphite fibers approximately 7–30 μm in diameter. Although fiber resistance varies directly with pressure change, this variation is generally not the variable used. Instead, the variation in contact area is used. As the pressure increases, more fibers are pressed together and the contact area increases. With more fibers in contact, the resistance decreases. For example, the resistance might be 2,000 Ω with no contact pressure, 1,000 Ω under a weight of 1 g, and 200 Ω at 5,000 g. Carbon fiber sensors are cheap but quite difficult to electrically connect. In one application, washers were formed from carbon fibers. The washer was placed on a bolt, which provided feedback to the robot on how tightly the bolt was turned.

Another research approach detects capacitance changes between conductive strips overlaying a rubber sheet. Changes in distance between the strips caused by a change in applied pressure can be detected by the resultant change in capacity. Capacitance changes are also being investigated to detect changes in temperature.

Ultrasonic transmissions can be used to translate compression, caused by pressure, into an electrical signal by measuring the thickness of an object. The transit time of an ultrasonic pulse through a known media, or reflected from the material's rear surface and back through the media, can be very accurately timed and the time can be translated to the amount of deformation. Since the amount of pressure needed to cause a specific amount of deformation is known, the force on the sensor surface can be calculated. Problems to be overcome include the effect of temperature, which will impact both the transit time and the amount of deformation, and the difficulty of obtaining good spatial resolution.

Ferroelectric polymers offer perhaps the greatest potential because they provide electrical signals from pressure and temperature changes, similar to the action of human skin. Biophysicists have determined that the response of human skin is analogous to the response of synthetic ferroelectric polymers. The polymer used most often is polyvinylidene fluoride (PVDF), which is available in thin sheets.

Pressure on a PVDF material produces a small electric charge across the film, which can be detected through electrical contacts. At the University of Pisa, Italy, researchers, using two layers of PVDF film separated by a layer of rubber to provide both insulation and a natural "give" to the surface, have developed a sensor with properties similar to the human skin. (This approach is the second illustrated in Fig. 5-3). Since the change in voltage produced by a temperature change is opposite to that produced by a pressure change, and since the separating rubber layer provides thermal insulation with a long time constant, a comparison of signals at different times from the two PVDF layers allows both temperature and pressure to be determined.

This film is very versatile. The outside sensor layer also provides a measure of surface texture if rubbed against the object. It has even been used to measure the pitch of a grating. Problems include the electronic difficulties inherent in charge amplification, the effect of pressure on one sensing point spilling over to adjacent points, and the lack of any signal during static conditions.

Motion Detection or Slip Sensing

Another approach to picking up fragile objects is *slip sensing*. Slip-sensing devices detect movement of the object due to gravity or some other force and, through feedback, can cause the gripper to be tightened slightly until the slippage stops. One technique uses an array to detect changes in the area receiving pressure. There are also methods based on translating linear motion into rotation, by having the object contact spheres that rotate under tangential forces. Some slip sensors operate similar to a record player, where a sapphire needle moves in conjunction with object motion, thus providing an electrical signal.

Workplace Sensing

The techniques described thus far have generally been placed on a gripper or built into a robot arm. It is also possible to have a freestanding array that the robot uses by placing an object on it. One such array, made by Lord Corporation, is shown in Figure 5-5. The robot arm is able to determine the shape, location, and orientation of the part through feedback from the array of sensors in the pad.

This tactile array has a 10 × 10 grid. It is possible to effectively obtain much higher resolution by using the robot to move the object around on the grid and comparing the results. In one application, the manipulator can check part orientation before mounting it into an assembly. A robot has even been programmed to turn over an object by dropping it on an edge so that it will fall upside down. Clever tricks like this can often be used in the workplace.

In another workplace application, a robot uses force sensing to weigh objects it is carrying. Accuracies of 1 part in 1,000 are possible. This system, developed by the Metler Corporation of Switzerland, uses precision force sensors to obtain an object's weight while the object is being moved. Note that there is no lost time, since the weight measurement is made while the object is being moved into position.

Figure 5-5. Workplace-mounted tactile display. (*Courtesy of Lord Corporation, Industrial Automation Division*)

5.3 SENSOR PROCESSING

Data from external sensors must be analyzed by a computer before they can be used. The first stage of processing—standardizing voltage levels and converting the analog data to digital form—is most often done at the sensor. This preprocessing produces a signal that can be transmitted over short distances to a computer.

A second requirement in sensor processing is usually calibration. Since most sensors are nonlinear and even have a zero shift bias built in, some type of standard may be necessary to provide the required accuracy. For example, the weight of the gripper should be removed so that only the weight of the object will be measured. The tactile coordinate system may need to be adjusted to match the gripper's own coordinate system. In optical approaches, the effect of ambient lighting may need to be considered. This level of processing may also be done at the sensor.

A third need in sensor processing is the extraction of parameters from the available array information. With tactile sensors, approaches include two-dimensional analysis, three-dimensional analysis, and force analysis. Most two-dimensional and three-dimensional analyses are based on algorithms developed originally for vision processing. The same types of parameters are often extracted, such as area (size), perimeter (function of shape), and major axis (orientation).

Simulations have been used in developing specific algorithms. In one simulation, a hand was provided with both tactile and force sensing.[3] The task is for the hand to orient itself with respect to the object in order to grip it. The difficult part is that no advance information on object size or shape was given to the computer. The technique studied used a gradient search procedure, with a threshold so that a certain level of confidence had to be obtained. The objective was for the robot to work with unknown objects, without using vision and with local tactile data only.

REFERENCES

1. Flynn, Anita M. *Redundant Sensors for Mobile Robot Navigation*, Report No. AI-TR-859. Cambridge, Mass.: MIT Artificial Intelligence Laboratory, 1985.
2. Datseris, P., and William Palm. Principles on the development of mechanical hands which can manipulate objects by means of active control. *ASME Journal of Mechanisms, Transmissions and Automation in Design*, Vol. 107, June 1985, pp. 147–156.
3. Fearing, R. S. Implementing a force strategy for object re-orientation. *Proc. 1986 IEEE International Conference on Robotics and Automation*. San Francisco, April 1986, pp. 96–102.

CHAPTER

6

VISION SYSTEMS

An accurate, fast, high-resolution vision system greatly improves the robot's versatility, speed, and accuracy for its complex tasks. For example, during welding the robot might have to follow curved seams, during assembly the robot often selects the appropriate part and installs it, and during inspection the robot verifies that previous manufacturing steps were performed correctly.

The earliest robot systems were used for hard automation and did not employ vision systems. In other words, parts had to be placed precisely, so part positioning fixtures were an expensive component of robot systems. A vision system lets the robot look around and find the part. Other sensors (e.g., contact sensing) could be used, but vision systems give the robot its greatest potential.

Vision systems provide information that is difficult, or impossible, to obtain in other ways [e.g., higher resolution, gray scales, color, and shape (three-dimensional)]. Their operating range is from a few inches to tens of feet, with either narrow-angle or wide-angle coverage, depending on both system needs and design. Humans have more experience and are more comfortable using vision, so we find it more natural, at least in concept, to develop programs and image processing techniques with vision systems. These image processing techniques have often been adapted for use with other sensors (e.g., tactile).

Although a few vision systems have been available for about ten years, they have not been widely used because of their limited capabilities and high cost. Not until 1985 were any but the simplest systems actually used in manufacturing. However, robot vision systems are now used more and more frequently because they allow new applications to be performed using robots. Most experts expect that within a very few years 25% of all robot systems will contain vision.

Vision systems are used in many areas other than robotics. Consequently, the general term *machine vision* is often used. The Automated Vision Association defines machine vision as "a device used for optical noncontact sensing to receive and interpret automatically an image of a real scene in order to obtain information and/or to control machines or processes." We use the terms *machine vision* and *vision systems* interchangeably to refer to systems in the visible spectrum that use a vision sensor with the appropriate image processing programming to locate the position, and in many cases the type, of an object. Photo cells may be used with conveyor belts to determine when an object has moved into a robot's range, but this is not a complete vision system (see chapter 4).

6.1 HUMAN VISION CONSIDERATIONS

Robot vision systems, though not always based on human vision, do have characteristics in common with it. In addition, human vision offers a point of comparison for any machine vision system, yet few engineers have much knowledge about human vision. A more complete discussion of these topics may be found in Poole.[1]

The human eye consists of rods and cones that are sensitive to light with wavelengths from 350 to 750 nm. Rods, being more sensitive to lower levels of illumination, allow us to see in very dim light. Their frequency response is also slightly shifted to the shorter wavelengths of light and are thus less sensitive to red. Cones allow us to distinguish colors, but since the cones are not as sensitive to light as rods are, the colors fade to shades of gray as darkness closes in. Because they are clustered in the central region of the eye, cones provide better central vision resolution. Figure 6-1 compares the responses of cones and rods to various wavelengths of light.

Spatial resolution, or visual acuity, measures the ability of the eye to see fine detail. The five types of visual acuity are minimum visible, minimum perceptible, minimum separable, vernier, and stereoscopic, depending upon how the acuity is being applied. For comparison with robotic vision systems, it is customary to consider the eye's resolution as 1 arc-minute.

The fovea, the region of sharpest vision, covers about 5° in diameter. Within this region are approximately 350,000 cones; the complete eye has perhaps 100 million rods and cones. Within the optic nerve 800,000 nerve fibers pass information to the brain. It appears that the central cones each have a single nerve cell connection (accounting for their higher visual acuity), whereas peripheral rods and cones must share nerve fibers.

Nerve impulses travel at the rate of 100 m/s and take approximately 0.5 ms to reach the brain. Changes in object light intensity seem to have no effect on the strength of these nerve impulses. Instead, different brightnesses affect the

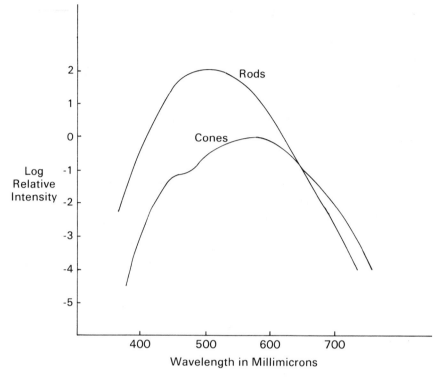

Figure 6-1. Eye sensitivity.

frequency of nerve signal recurrence. Frequency is also affected by the duration of the signal. This latter characteristic contributes to our ability to recognize motion. Motion detection is greatest in peripheral vision and seems to depend on the length of time the stimulus affects a rod and on changes in the adjacent rods.

Parallel nerve connections from the eye to the brain provide very fast data transfer and enable the eye to form images. In comparison, most robotic image processing systems must collect and transmit their data sequentially. This method is time consuming and usually delays processing until all the necessary information has been transferred. Another difference is that humans essentially see on a continuous basis and have developed the ability to recognize motion, especially in peripheral vision. The machine vision system usually takes snapshots (or frames) so that it is not overwhelmed with data. The result is that objects are never seen in motion but they appear in different positions in successive frames. Although software methods of detecting motion do exist, the recognition of motion by hardware has generally been impossible. Dana Anderson,[2] of the University of Colorado, has developed an approach to circumvent this limitation. Called a *novelty filter*, this motion detection hardware system compares two images and reinforces any changes found between them.

Six attributes of human vision have been matched in machine vision: the spatial resolution attributes (x, y, and z) and the color attributes (hue, saturation, and intensity). The human eye and a machine vision system can both resolve detail in the x- and y-directions. The z-axis (depth), however, is more difficult to solve in machine systems. The eye/brain combination uses numerous clues to determine depth, most of which are still untapped by machine vision systems.

Human depth perception is quite good. The average observer can discern which of two small objects is closer to him or her at a distance of 18 in even if the objects are separated by only 1/4 in. This ability translates to a difference in subtended angle of only 10 arc-seconds. Figure 6-2 illustrates the slight difference in angle ($\Delta\beta$) found between objects separated by slightly different depths. Because of numerous sources of errors, robot vision systems (as distinguished from laser ranging systems) cannot judge relative depth well. The robot, however, can estimate absolute depth better than humans.

Machine image processing of hue (color; e.g., red), saturation (color strength; e.g., deep red), and intensity (image brightness; e.g., bright red) differs from human image processing. From a physical viewpoint, hue corre-

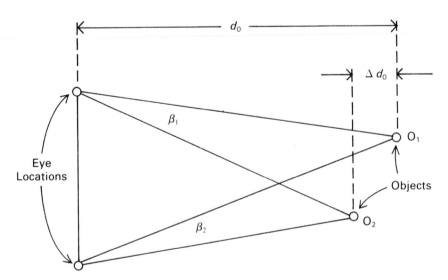

Figure 6-2. Stereo depth perception. (*Adapted from Harry H. Poole*, Fundamentals of Display Systems, *Washington, D.C.: Spartan Books, 1966, p. 277.*)

sponds to dominant wavelength, saturation to spread of wavelength, and intensity (luminance) to relative signal strength. Although some inspection systems use complementary colors to highlight markings, few machine vision systems actually recognize color. One example of a true color system is the Applied Intelligent System's Pixie-5000.

In addition to the lack of color information, monochrome systems do not separate color and saturation information from intensity. Therefore, a monochrome vision system cannot tell if a difference in intensity is caused by a difference in image brightness, image color, saturation, or a combination of these parameters.

Another difference is that the human eye, and by extension the robot eye, only needs high resolution in a small circle of central vision. But robot vision cannot automatically zero in on an object, so it must look in all areas of its visual field; therefore robot vision systems currently need equal resolution over the entire field.

6.2 MACHINE VISION APPROACHES

Robot vision systems are generally designed to answer one or more basic questions. Several of these questions, with the type of data output produced by the vision system, are shown in Table 6-1. Reliably answering these questions under any but controlled conditions has proven quite difficult and has been the subject of artificial intelligence studies for years. To answer these questions, some machine vision systems emulate various natural vision processes, while others use algorithms that have no natural counterpart. To understand the different approaches used in current systems, we break the vision task into three general steps: image acquisition, image analysis, and pattern recognition. We discuss these steps later.

Machine vision systems cannot uniquely represent and process all available data because of the computational problems and memory and time requirements imposed on the computer. Therefore, the system must compromise. Other problems include variations in lighting, part size, and part placement, and limitations in the dynamic ranges available in typical vision sensors.

Table 6-1. Vision System Capabilities

Question	Data Returned
1 Where is the object?	X, Y, range
2 What is its size?	Dimensions
3 What is it?	Identity code
4 Does it pass inspection?	Pass/failure code

Vision systems require specialized hardware and software.[3] It is possible to purchase just the required hardware with little or no vision application programming. In fact, few third-party image processing programs are available. A hardware-only approach is less expensive and can be more flexible for handling unusual vision system requirements. But since this approach requires image processing expertise, it is only of interest to users who wish to retain the responsibility for image interpretation, such as universities or system houses. The usual approach is to obtain a complete vision system, including computer system hardware and application software, from a vision system supplier. However, the user might still need custom programming for an application. Most major vision system suppliers specialize in providing software for only a few application areas.

Every machine vision system requires a sensor to convert the visual image into an electronic signal. Several types of video sensors are used, including vidicon cameras and solid-state sensors [a type of optical random-access memory (RAM)].[4] Many of these vision sensors were originally designed for other applications, such as television, so the signal must be processed to extract the visual image and remove synchronization information before the signal is sent to a computer for further processing. The computer then treats this digital signal as an array of dots (also called *pixels*) and processes this data to extract the desired information.

Image processing can be very time consuming. For typical scenes of 200,000 or more pixels, a machine vision system can take many seconds, even minutes, to analyze the complete scene and determine the action to be taken. The number of bits to be processed is quite large: For example, a system with a resolution of 512×512 pixels and 8 bits of intensity per pixel yields over 2 million bits to be processed.

If continuous images at a 30-Hz frame rate were being received, data bytes would be received at an 8-MHz rate. Few computers can accept inputs at these data rates and, in any case, there would be no time left to process the data. When higher-resolution systems, color systems, or multiple camera systems are considered, data handling requirements become astronomical.

Several methods can be used to reduce the amount of data handled and, therefore, the processing time. One approach is *binary vision,* which is used when only black-and-white information is processed (intensity variations and shades of gray are ignored). In binary vision a picture is converted to a binary image by *thresholding.* In thresholding, a brightness level is selected: All data with intensities equal to or higher than this level are considered white; all lower levels are considered black.

Another method of shortening processing time is to control object placement so that objects of interest cannot overlap in the image. Complicated algorithms to separate images are then unnecessary, and image processing time is reduced.

A third approach reduces data handling by processing only a small window of the actual data; that is, the object is located in a predefined field of view. For example, if the robot is looking for a fiduciary mark on a printed circuit board, the vision system can be told that the mark is always in the upper right corner.

A fourth approach takes a statistical sample of the data and bases decisions on this data sample. Unfortunately, all of these approaches ignore some of the available data and, in effect, produce a less robust system—processing time is saved, but some types of complex objects cannot be recognized.

6.3 IMAGE ACQUISITION

In image acquisition an image is obtained and digitized for further image processing. Although image acquisition is primarily a hardware function, software can be used to control light intensity, lens opening, focus, camera angle, synchronization, field of view, read times, and other factors. Image acquisition has four principal elements:

1. A light source, either controlled or ambient (scene illumination)
2. A lens system to focus reflected light from the object onto the image sensor (optics)
3. An image sensor, which converts the light image into a stored electrical image (image formation)
4. Electronics to read the sensed image from the image sensing element and, after preprocessing, transmit the image information to a computer for further processing

The vision system processor handles the remaining steps, image analysis and pattern recognition.

Scene Illumination

For the vision system to obtain an image, the scene must be illuminated. Ambient lighting is usually insufficient, so most systems use controlled lighting and, often, special lighting techniques. Controlled lighting also allows specialized image processing techniques to be used.

Table 6-2 lists ten types of lighting that can be used. Numerous light sources, including incandescent and fluorescent lamps, fiber-optic sources, arc lamps, strobe lights, lasers, and polarized light, abound. Figure 6-3 illustrates a fiber-optic system by Dolan Jenner for controlling illumination. Although most systems use light in the visible spectrum, infrared (IR) and ultraviolet light are also used.

Table 6-2. Types of Illumination

Lighting Direction	Function
Front	
Diffuse	Pick up features of the image
Specular	Detect surface defects .
Structured	Detect shape
Rear	
Silhouette	Binary signal representation
Transparent	Feature highlights
Collimated	Change light characteristics of object
Special	
Laser	Provide overlay lines or grid patterns
Beam splitter	Control light path
Part counting	Interrupt beam to indicate part present
Retroreflector	Object passes between the light source and the retroreflector; all light not being diffused by the object is returned to the sensor

Source: After M. P. Groover et al. *Industrial Robotics: Technology, Programming and Applications.* New York: McGraw-Hill, 1986, p. 167.

Figure 6-3. Fiber-optic illuminator. (*Courtesy of Dolan-Jenner Industries*)

126

Front lighting allows gray scale information to be obtained and thus resolve detail or surface features on the object. Side lighting helps bring out three-dimensional features. Back lighting supplies the least complex data, since it eliminates gray scales and provides the camera a silhouette of the object. It thus can only be used when gray scale information is not needed, and when a simple binary (black-and-white) image is sufficient. Figure 6-4 illustrates the

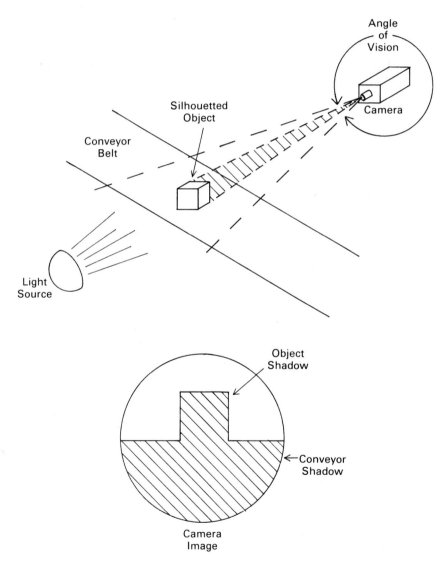

Figure 6-4. Backlighting.

application of back lighting, including an example of the type of image available. Silhouette robot vision can be used for property measurement, detection of missing parts, and simple inspection of final parts.

Diffuse lighting, which is similar to conventional lighting, is used when contour and surface characteristics are important. In diffuse lighting, the object is bathed with even light, allowing brightness variations in the object to be detected. By providing intensity data, or gray scales, this approach can be used for part identification, texture discrimination, and inspection of more complex parts.

With structured lighting, illustrated in Figure 6-5, the light is formed into a bright line by optical means or through a grating. The light source is kept to one side so that the fan beam will strike the conveyor belt obliquely. When an object, such as a box, moves under the fan beam, the resultant line on the top of the object is displaced toward the source of light. The vision system recog-

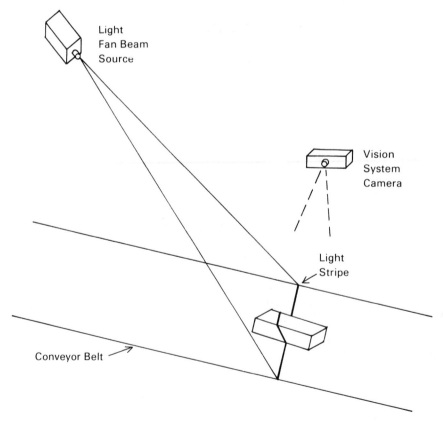

Figure 6-5. Structured lighting.

nizes the stripe displacement, and the robot is told an object is available. This system, still in use, was used to detect object presence on early systems such as Consight (a GM vision system).[5] With the addition of a method of scanning the light stripe to cover a horizontal area, as well as using the amount of displacement to tell height and shape, a three-dimensional image can be built up.

Many systems of this type use two light sources, with the light lines superimposed on the conveyor. This arrangement removes shadow effects that can interfere with determining the precise time an object crosses the beam. A similar approach uses directional lighting with a single light beam illuminating the object. This technique can be used to tell whether an object is present. If the beam is broken, an object is present. It can also be used to locate an object on a conveyor belt by a technique described later under ranging systems.

Optics

Another hardware area that relates to vision system performance is optics. All camera systems must have a lens to collect and focus the light on the light sensing element. There are three major factors to be considered in selecting lenses: (1) focal length, (2) light collecting ability, and (3) lens quality. In the following discussion, we assume that the task is to select a lens out of several candidates. We do not discuss factors involved in designing a complete lens, such as determining the number and type of lens elements required. Figure 6-6 illustrates the relationships among the object, the lens, and the image.

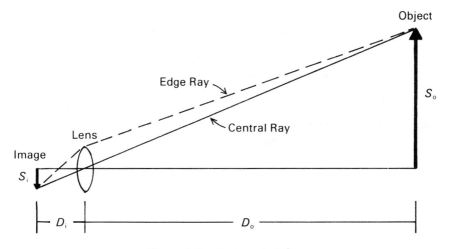

Figure 6-6. Optics principles.

To determine various optics parameters, we need some application-oriented information. This information includes the vision sensor being used, the size of the resultant light gathering area, the size of the object being looked for, and the size of the area to process for this object.

In most applications, image size S_i and object size S_o are known. The size of the image sensing element is known once a vision system is chosen, but the coverage area of the system depends on the application. Although in this type of application, image magnification calculations are not generally used, they are still convenient under some circumstances. Image magnification may be determined from

$$M = \frac{S_i}{S_o} \tag{6-1}$$

Image magnification is the inverse of lens power. If we know the focal length of the lens, we can determine image magnification from the lens focal length F and the object distance D_o:

$$M = \frac{F}{D_o - F} \tag{6-2}$$

There are several different formulas for focal length, some involving image distance. However, image distance is less accurately known in most systems than object distance or magnification. Thus, another useful formula relates focal length to magnification and object distance:

$$F = \frac{MD_o}{M + 1} \tag{6-3}$$

Light collecting ability is related to the lens aperture, or f number. Lenses with low f numbers are larger (for a given focal length) and can collect more light. These lenses are more expensive, however, and they also reduce the available depth of field (the area in which the object will remain in focus) and cause more aberrations and distortions.

Lens quality refers to a lens's freedom from various distortions. These distortions can come from several sources. One is diffraction, which affects different wavelengths of light differently, thus causing optical distortion, color fringing, and out-of-focus conditions. Other distortions are caused by monochromatic and chromatic (or color) aberrations.

Light Sensing Elements

Early vision systems employed vidicon cameras, which were bulky vacuum tube devices. Vidicons are also more sensitive to electromagnetic noise interference and require higher power. Their chief advantages are higher resolution and better light sensitivity. To reduce size, most current systems use solid-state cameras, based on charge-coupled device (CCD) or charge-injection device (CID) techniques, with a newer technique, charge-priming device (CPD), becoming available.

Solid-state cameras are smaller, more rugged, last longer, and have less inherent image distortion than vidicon cameras. They are also slightly more costly, but prices are coming down. The light-sensitive area on the chip is only $1/4$ in^2 and the smallest camera is currently about a 2-in cube. They are not damaged by intense light (although blooming does occur), and they do not exhibit memory lag, so they are free of smeared images from moving objects.

Since television was the earliest, and still is the largest, user of video cameras, it was natural that many vision systems adopted television standards. These standards cause problems when they are adapted for robotic vision. Television has a 4/3 aspect ratio, whereas machine vision algorithms need 1/1. Therefore, distance measurements at random orientations become difficult and time consuming.

Another problem is a nonlinearity in the response to light intensity, which is built into many cameras to compensate for nonlinearities in cathode ray tubes (CRTs). Gray scale linearity is determined by the formula

$$A \doteq KI^{\tau} \tag{6-4}$$

where the output amplitude A is some constant multiple of an exponential function of intensity I. Ideally, $\tau = 1$ for maximum linearity. In practice, though, often $\tau < 1$.

Another difficulty is in resolution figures. It is not always easy to compare information between television and machine vision systems. Television uses lines of resolution, and machine vision uses pixels. In standard television, scanning is done according to the Electronics Industries Association (EIA) RS-170 standard, with 480 lines divided into two fields. One factor in television systems that affects the definition of resolution is the correlation between the scanning lines and the image source (the Kell factor). Hence, television resolution cannot be directly related to the number of pixels in a machine vision system. However, television does have *ultimate picture elements*, which can be related to pixels. These elements have the smallest possible dimensions that picture elements can have, considering the bandwidth of the transmitted signal.

The total number of elements in a black-and-white picture is

$$n = \frac{2(\text{BW})P_V P_H}{f} \tag{6-5}$$

where P_V and P_H are the percentages of active horizontal and active vertical time, BW is the bandwidth, and f is the frame rate, both in hertz. When standard television is considered, the result is approximately 220,000 picture elements, with 480 pixels in the vertical direction and 458 in the horizontal.

Another difference between machine vision and television is the type of modulation used. Standard television images employ negative modulation (the lower the signal voltage, the greater the brightness), whereas most machine vision systems use positive modulation (i.e., a Gray scale of 0 refers to black, and 255 is the maximum white level).

Solid-state sensors have an available resolution of 250 × 250 pixels to 800 × 800 pixels, with a typical camera having 488 × 380 elements. Their sensitivity extends from 400 nm to 1050 nm, and often peaks in the near-infrared ($\approx$900 nm). Manufacturers thus often recommend operating the sensors with filters to eliminate IR wavelengths.

Both the CCD and CID chips use charge transfer techniques to capture an image. In a CCD camera light impinges on the optical equivalent of a random-access memory (RAM) chip. The light is absorbed in a silicon (photoconductive) substrate, with charge buildup proportional to the amount of light reaching the array. Once sufficient energy has been received to provide a picture, the charges are read out through built-in control registers. Some CCD chips use an interline charge transfer technique; others use a frame transfer approach, which is more flexible for varying the integration period.

A type of electronic iris can adjust the camera's sensitivity to light by controlling the amount of voltage across the elements and the length of time that the array is allowed to charge. The CID camera works on a similar principle. A CID chip is a metal-oxide semiconductor (MOS)-based device with multiple gates similar to a CCD. The video signal out is the result of a current pulse from the recombination of carriers. CIDs produce a better image (less distortion) and use a different readout technique than CCDs, which require separate scanning address circuits. CIDs are therefore more expensive than CCDs.

The principle difference between a CCD camera and a CID camera is the method of generating a video signal. In a CCD the charge must be shifted out of the array by switching gate voltages between the even and odd gates. The resultant flow of current creates the analog signal. In the CID, each element is discharged, and the resultant current is used directly (shifting is not required).

Linear models are also available. These models have a single row of light-sensitive elements and are used for line scanning applications. We can still obtain area pictures, if desired, from linear models by employing a moving mirror to scan the object or by moving the object past the camera (such as on a conveyor belt). This latter approach is used in many vision inspection systems, such as a system to inspect printed circuit boards (Fig. 6-7). Because only a single row of optical-sensitive elements is provided, these units give greater linear resolutions. For example, there are linear models with 256, 1,024, or 4,096 elements available.

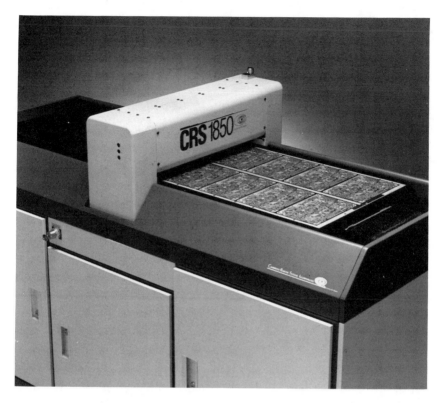

Figure 6-7. Machine vision inspection system. (*Courtesy of Cambridge Robotics Systems*)

The amount of charge stored is a linear function of both the object intensity and the length of time the camera is exposed to the light. If the intensity is insufficient for an image with good contrast, the exposure time can be increased, allowing more time to charge the elements. Although increasing exposure time increases system sensitivity, it reduces the number of images that can be processed each minute. It also increases the possibility of smearing, if the object or the camera is in motion. In high ambient lighting conditions, the charge can be read off more quickly, if desired, or the camera lens can be stopped down, thereby reducing the amount of light that falls on the pixels.

One example of an inexpensive camera system is the Micron Eye, available from Micron Technology. Figure 6-8 shows two of their cameras with associated circuit boards for installation in an IBM-PC. Their camera uses an IS32 optical RAM chip containing two sets of pixels, each 128×256. With special software it is possible to obtain 256×256 resolution in this system. Due to the mechanical arrangement of the pixels, a 128×128 readout is much easier to

Figure 6-8. CCD-based cameras with electronics. (*Courtesy of Micron Technology*)

obtain. Camera speed is also affected by the desired resolution, since each element must be read out sequentially, and thus higher resolutions require longer transmission times.

Preprocessing

Preprocessing can be defined as any data manipulation done on the visual image before image analysis. Preprocessing can be used for removing noise, digitizing the analog signal, selecting a small part of the image (windowing), and improving contrast. But its greatest contribution is reducing the amount of data (and bandwidth) to be processed by the vision processor. By taking snapshots of the data, narrowing in on a portion of the picture, and reducing the number of bits of intensity (by employing binary images), the preprocessor reduces the data work load.

Figure 6-9 illustrates the basic steps involved in processing a video image. On the preprocessing side (left), data are received from a camera, converted to the appropriate number of digital bits, and operated on for image restoration and noise removal. Then a complete frame of data is collected, assembled, and sent to the image processor. On the image processing side (right), the processor accepts the data, and performs such processing steps as image segmenta-

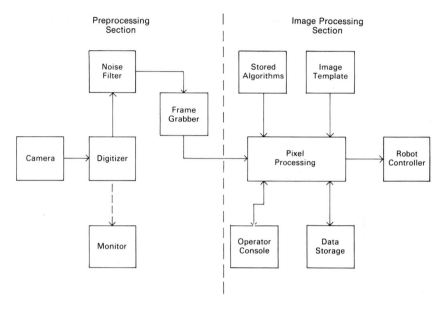

Figure 6-9. Vision system block diagram.

tion, parameter extraction, and image recognition. This latter step is often done through comparison of the processed image with a stored image template. Appropriate information about the object is then sent to an external monitor or a robot controller.

One function of the vision system electronics is to collect data from the vision sensor and prepare it for later processing. There is an enormous number of image points (as many as 1 million pixels) and only a limited time available (33 ms for a standard television frame) for collecting data. The speeds necessary require special-purpose processing techniques. Vision systems often operate at clock speeds higher than the computer speed, and special techniques, such as two-stage flash conversion, must be employed to perform image digitization in real time.

A flash converter uses $2^N - 1$ comparators to provide N bits of analog-to-digital data. For example, an 8-bit flash converter requires 255 comparators. Since all possible values are available at once, the conversion is very fast compared with other techniques, such as successive approximation. TRW has been a pioneer in this field, and offers 8-bit flash converters from $100 to $200.

Because of the speeds required, some companies are investigating parallel processing. Some parallel display processors have been available for about 12 years. In addition, parallel architectures and specialized very large scale integration (VLSI) chips have been developed, but most of them are limited to certain types of functions, such as point operations, convolutions, kernel representations, and functions related to the structure of the image grid. To be effective, other functions are also needed in parallel processing systems. Another problem with current parallel processors is that many have very limited storage (a few hundred bits), and some investigators are requesting at least 4,000 per pixel.

Quantization level refers to the number of bits set aside to handle intensity graduations. For example, 2^4 refers to 4 bits of quantization, allowing a total of 16 intensity levels to be portrayed. Studies on the human eye indicate that an observer can readily distinguish 256 gray scale increments (based on the Weber-Fechner fraction), which corresponds to a total of 8 bits of data quantization. Since a byte equals 8 bits, and since many instructions in a computer are made to handle byte manipulation and storage, it has been found convenient to provide an intensity quantization of 8 bits in many systems. Other systems restrict the digitization to 6 bits (64 levels), thereby reducing the amount of data that must be subsequently processed. Of course, many systems still use binary processing, in which case only 1 bit per pixel is required.

There are drawbacks to too much quantization. If more levels of gray scale are handled than necessary, equipment costs rise, processing time increases, minor scene lighting variations are magnified, and picture comprehension–pattern recognition capability will not be much improved. In many applications, it is preferable to use these extra bits to provide some type of color information, if possible.

An upper limit to the amount of quantization that can be employed comes from the available analog-to-digital (A/D) converters. Many are limited to 8 bits, although 10- and 12-bit converters are available. High accuracy is desirable in A/D converters because any errors in quantization will affect subsequent image recognition accuracy.

Even with an 8-bit quantization, all possible intensity values may not be represented in a picture. This lack of representation can occur due to low contrast, because only part of the picture (a window) is being examined, or due to brightness characteristics of the object. When any of these occur, the number of bits that must be processed can often be reduced.

Normal video digitization represents absolute values of the picture intensity. Another technique is also used to reduce the number of bits transmitted and stored, thus providing data in a different format. In this technique only the differences between adjacent samples, not their absolute values, are used for quantization. High digital values now represent image transition points, such as edges, which can help later processing steps.

Another function of preprocessing is image restoration, in which noise and blur, caused by image motion, are eliminated. It is even possible to correct for various system geometric distortions. A blurred image can be improved through convolution processing using a stored ideal image. Geometric distortions can be removed through calculations derived from the known geometry of the system and through inverse mapping functions with a known calibration grid pattern.

The Wiener filtering technique is one method of removing noise. Another approach to noise reduction is *medium filtering enhancement*. This approach is particularly applicable to situations requiring contrast enhancement. Contrast enhancement can be used with underexposed images that occur due to limited frame collection times or insufficient light.

Electronic circuitry may also be used for black-level adjustment or intensity standardization of the image to ensure that a given digital value always represents an image pixel of the same brightness. Electronics also performs *frame grabbing*, in which a single visual frame is selected from the available sequence of pictures and processed. Later another visual frame may be selected. This snapshot approach is used by most vision systems.

Because of processing time limitations, most vision systems extract a minimum amount of data. By reducing high-frequency content, some approaches reduce noise with little reduction in image understanding.

Run-length compression is another way of reducing data transmitted to the processor. In run-length compression, an intensity level and a number are transmitted. The number tells the processor how many pixels in a row have the same intensity level. For example, if 20 pixels all had intensity level 150, run-length compression would transmit 2 bytes, one representing the intensity (150) and the second the length (20), instead of the 20 bytes a normal system would transmit. This approach is practical when there are only a few levels on intensity quantization, since under these conditions the likelihood of high numbers of pixels with the same intensity level increases.

6.4 IMAGE ANALYSIS

Image analysis examines digitized data to locate and recognize an object within the image field. Different approaches can be used, but most image analysis techniques include segmentation, parameter extraction, and pattern recognition. Zimmerman[6] adds an extra first step, that of detecting an outline. Some investigators prefer to subdivide the process into low-level processing (bit level) and high-level processing (image level).

Figure 6-10 gives an overview of the image analysis process, showing the major steps in order of performance and various methods of accomplishing each step. Each step is carried out separately from the others. Current progress in image analysis has been helped by lower costs in computer hardware (especially in low-cost memory), a better understanding of human vision, and advances in the hardware available to generate the image.

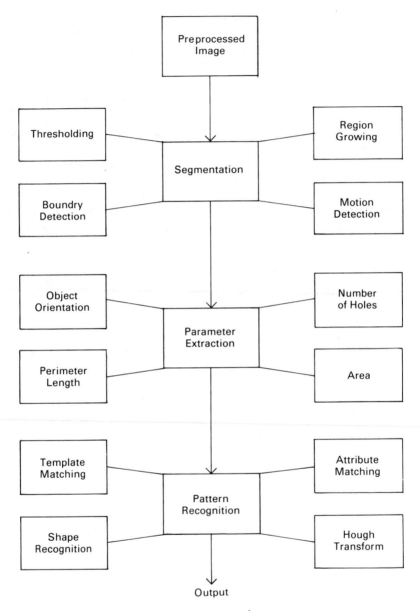

Figure 6-10. Image analysis steps.

Segmentation breaks a scene into several pieces or segments and allows the desired object to be isolated; if multiple objects of interest are present, segmentation separates them in the image. Parameter extraction then looks at segmented objects and determines key features about them, such as size, position, orientation, shape, intensity, color, and texture.

Pattern recognition attempts to match the observed features against stored criteria, thus allowing the objects to be identified. A major problem in this area is that slight differences often occur between objects of the same object class, due to slight dimensional differences (from tolerances in manufacturing), differences in perceived parameters (from limitations in the vision hardware), errors caused by changing shadows that affect the edge detectors, or even various software errors that propagate through the analysis.

To better understand the complexity of computer image analysis, we will look at an example of human image analysis. For example, most humans can easily follow any but the most difficult trail through a forest. The trail boundaries may be irregular and vary in width; portions of the trail could be overgrown with weeds; and occasional side paths, ending after only a few feet, may try to deceive us. However, it is impossible to program a computer to recognize such a path and follow it. Actually, a computer has trouble following a paved highway.*

When following this path, a human also performs the steps of segmentation, parameter extraction, and pattern recognition, although not consciously. For example, when concentrating on the path, the person can segment out that part of the image and pay little or no attention to flowers, trees, or anything else. To determine where the trail goes, the person can extract the edges of the path, even with breaks in them, and see the general direction in which he or she is going. Finally, a human recognizes what a path looks like and therefore can find it again after thrashing through some overgrown grass.

We should be able to see now why most machine vision tasks are much harder than they at first seem. Experienced personnel consistently underestimate by a factor of 3 the time it takes to code a vision algorithm.

Segmentation

Before the system can identify an object and act upon that knowledge, the object must be separated from the rest of the background. This process, called *segmentation,* is an essential first step in image analysis. Several methods of

---

*This is not an exaggeration. When the autonomous vehicle being built for the army by Martin-Marietta was first tested, it ran off a curved road because it thought a tree was the continuation of the edge of the road.

segmentation can be used.[7] One popular method is *thresholding,* because it is quite fast and can be used in many applications. Thresholding refers to selecting one level of image intensity as a threshold such that all of the desired object reproduces as white and all of the background (and clutter) appears black. If the computer is able to threshold perfectly, the result is a binary image in which the object can be quickly localized. If the same level of thresholding is applied over the entire image, the process is called *global thresholding.*

If the picture has noise, complex images, intensity variations, shadow effects, or other image degradations, global thresholding is insufficient and we must use *local thresholding.* In local thresholding, various regions of the picture have different thresholds set. One way to use local thresholding is to determine the average light intensity for each region of the picture and adjust the threshold in each area accordingly.

Another thresholding approach is based on which intensity level provides the maximum first derivative. (A first derivative is maximum where the image brightness changes the most, presumably at the object's edge). This brightness level is then used to threshold that area of the picture. Another technique, also found in human vision, being explored is *lateral inhibition.*

Histograms can also be used for thresholding or scene equalization. A histogram typically indicates two peaks: one showing the average brightness of the background, the other the average brightness of the object. Between these two histogram peaks is a valley, which represents a brightness level that few pixels in the image have, and which may indicate the transition between the object and the background. The intensity level corresponding to this valley is chosen for thresholding.

Binary systems require considerable optics and lighting skill if the resultant thresholding is to provide an image close to the original. However, because of shadowing, reflections, angular part placement, lighting constraints, and other factors, it is almost impossible in some applications to obtain a representative image. Under worst-case conditions, the image would be all white or all black. Even under reasonable conditions, part of the image often merges with the background, making part identification, position, or orientation extremely difficult.

A second approach to segmentation is *boundary detection.* Two types of algorithms are commonly used, one for edge detection and one for correlation. If the boundaries of an object (the edges) can be determined, object location has been defined. However, edge detection generally requires complex calculations. Edge detection can be accomplished by using small (3×3) templates that operate on each pixel of the image, thus providing the intensity and direction of image gradients.

Another method of boundary detection is the use of gradient operators (which involves partial derivatives in two directions). A convolution mask, such as the Sobel operator, is often used. The mask cannot be used alone, but requires another step, such as linking. The Sobel operator is one example of a template, but many other types of operators are available. Chapter 11 looks at some of these more advanced approaches.

An example of a 3×3 mask is

a	b	c
d	e	f
g	h	i

where the Sobel operator is defined as

$$\text{Sobel} = \sqrt{\text{grad } X^2 + \text{grad } Y^2}$$

and

$$\text{grad } X \approx (g + 2h + i) - (a + 2b + c)$$

$$\text{grad } Y \approx (c + 2f + i) - (a + 2d + g)$$

Grad Y represents the vertical edges, grad X the horizontal edges. For this operator, maximum weight is given to the middle operators b, d, f, and h, and no weight is given to the center value e. In addition to the values shown, other values for the convolution mask can be used.

The Sobel operator adjusts each pixel's brightness level with an average brightness level from some set of surrounding pixels. Using the Sobel operator requires the process to be repeated for each pixel, thereby requiring extensive computations.

Some types of edge detectors are severely affected by noise in the picture They can also provide poor directional correlation information since they are limited to eight directions.

Various investigators have found the difference of Gaussian (DOG) function to be better for complex images. In this approach, a representation of an image is smoothed with a Gaussian filter, which limits it in the frequency and spatial domains. Zero crossings are then extracted by using a Laplacian operator. The DOG function can be a very powerful edge-finding method. A related approach uses multiple image scales with different half-power widths for each scale.

Another method of edge detection is based only on zero-crossing detection. This technique, developed at MIT, involves fewer computations.

Boundary tracking can also be used for segmenting. In boundary tracking, an edge-following procedure is used to detect and track the perimeter of an object. Heuristic techniques can often be used to fill in any gaps in lines found before image recognition. We have to be careful with this approach in an inspection application because the gap could be the failure we were searching for.

A third approach to segmentation is *region growing*. In this approach, pixels are grouped together into regions based on segmentation. For example, thresholding could indicate several objects in an image, and region-growing techniques (such as a run test) could be used to allow regions containing each object to be separately identified.

Another method of region growing is the *split-and-merge* procedure. The image is split into four parts, and the areas are examined for homogeneity. If the areas are homogeneous, the two images are merged. If not, the areas are split again, and the procedure is continued.

Refer to Figure 6-11 for an example of split and merge. After the first split, b and d are found to be homogeneous (they have the same average intensity) and a and c are found to differ from each other and from b and d. After this first pass, b and d are connected (merged), and a and c are split again. After three

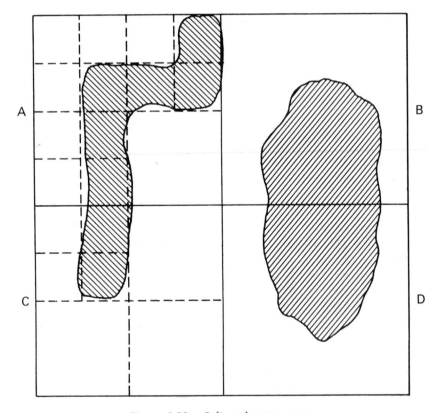

Figure 6-11. Split-and-merge steps.

split-and-merge operations, we find that six elements within area a have merged, along with two elements from area c, to form the part shape we were looking for. Note that once b and d are considered homogeneous, no more processing time is spent on them. This procedure is good, but if many splits and merges are needed, it can be too slow.

Motion can also be used to segment the image. Motion is a key for detecting certain types of objects. To use motion, we subtract the first image from the second. Any data remaining must then be due to image motion.

Parameter Extraction

Parameter extraction identifies key features for further processing. Features that may be extracted include

object area	diameter
gray level	ellipse eccentricity
perimeter length	aspect ratio (D/H)
size of minimum enclosing	compactness (perimeter^2/area)
rectangle	number of holes
orientation	principal axis
center of the object	various parameter ratios

Various approaches are used to obtain this information. One obvious method is to locate a fiduciary mark and use it to identify, orient, and position parts. A second approach, applicable to split-and-merge segmentation, allows position and shape data to be obtained directly from the final pattern of the pixels.

Andersson[8] described a system for extracting parameters from gray scale images by using a custom VLSI chip. This chip determines the object's area, center of gravity, orientation, and size. It was actually tested in a robot to track and catch Ping-Pong balls.

Parts orientation is often very important. In one technique for obtaining orientation, an ellipse having the same area as the segmented image is computed. The major axis of the ellipse defines the object orientation. Another method connects three identified points on the object. In addition, light intensity distribution and shading can sometimes be used.

Another important parameter is shape. Shape discrimination can be contour-based, region-based, or even use smoothed local symmetries. The first two methods offer the most promise. Many objects have simple shapes that can easily be found, oriented, and identified. Segmentation can give shape information, especially the split-and-merge method. A second approach is modeling, in which small image pieces are recognized and put together to build up a larger, more complex image. A third technique is image organization (a hierarchical approach)—once piece A is recognized, it is easier to find piece B, which is attached to it.

Another approach is based on the difference of Gaussian filters for edge detection and object determination. Still another fits quadratic and higher-level equations to images. Some investigators are combining vision data with other types of sensor images, such as ultrasonics and tactile. Known as *sensor fusion*, this combination approach probably offers the best long-term potential for the most complicated objects.

Image texture is often very important for both object recognition and object orientation. Of the approaches suggested to obtain texture information,[9] fractal representation is perhaps the most interesting. This approach is based upon features of the object that are independent of scale, at least over a reasonable range. Fractal methods produce results similar to human perception of texture.

There is an important difference between parameter extraction and image recognition. Parameter extraction is based on image descriptors and operates directly on the image. It then requires the recognition system to see various relationships within the extracted parameters. The result is that the recognition process does not operate on the image directly but on a representation of the image, which simplifies the computation but does not consider all available information.

Image Interpretation

The final step in robot vision is image interpretation, which consists of matching the image with a stored model (pattern recognition), deciding on the basis of the match whether the match is successful, and converting this decision into a form suitable for use by the robot control system.

Many successful robot vision systems just find the part and do not recognize it. For example, if the robot is to remove boxes from a conveyor belt as they go by, the robot need not recognize the box; it just needs to known where it is. In this case it can be assumed that extraneous items will not be on the conveyor belt. However, this section discusses recognition of a specific type of object from a collection of different objects, the fundamental pattern recognition process.

In the most complex image analyzers, three levels of processing may be used. At the lowest level, the processor is handling raw images, and its job is to detect the edges and handle some of the stereo processing functions (if present). At the intermediate level, the processor builds an abstract image from the local descriptions passed on by low-level actions. At this level, the processor determines relations and groupings.

At the highest level, the processor converts the middle-level information into a recognized image. At this level, it is only working with abstract symbology. In many applications, the processor uses knowledge about the object and the application to help recognize the image, whereas at lower levels it was concentrating on local data extraction. See section 6.6 for an example of this multiple-level approach to processing data.

Recognition algorithms can be based on the presence or absence of identifying marks or holes, and on object shape, size, and, in some applications, location. Approaches include template matching, attribute matching, heuristic approaches, and syntactic and semantic techniques. Gray scale, color, and three dimensions, though possible, are used less often because of equipment costs and processing time considerations.

Template matching is particularly useful when the system needs to locate a specific feature in an image, such as a cross being used as a fiduciary mark. Most template matching systems are based on some type of image correlation. A correlation system produces a correlation coefficient to indicate how well the image matches the model.

Many types of correlation approaches can be used for this purpose, but some suffer from errors caused by lighting variations (causing correlation systems to show no match) or rotational errors. The most common method is binary correlation, which is less computationally intensive than other methods but can generally only be used on sharply defined images. Cross-correlation is another frequently used approach. Rotational errors can sometimes be removed if object orientation or principle axis determination has been determined.

Normalized correlation produces better results than cross correlation when the image is degraded. Normalized correlation has been used as part of a stereo ranging system (where it is important that measurements be made from the same point) and in tracking the motion of autonomous robots. Recently, Cognex corporation has adapted a normalized correlation method for its model 2000 vision system.

Another approach to image recognition is attribute matching. The attributes measured during parameter extraction are matched and compared with attributes stored for the model object. The system uses functions such as center of mass, image area and number of holes to make a decision. This technique requires a classification strategy. Ideally, available parameters should be independent of rotation, scale/size, and certainly location. This calculation is computationally easy and thus takes less computer time than most other approaches. The biggest problem with using it is in segmenting the image to determine which parts of the image belong to the object.

A heuristic approach applies intuition and experience to the problem. For example, such a simple concept as separate locations can be used. In this technique, if the object is on one side of a predefined boundary, it is one type object (i.e., a bottle of ginger ale). If it is on the other side, it is a different object (i.e., a bottle of cola). This is a key approach used in practice in structured environments, but it obviously cannot be used in unstructured situations. Other heuristic approaches can be used in more unstructured environments; for example, the robot might look for a large difference in size between the desired object and competing objects.

A more theoretical approach that uses syntactic and semantic methods has been developed. In the syntactic step, a set of primitives (small subsets of the object) is used to build the object. The semantic step then checks the object to determine if it is reasonable. Let us take as an example the recognition of a large bolt. The image is split into a series of primitives or subelements and then built into the final pattern through a set of rules. In this example, the primitives can be short arcs, tee's, and angles, all individually recognized. A series of the short arcs are matched to the lines found at the head of the bolt, other subelements then complete the head, and, finally, the syntactic processor has recognized the bolt.

Feature weighting uses the principle that some parameters are more important than others for recognizing objects. After the parameters are measured, weighting values are applied to each parameter. The resulting values are then used to make the final image determination. For example, the shape of a pair of pliers can be weighted more heavily than its size for purposes of recognition. Once the object is identified, normal template matching could be used to see if the object was within tolerances for inspection purposes.

What happens after an object is recognized? The robot receives information about the location, orientation, and type of object, but the robot computer must determine the appropriate action. For example, if an object was determined to have the wrong orientation, the robot could be programmed to reorient it. In an inspection system, the robot might reject a bad printed circuit board. And in a materials handling application, the robot decides where to place an object based on what kind of object it is.

6.5 SUMMARY OF THE STATE OF THE ART

Vision offers difficult challenges due to the high level of computer processing required and a lack of good recognition algorithms currently available. However, machine vision systems have made major advances in the last several years and thus have opened up new application areas.

In designing a vision system, engineers often make simplifying assumptions or overlook problems that might limit the applicability of the system, such as shadows on the image, the effect of surface texture and shading, unwanted light reflections, and object movement during the analysis. The result can easily be an object with a viewing angle, orientation, apparent shape, and apparent size different from data stored as the template. In this case we may easily recognize the object as a telephone (for example), while the computer is unable to match any stored template.

Changing light conditions are a constant problem in any system. Light intensity reaching the object may change as the object moves along the conveyor belt, as the day passes and external lighting dims, or because a shadow covered the object when someone passed by it. Therefore a vision system must offer robust performance under changing lighting or other environmental constraints.

The needs imposed on computer vision obviously depend upon the application. Some systems can operate with gross silhouette-type data (black-and-white), but most require levels of gray for processing, which increases the number of bits to be stored and the data processing time. During an inspection application, for example, the vision system is allowed more time, perhaps as much as 10 s. However, for robot control during production, time is more critical, so a processing time of less than 1 s is usually demanded. Systems limited to detection of simple parts can recognize 2 to 10 parts/s, and some have even recognized 15 parts/s. In many cases, the system can be set to trade off between accuracy and processing speed.

Here are some of the characteristics of vision systems in 1983:

1. Maximum resolution was typically 512×512.
2. Almost all systems used binary processing.
3. Recognition speed ranged from a few tenths of a second to a few seconds.
4. The level of sophistication was very low, limiting potential applications.

Now note the advances that have occurred:

1. Resolution is $1{,}024 \times 1{,}024$ (2,048 in linear cameras).
2. Many systems use 256 gray scale levels (8 bits).
3. Some systems can handle 256 different colors (8 bits)
4. Speed is less than 0.1 s for the simplest recognition tasks, 1 s for most tasks.
5. Elaborate assembly and inspection tasks can now be supported.

An important, additional consideration in any system is cost. As expected, stereo and color systems have higher cost, lower capabilities (especially speed), and require larger system memories. CCD cameras are much less expensive now, even though they have increased resolution and reduced overall size. However, their design is still oriented toward television rather than robotics applications, few improvements have been made in optics and scene illumination. Dirty conveyor belts still cause problems for vision systems.

New work is now being carried out in analog rather than digital areas. For example, if a CCD sensor is exposed to light and an image is formed on it, it is possible to let the sensor slowly discharge, thus reading its values several times during the discharge. This design can supply several different values for the Gaussian convolutions some systems need for processing, and it can help reduce the bandwidth between the sensor and the processor.

6.6 APPLICATIONS AND AVAILABLE SYSTEMS

Machine vision systems are finding greater acceptance in the marketplace, particularly in robotics applications. As discussed in chapter 1, by 1990 it is estimated that 25% of all robots sold will include vision systems. But three steps will be necessary: (1) lower cost (few applications can justify the high costs of some systems just to add vision); (2) faster response time; (3) more application-oriented software, to allow the vision system to be used in a wide range of practical applications, not just laboratory conditions. Progress in all three areas is taking place, and vision systems are now being used in such diverse applications as label inspection, reading characters, seam following, robotic grinding, complex assembly, and final inspection. In the agriculture field, vision uses include harvesting oranges; sorting cucumbers; orienting corn ears; finding defects in tomatoes; checking the color of bell peppers; examining apples for bruises; inspecting potato chips; and identifying fish.

One of the earliest robot vision systems, developed for General Motors, was Consight, a vision-controlled robot developed in 1979 to retrieve parts from conveyor belts. Ten years later, the largest current users of robot vision systems (and all types of robots) are still automotive companies. They use robot vision systems in many areas, with their major use in small parts inspection during manufacturing.

Vision systems are making inroads in many other industries, including electronics (PC board inspection), health care, food and beverage, glass, chemicals, paper products, and the postal service. Typical vision processing systems are used for parts orientation, parts alignment, feature recognition, noncontact measurements, and seam tracking. In appliance manufacturing, vision systems perform surface flow analysis—that is, they find foreign material, scratches, dents, and cracks from the sharp contrast with the surface paint.

Four examples of the currently available vision systems include: (1) Cognex Corporation (reads most characters that are legible to humans), (2) Automatix, Inc. (follows seams in arc welding operations), (3) Copperweld Robotics (assists in complex assembly operations), and (4) Honeywell (serves as final inspection stations).

Figure 6-12 shows a Hitachi HV/R-1 vision system that can recognize shapes, locations, and part orientation. It is shown in conjunction with an assembly robot that uses the information to locate parts and assemble them

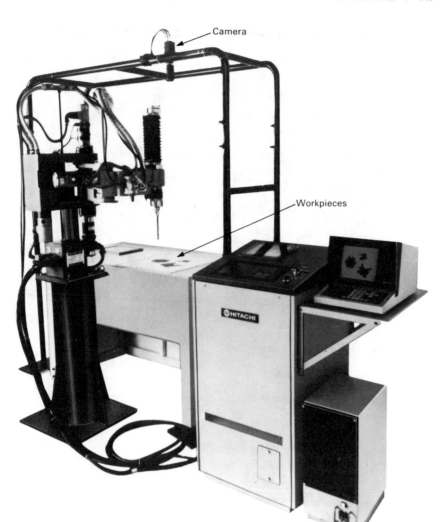

Figure 6-12. Robot with vision system. (*Courtesy of Hitachi America, Ltd*)

into a unit. Figure 6-13 is an actual error display from a Cambridge Robotics CRS 1850 vision system. This system has found two errors and displayed them. X-axis movement by the camera determined one coordinate of the errors and PC board frame movement provided the other coordinate.

Universities have cooperated with industry to solve practical vision problems. For example, an early difficult task addressed by researchers was bin picking: the robot had to reach into a bin and remove a single part from among

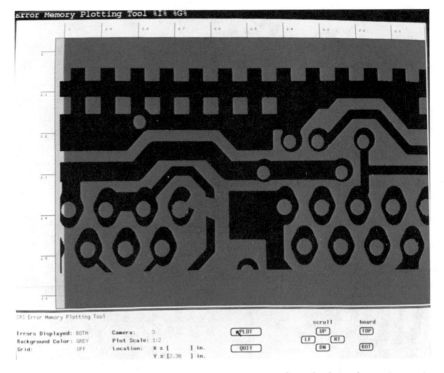

Figure 6-13. Inspection system display. (*Courtesy of Cambridge Robotics Systems*)

a jumble of parts. The task was even more difficult if the robot had to choose from several part types kept in the same bin. Researchers at the University of Rhode Island[10] and other universities have contributed to the development of vision systems that can perform this complex task.

One successful area of robot vision systems is inspection. Inspection applications have four basic steps: find the part, recognize the part, inspect the part, and report the result. The user typically sets an accept/reject threshold, depending on the quality level desired. Most inspection systems can also keep track of reject statistics to allow tracking of production defects. Some inspection systems are designed to work quickly by concentrating their attention on the few areas of the part where failures are most likely to occur; however, other (rare) errors could be missed.

In a battery inspection task, a typical application, the vision system must determine that there is a shoulder at the positive battery end, that there are both positive and negative tips, and that there is a two-color marking on the battery wrapper. The battery inspection system runs at 450 parts/min and checks color registration, correct length and width, and correct formation of positive and negative tips. Measurements are accurate to about 0.015 in. The batteries move on a conveyor belt, and one camera looks at each battery in turn. Misregistration of color is tested first, since no further tests are necessary if there is an error here, and 80% of the errors occur here. Next the shoulder is checked. Then the positive tip is inspected, and finally the negative tip. At any reject point, a blower is activated to force the battery from the line.

Robots can also inspect semiconductor leads for coplanarity, skew, and diameter to dimensional tolerances of 0.25 mil. A final example is the Gemini system developed by GEI. This system inspects packages to make sure that they are properly closed and that labels have been properly placed. It can also measure distance.

Still other applications include the Partracker system by Automation Inc. This system locates a part when the object's position is not precisely known and then performs the necessary welding or assembly function. For example, the location of automobile seams can vary along an assembly line by as much as 2 in. Partracker enables the robot to locate the seam and braze the gap.

Octek of Burlington, Massachusetts has developed a system that classifies fresh fish. Currently in the prototype stage, the fish monitoring system (FMS) applies vision processing to fish as they move along a conveyor line. The system uses back lighting and a binary image to identify the fish type and to calculate its length and width, from which weight approximations are made. The fish are categorized as cod, haddock/halibut, flounder, redfish, catfish, and other. Statistics are kept of the total quantity and weight of each class of fish on board. A related system by Opcon of Everett, Washington, uses line imaging techniques.

The RV-200 is a vision system manufactured by NACL, a joint venture of NEC and Hughes Aircraft. This system allows up to four TV cameras as inputs, and can recognize objects by their shapes. Thus it can be used for part identification and part positioning. Using the 8086 and 8087 microprocessor chips and a TI-22A CCD camera, the system has 256 (H) $\times$ 240 (V) pixels and 64 gray levels. A special programming language, ARVOL, has been developed for applications programming.

One interesting application, still in the research stage,[11] is the use of vision to determine the current position of a mobile robot.

6.7 RANGING TECHNIQUES

Many different approaches have been developed to determine range to a target.[12] In addition to techniques covered in the next two chapters, four visual approaches can be used: stereo imaging, camera angles, stadiometric, and light stripe. Human vision can estimate depth (range) by using one of the first three approaches, the most natural being stereo imaging (Fig. 6-14). Stereo imaging

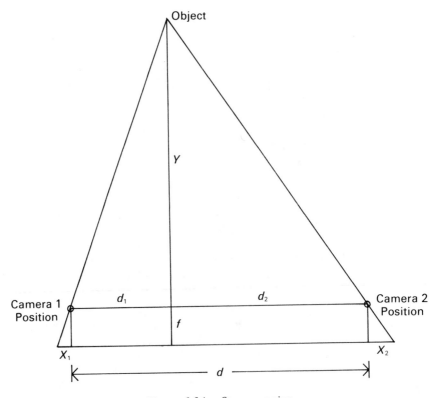

Figure 6-14. Stereo ranging.

determines the range to an object when two images are available, either through two cameras, or through the movement of a single camera. In this approach, it is assumed that the two sensor positions are located in a plane parallel to the image plane. By similar triangles the range Y can be obtained. Since $d_1 = d - d_2$,

$$\frac{Y + f}{f} = \frac{d - d_2 + x_1}{x_1} = \frac{d_2 + x_2}{x_2} \tag{6-6}$$

Rearranging and simplifying the last two terms gives

$$\frac{d}{x_1} = \frac{d_2}{x_2} + \frac{d_2}{x_1} \tag{6-7}$$

Solving for d_2, we get

$$d_2 = \frac{dx_2}{x_1 + x_2} \tag{6-8}$$

Substituting in (6-6) and simplifying then gives

$$Y = \frac{df}{x_1 + x_2} \tag{6-9}$$

In this formula, x_1 and x_2 are variables that determine the object range, and f and D are held as constants.

It is also possible to use a known range (R) to calibrate the system. Equation (6-9) can be rearranged to use this known range:

$$Y = \frac{R(R_1 + R_2)}{x_1 + x_2} \tag{6-10}$$

An advantage of this equation is that the calculations are much easier. Because of the ratio of two sums, absolute dimensions are not necessary and pixel positions can be used directly. In this equation, R_1 and R_2 correspond to the values of x_1 and x_2 for the known range. Once values for a standard range are determined, distance to a specific target (for example, a fiduciary mark) can be computed from the corresponding pixel positions (x_1, x_2) of the left and right cameras. Equation (6-9) holds for three-dimensional data as well, but x_1 and x_2 are then vector distances. Note that if the object point being viewed is located on either side of the two image points, the difference between the two offset distances (rather than the sum) must be used.

Depth accuracy is limited by the number of pixels available and by nonlinearities in the image. Accuracy is not as good for objects not located between the two cameras, since these calculations require the use of differences, which, by their very nature, reduce available accuracy.

Finding the same point in each picture is necessary to any stereo ranging system and can be quite difficult. Known as the *pixel correspondence problem*, there are two basic approaches to solving it. One approach searches for a characteristic pattern (the fiduciary mark approach); the other uses correlation techniques within a moving window.

Another approach to ranging is shown in Figure 6-15. In this case, the angles of the two cameras are automatically adjusted until some recognizable point on the object (such as a cross hair) is centered in the picture. Then straightforward trigonometry gives the range to the object as $d_1 \tan \theta_1$ or $d_2 \tan \theta_2$. Since the sum of d_1 and d_2 is known (d) and both θ_1 and θ_2 can be measured from the camera tilt position, the two triangles can be solved for Y as follows:

$$Y = \frac{D \sin \theta_1 \sin \theta_2}{\sin(\theta_1 + \theta_2)} \tag{6-11}$$

This approach is seldom used because it requires both cameras to move, thus increasing the system cost, complexity, and time required to determine the range. A modified approach using a laser to create the cross hair has also been employed.

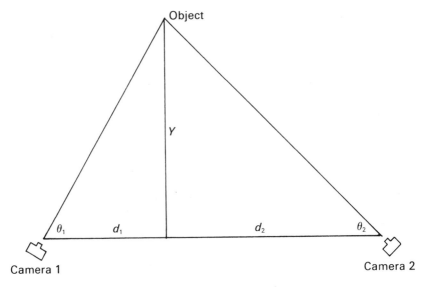

Figure 6-15. Camera angle technique.

In stadiometric ranging, the calculation is based on knowing the object width (or height) in advance. Then the angle subtended by the object (α in Fig. 6-16) is the only value needed to determine the range. When the vision system measures this angle, the range is computed from the expression

$$R = \frac{W}{2 \tan \alpha/2} \tag{6-12}$$

If $\alpha < 10°$, then $\tan \alpha/2 \approx \sin \alpha/2$ and the sine function may be substituted in the formula. Note that the object must be symmetrical, or else the object orientation must be known. These constraints do not generally pose a problem in the factory.

Range information can also be obtained through the approach shown in Figure 6-17. The light source projects a vertical stripe of light. This light stripe will appear on the wall behind the conveyor at position Y_1, which represents no object present. When an object intercepts the light stripe, it will be seen by the camera displaced to the side by an amount proportional to the depth position of the object on the conveyor (Y). The system can then calculate the range. If ΔX is the displacement of the stripe, the depth is

$$Y = \Delta X \frac{Y_1}{X_1} \tag{6-13}$$

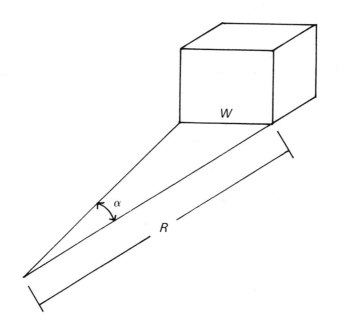

Figure 6-16. Stadiometric ranging.

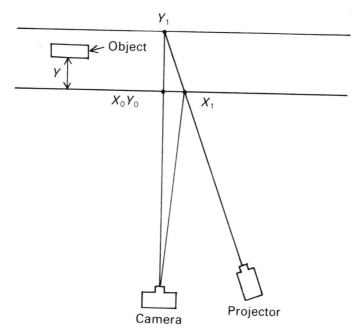

Figure 6-17. Light stripe ranging.

REFERENCES

1. Poole, Harry H. *Fundamentals of Display Systems*. Washington, D.C.: Spartan Books, 1966, pp. 276–285.
2. *EE Times*. Oct. 20, 1986, p. 33.
3. *Machine Vision Systems: A Summary and Forecast*. Lake Geneva, Wisc.: Tech Tran Consultants, 1985.
4. Kell, R. E. Survey of off-the-shelf imaging systems. *Proc. 3rd Annual Vision Conference*, 1984.
5. Holland, S. W., L. Rossol, and M. R. Ward. Consight-I: A vision controlled robot system for transferring parts from belt conveyors. *Computer Vision and Sensor Based Robots*, George G. Dodd and Lothar Rossol, eds. New York: Plenum Publishing, 1979.
6. Zimmerman, N., G. J. Van Boven, and A. Oosteline. Overview of industrial vision systems. *Industrial Applications of Image Analysis*. Pijnacker, Netherlands: DEB Publishers, 1983, pp. 203–229.
7. "Robot Sensing and Intelligence," IEEE short course presented by satellite transmission in November, 1983. Prepared by the Institute of Electrical and Electronics Engineers, Parsippany, N.J.
8. Andersson, Russel L. Real-time gray-scale video processing using a movement-generating chip. *IEEE Journal of Robotics and Automation*, Vol. RA-1, No. 2, June 1985, pp. 79–85.
9. Kuklinski, Walter B. Image Processing Considerations in the Development of Computer Vision Systems. Mini-tutorial presented at Electro 86, IEEE, 1986.
10. Kelly, R. B. Binary and gray scale robot vision. *Proc. SPIE, Robotics and Robot Sensing Systems*. San Diego, August 1983, pp. 27–37.
11. Drake, K. C., E. S. McVey, and R. M. Iñigo. Experimental position and ranging results for a mobile robot. *IEEE Journal of Robotics and Automation*, Vol. RA-3, No. 1, Feb. 1987, pp. 31–42.
12. Orrock, J. E., J. H. Garfunkel, and B. A. Owen. An integrated vision range sensor. *Robot Sensors*, Vol. 1: *Vision*, Alan Pugh, ed. London: IFS Publications Ltd., 1986.

ULTRASONIC SYSTEMS

Installations using several coordinated robot arms performing a single task are becoming more common. Interest in mobile robot systems is also increasing. Both applications require an inexpensive but effective short-range collision-avoidance system that does not rely on object contact. A collision-avoidance system must provide sufficient awareness of the robot's environment to prevent impacts while the robot is in motion. Therefore, accurate range information is very important. This range data can also be used for other functions, such as navigation or object recognition.

Ultrasonic systems are based on the transmission and reception of sound at frequencies above the range of human hearing. They offer important advantages in ranging, and most of their techniques have not been fully investigated. Compared to vision, ultrasonic systems are less expensive, faster, and more accurate over medium distances (a few inches to tens of feet), and they provide object parameters not otherwise easily available.

Unfortunately, much less work has been done in applying ultrasonics to robotics than in applying vision. Certain specialized ultrasonic techniques have received little research attention. One reason is that people are more experienced in working with and understanding visual images. A second reason may be that there is little ultrasonic information available in robotics literature, and what there is covers only portions of the topic. This chapter gives an overview of all areas of ultrasonics important to robotics and indicates some promising areas for future investigative work.

7.1 SONAR FUNDAMENTALS

It has long been recognized that ultrasonic systems may be used to determine range and bearing (direction) to a target. This technique is the basis of sonar (sound navigation and ranging), developed originally for underwater applications where conventional radar systems could not operate. However, sonar techniques can also be used with air as the medium. In this area, nature is ahead of us, since bats have been employing ultrasonic principles for target tracking, collision detection, and navigation for centuries.

With careful design (including compensation for air temperature), ultrasonic ranging is possible with accuracies of 1/10 in at 10 ft (<0.1%). With array processing and other techniques, azimuth resolutions as small as 1° or less are possible. Figure 7-1 illustrates the basic principle behind sonar or ultrasonic ranging—the reflection of sound energy from objects in the beam path. When a sound transmission source sends out a pulse of acoustical

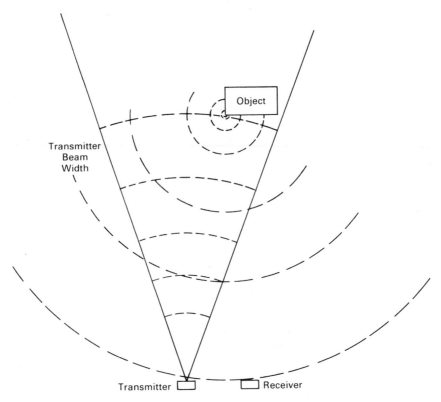

Figure 7-1. Basic sonar principles.

energy, the resulting acoustical wave travels along a path defined by the beam width of the transmitting element. A portion of this energy is subsequently reflected back from any obstacle (target) that the sound waves reach. This returned energy can be detected by a receiving element, and the delay in time between the transmitted pulse and the received pulse can be measured. This delay time is in direct proportion to the range of the target and allows the exact range to be calculated. Transmitting a pulse of energy is repeated each time a range measurement is desired. Although the principle works with sounds of any wavelength, the best frequencies for most ultrasonic ranging systems are 40 to 80 kHz. As discussed later, this frequency range seems to give the best trade-off between resolution and longer ranges.

Information other than range is also available from an ultrasonic system, including bearing (object direction), object velocity, and parameters that can be used in object recognition or shape detection. This information requires more sophisticated techniques, many of which are discussed in this chapter.

Central to any ultrasonic ranging system is the speed of sound in air. This subject has not received complete coverage in most published papers on ultrasonic systems, and even complete acoustics textbooks (such as Kinsler's[1]) leave out some important topics for robot applications. Many published formulas are not accurate enough for robotic ranging applications, and some robotics papers have, unfortunately, given erroneous information. Hence, ultrasonic theory as applied to robotic ranging systems is difficult to locate.

7.2 THEORETICAL ACOUSTICS

The study of ultrasonics requires information from many branches of physics. The basic laws of sound propagation are derived from four areas within physics. Hydrodynamics provides the fundamental equations of sound flow. Statistical mechanics relates the macroscopic behavior of physical systems (in this case the air) composed of numerous molecules to the laws of forces that govern intermolecular relationships. Thermodynamics provides the fundamental conservation principles around which all sound interactions are based. Meteorology provides some of the necessary information on related effects like humidity, altitude, and attenuation.

Velocity of Sound

Sound energy propagates as waves through a medium. The velocity of these waves depends on several characteristics of the medium. For example, sound travels much faster underwater than it does in air, and even in air sound velocity varies as the composition of the air changes.

The atmosphere is a mixture of gases, most of which obey the equations of state for ideal gases fairly closely. According to Dalton's law, the sum of the partial pressures of gases in a mixture is equal to the total pressure of the mixture. In addition, the pressure of any individual gas is related to its molecular weight. Consequently, we can base equations on ideal gas equations if we include in these equations the molecular weight of the mixture of gases in the atmosphere.

The mechanical disturbance of local gas molecules by a source of sound produces the original sound wave. Sound propagation involves the transfer of this vibrational energy from one point in space to another. As such, it is based upon the relationship of five variables at one point in space and time to the same five variables at a different point in space and time. These five variables include two thermodynamics variables (pressure P and density ρ) and three kinematics variables (the three components of the velocity field).

The propagation of sound must satisfy the principles of conservation of mass, momentum, and energy. Since sound propagation relates the changes of state from one point to those of its neighbors, it is possible to develop time-dependent solutions for the resulting five mathematical equations.

Sound transmission is the propagation of small, local pressure disturbances on the underlying atmospheric pressure and density. Thus, the resultant time-varying pressure is the sum of the original pressure and small sound pressures (disturbances) imposed upon it. Historically, Euler developed the appropriate equations of motion for ideal fluids (including gases), based on equations of continuity (from kinematics) and on Newtonian laws (dynamics). These equations lead to the ideal gas equation of state, discussed later. Through mathematical treatment of the appropriate energy equations, they provide a formula for determining the velocity of sound.

Atmospheric sound produces longitudinal waves; that is, the density changes produced are in the same direction as the wave propagation. For conservation of mass to hold, this change in density per unit time must be equal in magnitude (but opposite in sign) to the scalar product of the density and the divergence of the Eulerian (pressure) velocity:

$$\frac{d\rho}{dt} = -\rho \text{ div } v \tag{7-1}$$

From this equation and the conservation of energy equation, the divergence of v can be expressed as a function of sound propagation velocity c:

$$\text{div } v = \frac{1}{\rho c^2} \frac{\partial P}{\partial t} \tag{7-2}$$

Assuming that there is no rotational motion and no imposition of external forces (both valid for standard sound propagation in the atmosphere), and recalling that for a perfect gas pressure is an exponential function of density, we have

$$P = A(S)\rho^\tau \tag{7-3}$$

where τ is the ratio of the specific heat at constant pressure (C_p) to the specific heat at constant volume (C_v). From these three equations, we can derive the basic equation for the velocity of sound:

$$c^2 = \frac{\tau P}{\rho} \tag{7-4}$$

For diatomic gas molecules (the type which make up the bulk of the atmosphere), $\tau = 1.4$; for monotomic molecules (such as neon), $\tau = 1.0$.

To rearrange equation (7-4) into a form that includes temperature, we start with the equation of state for a perfect gas:

$$\frac{RT}{PV} = 1 \tag{7-5}$$

where R is the molal gas constant, and T, P, and V are the temperature, pressure, and volume of the gas, again referring to molal quantities. Rearranging equation (7-5) to solve for P, we have

$$P = \frac{RT}{V} \tag{7-6}$$

Substituting for P in equation (7-4) gives

$$c^2 = \frac{\tau RT}{V\rho} \tag{7-7}$$

Note that density ρ is defined as mass per unit volume:

$$\rho = \frac{m}{V} \tag{7-8}$$

If the volume is 1 mole to make it consistent with the terms in equation (7-7), the mass m can be expressed in gram molecular weight. Substituting (7-8) into (7-7) gives

$$c^2 = \frac{\tau RT}{m} \tag{7-9}$$

or

$$c = \sqrt{\frac{\tau RT}{m}} \tag{7-10}$$

Using $\tau = 1.4$, $R = 8.3170 \times 10^7$ erg/mole-deg, and $m = 28.97$ g/mole, and taking the square root gives

$$c = 2{,}005\sqrt{T} \quad (\text{cm/s}) \tag{7-11}$$

(1 erg/g $= 1$ cm^2/s^2).

If T is expressed in degrees Celsius rather than kelvins and c is converted from cm/s to m/s, we have

$$c_m = 20.05\sqrt{T_c + 273.16} \tag{7-12}$$

When this equation is modified to use the Fahrenheit scale, with speed in feet per second, the result is

$$c_f = 49.03\sqrt{T_f + 459.68} \tag{7-13}$$

At normal room temperature ($20°$C, $68°$F), the values are

$$c_m = 343.3 \text{ m/s}, \quad c_f = 1126.3 \text{ ft/s}$$

Note that equations (7-12) and (7-13) appear to include three constants and one variable, temperature. This overlooks the many simplifying assumptions that have been made. Although accurate enough for many purposes, additional factors must be considered when high accuracy is required. As shown, equations (7-12) and (7-13) are only valid to within 1% for most conditions. The effects of humidity and altitude on velocity have both been neglected.

If 0.1% or better accuracy is desired, the effect of humidity must be considered. In equation (7-5), the equation of state assumed a perfect *dry* gas. To include the effects of humidity, the value of T in that equation (which also affects subsequent equations) must be replaced by T_v, the virtual temperature, computed as

$$T_v = \frac{T}{1 - 0.379(P_w/P_t)} \tag{7-14}$$

where P_w is the partial pressure of the water vapor, P_t is the total atmospheric pressure, and T and T_v are in kelvins. Substituting in equation (7-12), then, we obtain

$$c_m = 20.05 \sqrt{\frac{T_c + 273.16}{1 - 0.379(P_w/P_t)}} \tag{7-15}$$

The partial pressure of water vapor can be obtained from relative humidity and temperature information. If relative humidity (RH) in percent is known, we have

$$P_w = \frac{P_s(\text{RH})}{100} \tag{7-16}$$

where P_s, the saturation pressure of air, is a nonlinear function of temperature with values of 0.08854 at 32°F, 0.25630 at 60°F, and 0.5069 at 80°F; P_t, the total pressure of the air, also varies but can usually be taken as 14.696 psi at 32°F.

For highest accuracies, equations (7-14)–(7-16) should be used. However, if accuracies to within 0.1% are acceptable, a simplifying equation can be used for temperatures from -20 to $+110°F$:

$$c_f' = c_f + RH \left[0.33 + 10 \left(\frac{T_f}{100} \right)^3 \right] \tag{7-17}$$

where RH is the relative humidity expressed as a decimal. For example, at a temperature of 79.7°F (26.5°C), RH $= 0.37$ (37%), and $P_w/P_t = 0.0126$. The velocity computed through equation (7-13) is 1138.7 ft/s. However, the actual velocity is 1141.4, and thus equation (7-13) produces a 0.24% error. When equation (7-17) is used to correct for humidity, the computed velocity is 1140.4 ft/s, which reduces the overall error to 0.088%.

Absolute accuracy may not be required in all ranging applications. In some cases, relative ranges can be used, thus removing the effect of errors in humidity or air temperature. It is also possible to compute an accurate sound velocity on location by ranging to a known reference point. With a known distance, the actual sound velocity can be computed and used to determine range to other objects. The resultant accuracy would be no better than about twice the error in range measurement of the reference point and would have to be updated with temperature changes or major changes in humidity.

A correction could also be necessary for altitude, if the system is not operating near sea level. This correction is necessary since we used a figure of 28.97 g/mole for the molar mass of the atmosphere in equation (7-9). Air density decreases with higher altitudes, as does the effect of humidity. Fortunately, the velocity change due to this factor is less than 0.01% for low altitudes and is usually ignored.

Sound Reflection

Sound energy must be returned to a receiver in order to be detected. In everyday sound (such as a concert heard over the radio), energy is radiated over a wide angle. In addition, this sound energy reflects off walls and diffracts around door openings, thus allowing it to reach the ear of a listener who may be in another room or who may be on the other side of a large object.

When sound energy is used for ranging purposes, only the direct path of the sound is usually of interest, since this path is used in the direct calculation of range. (There are some advanced applications in which sound traveling over an indirect path may be useful. Some of these applications are discussed in section 7.4.) Understanding this direct path and the amount of energy returned involves several considerations.

When ultrasonic energy impinges on a target, boundary level effects cause some of the energy to be reflected and some to be absorbed. The amount absorbed is a function of the density of the material, the angle of incidence, and the speed of sound, among other factors.

If a plane sound wave (S_i) is incident on a plane surface at an angle θ from the normal, a reflected wave is produced with its amplitude a function of the type of material. The reflected intensity S_r is

$$S_r = S_i, \left| \frac{Z \cos \theta - \rho c}{Z \cos \theta + \rho c} \right|^2 \tag{7-18}$$

where Z is the specific acoustic impedance of the reflecting material, and ρ and c are the density of air and the velocity of sound in air. Note that the product ρc is approximately equal to 44 (metric units). Under conditions of reflection from a solid object into air, Z is always much larger than ρc. Thus almost all of the energy is reflected, with only a small amount absorbed by the object.

Also of significance is the directivity of the response. Ray tracing is just as applicable in sound wave propagation as in light wave propagation. With light waves, our eye can see objects because the incident light energy is scattered in all directions by most objects, which means that some energy will reach our eye, regardless of the angle of the object to us or to the light source. This scattering occurs because the roughness of an object's surface is large compared to the wavelength of light (approximately 550 nm or 0.000055 cm). Only with very smooth surfaces (such as a mirror) does the reflectivity become highly directional for light rays.

Ultrasonic energy has wavelengths 10,000 times as large as light (≈ 0.25 in). Therefore, ultrasonic waves find almost all large flat surfaces reflective in nature. With little scattering, the amount of returned energy strongly depends on the incident angle of the sound energy. As surface roughness increases, an increasing amount of energy will be reflected back at varying angles. With irregularities in the surface, such as a picture on a wall, the edge of a door, light switches, and so on, the discontinuities will reflect some energy over a large angle. Thus, even with smooth surfaces, any edges present will reflect energy back to the transmitter. Figure 7-2 illustrates ultrasonic energy impinging on a large wooden crate and illustrates the directivity of the reflection. Due to its placement angle, only sound energy reflecting from the crate edge can be detected by the transducer. The rest reflects in various directions.

This scattering has an important effect. Nearby objects, even if detected by the ranging system, often reflect energy in other directions, where it may encounter another object and then be returned to the transducer. Because of

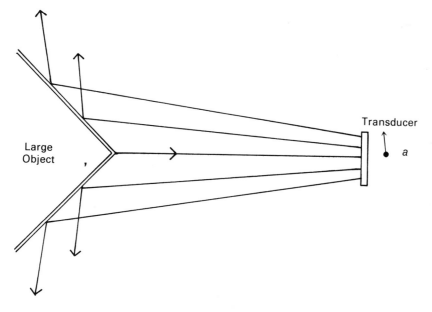

Transducer

Large Object

Figure 7-2. Ultrasonic beam reflections.

the longer path lengths, the arrival time will be later than for the energy returned by the direct path. If the ranging system was designed to accept several returns in order to detect range to more than one object, it would report a false location for this reflected return. Therefore, most ultrasonic systems reject any return after the first.

It is also difficult to detect very small (point size) objects. Most of the sound energy tends to be reflected from objects that are larger than a wavelength, whereas most of the sound energy tends to go around objects that are smaller than a wavelength. In addition, for objects smaller than a beam width, the amount of returned energy is also proportional to the cross section of the object. The combination of these two factors makes it extremely difficult to detect very small objects, such as a wire.

Ultrasonic Attenuation

Except under near-field conditions, the intensity of an ultrasonic wave decreases with the distance from the source. (Near fields extend approximately to a distance of $a^2/4\lambda$, which represents only a few inches.) This decrease is caused by geometric factors and by energy absorption of the atmosphere. Geometrically, energy spreads out as it is radiated, and the amount of energy thus incident on a constant-size object smaller than a beam width will decrease as a function of $1/d^2$, where d is the distance from the transducer to the object.

Most of the incident energy is reflected, and it is this energy we wish to retrieve for a ranging system. Energy returned from rays reaching a surface normal to the wave will continue to diverge after reflection, with the result that two-way range must be considered. Therefore, the energy from standard reflection decreases as a function of $1/4d^2$. Since the edges of objects scatter the sound wave, the returned echo strength from edges again decreases with distance (approximately a $1/d^2$ function). In summary, the amount of loss due to geometric scattering depends on the surface characteristics of the object, and will range from $1/4d^2$ to $1/d^4$, depending on the reflecting surface.

Ultrasonic energy is also absorbed by the air due to viscosity, heat conduction, and molecular-vibration-based absorption mechanisms (particularly those due to water vapor). These losses are usually considered together as an absorption factor. The actual loss varies with temperature, sound frequency, and relative humidity. Although various theoretical formulas have been developed, none are accurate enough to predict this absorption, so measurements under actual conditions must be made.

Frederick[2] ran a series of tests to determine the attenuation of ultrasonic energy due to energy absorption. He found the attenuation (in dB/ft) to be 0.2 at 20 kHz, 0.5 at 40 kHz, and 0.8 at 60 kHz, under conditions of 760 mmHg barometric pressure, 26.5°C temperature, and 37% relative humidity. His actual result was much larger than the classical (Stokes) attenuation figures of 0.015 dB/ft at 20 kHz and 0.13 dB/ft at 60 kHz. Frederick's tests indicated that attenuation increases with temperature and frequency and seems to peak at medium levels of relative humidity, at least at higher room ambients (Table 7-1).

The effect of energy absorption on ultrasonic systems is a trade-off in range and frequency. One can compute either the maximum range that can be obtained with a given frequency or the maximum frequency that can be used at a given range.

Table 7-1. Attenuation as a Function of Humidity

Temperature (°C)	Relative Humidity (%)				
	10	30	50	70	90
0	5.92	6.85	7.98	9.21	10.47
20	8.18	15.08	20.2	22.0	21.4
40	19.5	27.2	20.5	16.1	13.5

Table shows loss in dB for each 10 m of travel through the air at various temperatures and humidities.

Limitations

There are upper limits to both sound frequency and amplitude. A gas can no longer conduct sound when the intensity is so large that the refractory portion of the wave becomes zero (cavitation would be produced) or when the wavelength is about equal to the mean free path of the molecules. Thus, the maximum amplitude of acoustic energy propagated in the air is limited to 190 dB, although major waveshape distortions would occur far below this intensity level. The maximum frequency limit is about 1,000 MHz. Both of these limits would never be approached in actual systems, because the attenuation in air at high frequencies and the potential damage caused by high sound intensities would prevent reaching these limits. Sound frequencies as high as 500 MHz have been generated.

Note that sound intensity is usually expressed on a log scale, with the reference intensity taken as

$$S_0 = 10^{-16} \, W/cm^2 \tag{7-19}$$

which corresponds approximately to the lower limit of hearing by humans.

There is also a theoretical upper limit on discrimination between two objects at slightly different ranges. In any ranging system, this limit is defined by the Rayleigh[3] resolution criteria, which says that two objects are resolvable in range if they are spaced at least 2λ apart. At 50 kHz, this would be 13.5 mm (0.5 in).

7.3 PRACTICAL CONSIDERATIONS

Range Calculations

Once velocity under current operating conditions is known, range in an active sonar system may be determined from the following range equation:

$$R = \frac{ct}{2} \tag{7-20}$$

where t is in seconds, c is the speed of sound (in either meters per second or feet per second), and R is the one-way range in compatible units. The measured range is slant range to the nearest portion of the object that reflects energy back to the receiver. This point is important because it is not always clear to a ranging system which spot on the object reflected the energy.

Although the basic ranging formula is quite accurate, several factors should be kept in mind when considering the accuracy of the result. One factor is error in the velocity of sound. Obviously, the accuracy of R can be no more accurate than c. As discussed in section 7.2, several sources of errors, such as temperature and humidity, contribute to this error. For example, if a $10°$ temperature differential is not considered, the computed range will be in error by about 1%. The minimum error contributed by sound velocity errors is seldom less than 0.05%.

Other errors develop over the value of time t. Uncertainty about the timing of the transmitted pulse is one possibility, but it can usually be eliminated. Since many ultrasonic transducers need high-voltage pulses, which typically come from a source of stored energy in the circuit, there are variable delays in the time a pulse is actually transmitted compared with the time a signal to transmit was given. This potential transmission timing error can be significantly reduced if a signal is taken from the transmitter circuit itself and used for timing control.

A much more significant error is associated with the received pulse time. Since the delay in time of reception is the value being measured, the normal technique is to send out a pulse of sonar energy and time the difference between the start of the transmitted pulse to the start of the received echo. This pulse of energy is created by transmitting a number of cycles of the selected ultrasonic frequency. Pulse duration (t_p) depends on the number of cycles and its frequency:

$$t_p = \frac{n}{f} \qquad (7\text{-}21)$$

where f is the sound frequency in hertz and n is the number of cycles in the pulse. Thus, if 25 cycles are transmitted at a frequency of 50 kHz, the pulse duration is 0.5 ms.

It is possible to determine reasonably accurately the time that the pulse was transmitted and when a return was received. The difficult part is determining which of the transmitted cycles was the first one received. Thus, if 25 cycles of 2 μs wide (50 kHz) are transmitted and the signal is received in 20 ms, at first glance it would appear that the target range was 11.263 ft (normal room conditions). However, if the fifteenth pulse was the first one to exceed the threshold of the receiving system, the actual time between transmission and reception must be reduced by 30 μs. The actual distance is then 11.246 ft, a difference of 0.15%, or a range error of 0.2 in.

Since most receiving systems have an integration effect built in, requiring three or more pulses to be received before indicating the presence of a target, the actual received pulse from a timing standpoint must be at least the third pulse. This integration is to eliminate noise spikes from triggering the receiver. For highest accuracy, received time must be reduced by an amount corresponding to which pulse triggered the system. If the system uses a chirp pulse (one with a series of different frequencies) to enhance detection, it may be

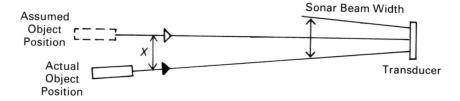

Figure 7-3. Effect of beam width.

even less certain in these systems which pulse was detected first. This topic is discussed further in section 7.4.

Finally, errors arise due to geometrical effects. Equation (7-20) assumed that the echo is directly returned from the target (secondary echoes are discussed later) and that the sonar beam width is negligible. One effect of practical beam widths can be seen in Figure 7-3, which illustrates the case where the object is off center but normal to the transmitted beam. The slant range R as computed is correct, but the value of the X-component of range is wrong. That is, the computer would assume the object was along its center line ($X = 0$) rather than having a horizontal displacement X. The actual displacement is a function of the angle to the object:

$$X = R \sin \alpha \qquad (7-22)$$

For a 20° beam width (or $\pm 10°$), the X-component could be wrong by as much as 17% of the total range, or 1 ft at 6 ft. The Y-component of the range is also slightly affected, but the error is less than 0.2%.

A similar situation occurs for objects at an angle to the sonar beam. With the geometry shown in Figure 7-4, the range received will be to the closest point on the object, but the system may assume the object is positioned as shown by the dotted lines.

The minimum range that can be detected with a separate receiving element depends on the length of the transmitted pulse, since, in general, the receiver must be shut down during this time to prevent the system being triggered on the transmitted energy pulse. Thus, minimum range is

$$R_{min} = \frac{ct_p}{2} \qquad (7-23)$$

where t_p is pulse duration. For example, with a 0.5-ms pulse, the minimum range is 3.4 in.

When the same transducer is used for both transmission and reception, additional time is needed to allow ringing in the transducer to settle. This increases the minimum range by about 50%.

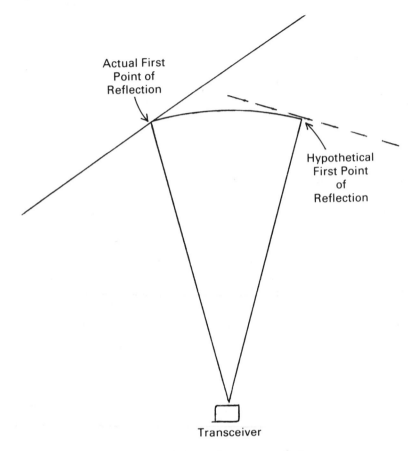

Figure 7-4. Target placement confusion.

Wavelength

Sound energy is propagated in waves that have a characteristic wavelength. Wavelengths may be calculated as follows:

$$\lambda = \frac{c}{f} \tag{7-24}$$

where λ is the wavelength in the same distance units that c is expressed in, and f is the frequency in hertz. The wavelength of sound within the midrange of human hearing (200 Hz to 1 kHz) is between 1 and 5 ft. The wavelength at typical robotics ultrasonic frequencies (40–80 kHz) is approximately 1/6 to 1/3 in. Frequencies in this range are also employed by most bats.

In addition to wavelength, a wave number K is often used, especially in calculations involving beam widths:

$$K = \frac{2\pi}{\lambda}$$

(7-25)

Horizontal Resolution

Sound energy radiated from any transducer has beam width. In essence, there is a cone of energy leaving the transmitter, and sound waves will impact all objects within that cone. The solid angle subtended by the cone defines the beam width of the sound. In actual systems, the energy does not drop off sharply but tapers off. Beam width in an actual system is defined as the point where the intensity is 3 dB down (the amplitude is 0.707 of peak value).

Beam width is affected by the frequency of the sound (higher frequencies have a narrower beam width) and by the size of the transmitting elements (because of phase interference, larger elements produce narrower beam widths). The azimuth resolution (along a plane perpendicular to the sound wave) depends on the number, spacing, and phasing of the transducers. For a single transducer, beam width depends on the size of the transducer and the wavelength of the ultrasonic energy transmitted.

Most transducers are circular and can be treated as a circular disk set in an infinite baffle. Maslin[4] showed that the beam width (on each side of center) is approximately equal to

$$\alpha = 2 \sin^{-1} \frac{1.62}{Ka}$$

(7-26)

where K is the wave number $(2\pi/\lambda)$ and a is the disk radius (in the same units as λ). One commercially available transducer (from Polaroid) is 1.5 in in diameter (0.75 in radius) and operates most efficiently at 49 kHz. With $\lambda = 0.275$ in and $K = 22.8$, $\alpha = 2 \sin^{-1}(0.095) \approx 11°$. Since the beam is symmetrical, the total beam width is $22°$.

It is often assumed that beam width is independent of range, but in one series of tests conducted by the author, the effective beam width from a typical transducer increased about 10% for every doubling of distance (from 2 ft to 20 ft).

7.4 ADVANCED CONSIDERATIONS

Arrays

One way to control the direction of sound transmission and/or to improve azimuth resolution is by using multiple transmitters or receivers. If control

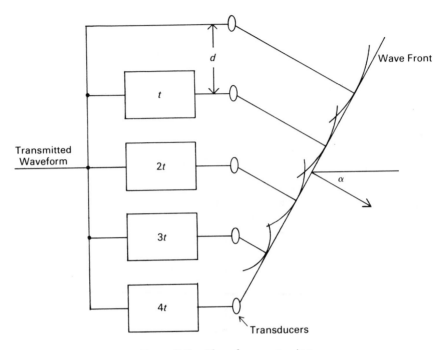

Figure 7-5. Phased array steering.

over the timing (or phase) of the signals is employed at the transmitting or receiving elements, the technique is called a *phased array*. Phased array systems were first developed for radars to decrease the time it took to scan a radar beam when it looked for a target. It was then adapted to sonar systems for a similar purpose and for focusing a beam to a small area.

Steering of the beam is accomplished through control over the phase shifts given to the transmitted signal. For example, Figure 7-5 shows a beam aimed to one side by putting in delays of varying lengths into the signal path to each transducer. Since electronic circuitry can be used to control the amount of delay, the beam can be electronically steered over a wide angle. Although the illustration shows a two-dimensional (2D) array, three-dimensional (3D) arrays work under similar principles. Five elements are shown in the illustration, but the approach works for any number: the greater the number, the narrower the resultant beam.

In most cases, we know the array steering angle, and it determines the amount of delay between elements. The formula is

$$t = \frac{d \sin \alpha}{c} \tag{7-27}$$

where t is the required delay time between pulses sent out through adjacent elements, d is the distance between elements, α is the angle that the beam is to be deflected, and c is the speed of sound. For example, let $d = 1.75$ and $\alpha = 10°$. Then the required delay is 22.7 μs between each of the elements.

If the purpose of an array is to focus the beam rather than to steer it, two approaches can be used. For an array of transducers, a simple Gaussian taper, which provides higher energy in the center of the beam, will focus the beam to approximately 40% of the array diameter.

Tighter focusing of the beam is possible through phased array techniques. By producing a more narrow beam of energy, the system will have better azimuth resolution. In phased array focusing, pulse transmission from the center of the array has a controlled delay, compared to the edges, resulting in the equivalent of an acoustical lens focusing the beam energy. Depending on the number of array elements, pulse characteristics, and time delays used, the beam can be focused down to 4λ (about 1 in).

Figure 7-6 shows how an array can be focused to a point. The calculation of the amount of delay is more complicated in this case, and the value of N (the amount of delay for focusing at a point) cannot be computed from equation

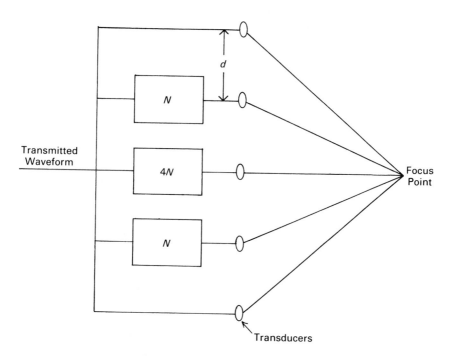

Figure 7-6. Phased array focus.

(7-27). In the figure the array is to be focused at a point 5 ft away, and N is approximately 2 μs. Note, however, that this focusing is only at a specific range, with resolution at nearer and further distances much larger. Over reasonable distances, phased arrays always provide beam widths that are appreciably smaller than single transducers.

Phased array techniques using a single transmitter can also be applied to the receiving element. In this case the transmitter will illuminate an area with ultrasonic energy, and the receiving array can focus on various portions of this area.

Two-Receiver Systems

In beam splitting, two (or more) ultrasonic range measurements are made to the same target. The second may be from the same transducer after it has been rotated one half beam width, or from a second transducer with an overlapping beam pattern, as shown in Figure 7-7. In either case, if the object is seen in

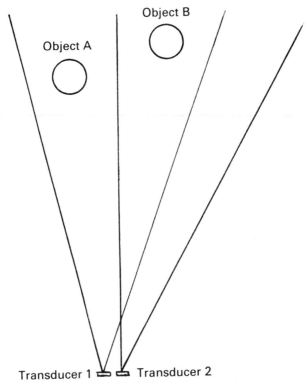

Figure 7-7.   Beam splitting.

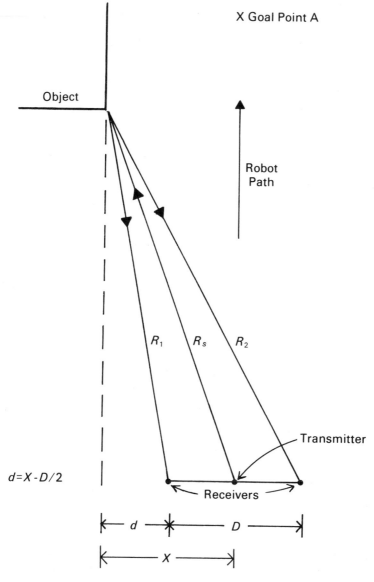

Figure 7-8. Difference of arrival time.

both measurements (such as object B), it must be located in the overlapping parts of the beams. If it is seen in only one (such as object A), it must be in the nonoverlapped portion of the beam. The net result is to improve azimuth resolution by a factor of 2 or more.

Another suggested approach is to use the difference in arrival time of two ultrasonic returns to detect more accurately the angle to the target. With this approach, equivalent angular accuracies as low as 1° can be obtained. This technique will also improve azimuth resolution. Figure 7-8 shows a mobile robot that must travel to a goal point (A). The robot has one transmitting and two receiving sensors. To make sure it will clear the obstacle shown, the robot must determine the horizontal distance to the object (X) and compare it to half of its own width. Horizontal clearance may be determined as follows. Two receivers are located a distance D in apart at the front of the robot. A sonar transmitter is mounted between them in the center. Sonar signals emitted from the transmitter will be reflected from the corner of the object. A typical value for D might be 18 in. Separate signals are received by each sonar receiver. Knowing the transmission time and the time of each reception, the robot can accurately determine separate ranges (R_1, R_2) from each receiver to the object. With R_1, R_2 and D known, the slant range R_s to the object and the horizontal distance X can be computed:

$$R_s \approx \frac{R_1 + R_2}{2} \tag{7-28}$$

$$X \approx \frac{R_2^2}{2D} - \frac{R_1^2}{2D} \tag{7-29}$$

Both formulas, although approximate, have only a small error. In this example, we will assume that range can be measured to 0.3%. For the area of interest, along the side of the robot, we can calculate the worst-case error. This error occurs when one range measurement has an error of +0.3% and the other an error of −0.3%. This condition produces a 10% error in X. Since clearance is measured to half the robot's width, for a robot 30-in wide the added clearance requirement is only 1.5 in, which is equivalent to having a system with a horizontal beam width of about 1°.

Chirp Systems/Frequency Selection

In standard systems, the choice of frequency for the ultrasonic system is based upon at least three factors: attenuation through the air, surface reflectivity, and environmental noise. Environmental noise, if present at ultrasonic frequencies, can interfere with any measurements and can affect the choice of frequency. Attenuation through the air is worse at higher frequencies. This factor must be considered if ranges beyond 10 ft are needed, although frequencies up to 50 kHz are usable to ranges of perhaps 50 ft.

Surface reflectivity can affect whether a target is seen. The amount of energy returned is a function of surface reflectivity. This dependence, along with the frequency selected, produces a variation in the amount of energy reflected, the amount absorbed by the object, and the amount of reflective energy scattering that occurs. Even slight differences in ultrasonic frequencies can mean the difference between seeing a target and missing it.

Therefore many systems employ several frequencies in the ranging system so that if one frequency is not properly reflected back another might be. If the transmitted frequency is changed smoothly over a range of frequencies, the resultant pulse is referred to as a *chirp* (due to the sound it would produce if it were in the audible spectrum). Another approach, called *pseudochirp,* uses four to eight fixed frequencies and divides the available transmission time among them all. Polaroid uses this technique in its autofocusing camera to improve ranging ability.[5]

A chirp system increases range error and minimum ranges that are present in the system. Since multiple frequencies are transmitted, more total cycles must be accommodated. Thus the length of the transmission and the minimum range of the system are increased.

Also, since the very purpose of using multiple frequencies is to increase the possibility of a return, it will be difficult (almost impossible) to determine which frequency triggered the receiver and therefore what the actual round-trip time was. One method is to assume that the round-trip time should be shortened by a value equal to one fourth of the pulse width.

Some systems can measure the frequency of the incoming pulse. A system of this type will at least be able to determine in which portion of the transmitted pulse the first detection occurred.

It is also possible to use a chirp system to measure range without making time measurements or being concerned about occasional missed pulses. If the ultrasonic transmission starts at one frequency and is then linearly reduced in frequency until a return is detected, the returned frequency can be compared with the current frequency of the oscillator. Since the oscillator is now at a much lower frequency, the difference in the frequency is a function of the distance to the target (the greater the distance, the greater the difference frequency). This difference frequency is independent of which cycle is detected, and can be converted to a range measurement.

Gain Control

Attenuation is a function of distance, and the length of time before a return is received is also a function of the distance to the object. Thus, it is possible to increase the gain of the receiving circuitry as a function of time. If this approach is taken, nearby objects (which produce the strongest return) will have a lower gain for detection, and more distant objects will have a greater gain for their weaker returns. Noise in the environment will receive little gain while the system is looking for nearby objects and will pose little problem during this time. At larger ranges, with increased gains, noise will once again become a problem.

Received Pulse Width and Amplitude

Point-size objects will return a signal of small width (few pulses). The larger the object's size and the more beam area it covers, the longer the time that returns will be received. Thus the number of cycles received (the wider the return pulse) gives a rough indication of the size of the object.

Amplitude varies with range, but it also varies with the type of material being illuminated and whether the return is from an edge or from a flat object. In some applications the amount of energy expected to be returned at the specific range can be predicted. The expected value can then be compared to the actual amount received, and surface characteristics can be estimated.

Motion Detection

When the source and the target have a relative velocity between them, a Doppler effect comes into play. The *Doppler effect* is an apparent frequency change in the signal caused by motion. Figure 7-9 illustrates this effect when the velocity is along the X-axis. The sound wave will change in frequency as a function of the ratio of the relative velocity v to the speed of sound c. With β equal to the ratio of v/c,

$$f' \approx f(1 + \beta \cos \theta) \tag{7-30}$$

where θ is the angle between the direction of travel and the direction of the source. In the figure, $\cos \theta = x/r$ for object O_1, and $\cos \theta = 1$ for object O_2.

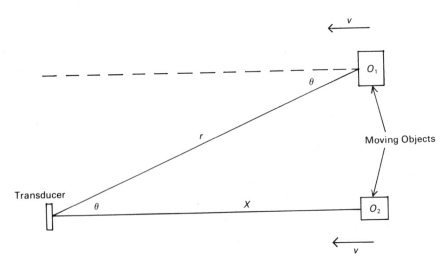

Figure 7-9. Doppler effect.

Although equation (7-30) is only approximate, at velocities under 100 ft/s (114 km/hr), the error is less than 1%. If the system can detect a change in frequency, it can be used to detect the presence of motion and to determine the magnitude and direction of this motion.

Second Echoes

We have been discussing the first received echo, which can usually be accurately detected and used in range measurement. It is also possible to look at the second echo and determine range to a second target. The difficulty is in determining whether this second echo is a direct echo from a second object, or a second, indirect echo from the first target. The latter can occur if some of the energy received by the first target is reflected in a different direction, strikes a wall or other object, and then is returned to the transducer. This added path length would delay the arrival of this part of the energy and give a false second echo.

Nevertheless, second echo detection can be used under some circumstances. For example, you may know that there is no nearby object to return a false echo from the first target before the echo from a second object has reached the transducer. Or if you know the approximate range to the second object, you could ignore all returns not within that range window. This approach is particularly good for mapping, where you have some information about the area (either a priori or through earlier range mapping). Now you know the approximate distance to objects and are only trying to determine the robot's position with respect to each object.

Another application of a second echo might be to ensure the absence, rather than presence, of an object at the indicated range. For example, if you are approaching a known door location, you can tell that the door is open if no return is received at that range (whether the return is the first, second, or third echo). If there is a return at the right range, and it might be a false echo, you would not be sure whether the door was open (it was a false echo) or the door was closed (it was the door echo). However, if there is no echo at that range, you can be sure the door is open.

7.5 ULTRASONICS IN BATS

Studies have been done on bat collision-avoidance systems, which also use ultrasonic techniques. Some of this work is also applicable to robotics, at least in providing comparative results and in defining some sonar techniques (such as sonar harmonic pulsing) that are not currently being used in ranging systems. There is no current system employing some of these "natural" principles to detect and avoid obstacles, so we briefly describe ultrasonic ranging by bats.

Bats use a triangulation method to determine angles. In this approach, a chirp, emitted from the bat's mouth, is picked up at slightly different times by the bat's two ears. Phase differences between these two received signals allow the bat to detect both range and bearing (horizontal angle) to the target.

Studies of echo location by bats indicate the bat's ranging accuracies and serve as an indication of the accuracy possible for sonar object detection. Masters[6] determined horizontal resolution to be 1.6° for moving targets and 1.5° for stationary targets. These values are compatible with the author's studies of systems that use a single transmitter and two receivers. It should certainly be possible to match, if not surpass, the angular resolution of bat sonar, although many current systems cannot.

Lawrence and Simmons[7] have determined that although bats ears are in a horizontal plane they can still locate objects within 3° vertically. Fuzessery and Pollack[8] have shown that one species of bat uses three harmonics to determine vertical target positioning. Harmonic analysis using three separate harmonic-specific receivers could therefore be considered as one method to improve angular resolution. The use of harmonic frequencies is an avenue for study not previously utilized. Previous ultrasonic systems have been limited to single- and swept-frequency systems. Still another direction for investigation is suggested by Simmons,[9] whose research indicates that bats use phase information to refine acoustic imaging.

7.6 SYSTEM CONSIDERATIONS

Transducers

A *transducer* is a device used to convert an electrical signal into sound energy, and vice versa, and is a key element in any ultrasonic system. The standard ranging system uses separate elements for transmitting and receiving energy. It is also possible for the system to use the same element for both transmitting and receiving. A single element requires less circuitry and is less expensive. However, separate elements eliminate transmitter ringing as a factor and thus allow shorter ranges to be detected. Separate elements are also usually desirable in phased-array-based systems.

Three types of transducers are available for the transmission of ultrasonic energy in air: piezoelectric, electrostatic, and magnetostrictive. Piezoelectric devices change their length under an applied electric voltage. They are typically made from quartz, adenosine diphosphate (ADP), barium titanate, or ceramics.

Magnetostrictive devices change length under an applied magnetic field. They require a bias field to allow movement in both directions, this bias field changes the device length slightly. Then if an AC magnetic field is impressed on the device, its dimensions can change in both directions. This change is then coupled to the air and produces the desired ultrasonic energy. A formula for the change in dimension of magnetostrictive transducers is

$$\frac{\Delta l}{l} = S_0 = CB_0^2 \tag{7-31}$$

where l is the length of the material, Δl is the change in length, S_0 is the strain in the material, C is a constant, and B_0 is the flux density.

Electrostatic transducers move a thin element under the field produced by an electrostatic charge. In a typical electrostatic transducer, a thin plastic foil

Figure 7-10. Typical ultrasonic transducers. (*Courtesy of Polaroid Corporation*)

with a gold conductive coating is placed over an aluminum backplate, thus forming a capacitor. A DC bias voltage is applied to this capacitor, and thus an electrostatic force is produced. An AC signal voltage is then used to vibrate the transducer according to the applied frequency. This vibration sends appropriate sound waves through the air.

Polaroid uses this method in its transducer. Figure 7-10 shows several Polaroid instrument-grade transducers, along with a small circuit board designed to produce the necessary DC bias and coupling the desired AC pulse to the transducer.

Transducer dimensions affect the power and frequency transmitted. If the transducer dimension is equal to $\lambda/2$, the transducer will resonate at that frequency, and the resultant energy transfer will be much greater. Because of characteristics in the transducers, piezoelectric transducers have a much higher Q. In other words, they produce more acoustical energy for the same input signal, but they also are much more sensitive to frequency. Since they have less band width, frequency control is more critical.

Trade-offs and System Considerations

Before discussing trade-offs, we review some data about ultrasonic systems. Most ranging systems provide only limited information, such as range data to the closest target, and report only limited angular resolution. But ultrasonics is not so limited. Processing techniques exist that can provide much other information. For example, both acoustical beam focusing and phased array techniques have been employed[10] to improve the beam's angular resolution at the expense of reducing the field of view. These techniques require some type of scanning to cover the original field, thereby increasing complexity and requiring more time. On the other hand, both experimental work and the results of bat studies indicate that a two-receiver system can pinpoint objects while retaining a wide field of view. Other information that can be extracted from a sonar signal includes difference in range between two sonar receivers, variations in amplitude of signals at a known range, length of the returned signal pulse, presence of multiple echoes, use of diffraction effects, and processing second echoes.

The design of any sonar system involves trade-offs. The higher the selected frequency, the better the range resolution, the narrower the beam width, and the less likely it is to miss small targets (*i.e.,* go around them); however, the lower the frequency, the greater the range in air (because of less absorption) and the larger the system field of view. Another trade-off is in the number of pulses transmitted. The more ultrasonic pulses that are transmitted, the greater the likelihood of detecting a target return (because of the buildup of coherent return energy), but the resulting range resolution is poorer and system minimum range is increased.

The length of the pulse chosen is typically from 0.5 to 1.0 m. If the pulse is much longer, errors in received pulse timing will increase. If the pulse is much shorter, it may not contain enough energy (which is proportional to pulse length) to reliably detect all targets. See the excellent article by Everett,[11] which discusses in some depth many of the problems and limitations found in sonar systems.

Collision Avoidance

The task of collision avoidance for a mobile robot involves determining the distance of objects from the robot and the relationship of those objects to the robot and its projected path. This calculation must often be made under the constraint that the robot is moving. Thus, the robot must take measures before some critical distance (R) to avoid colliding with the object.

The critical distance is a function of the robot's radius and speed, beam angle, detection sampling rate, object size, and measurement errors. Figure 7-11 illustrates the geometrical relationship of several of these variables. Assume that the robot is traveling in the same direction as the center of the sonar beam, that angular data are known only to the sonar beam width, and that a sonar reading taken reveals an object. This reading does not reveal if the object is within the robot's projected path. That determination can be made only when the robot gets to the critical range R, which is the intersection of the sonar beam and the projected path of the robot. At this range, if the object is still seen the robot will hit it. The further an object is outside of the horizontal of the projected path, the greater the range at which it will cease being observed. Those objects on the outside edge of the path will cease being observed precisely at range R, while objects extending into the path are still visible, thus implying a collision.

It is also important to consider that the sonar measurements are only taken periodically, so the critical range becomes a band and the length of the band is the distance that the robot will travel at its speed between observations.

There are many potential errors. The foregoing discussion assumes that the object has some reflecting point that has the same height as the sonar transducer, that the sonar beam angle can be determined with a high degree of accuracy, that there is no error in the range measurements returned by the sonar, and that the sonar and direction of robot movement are completely aligned. If these assumptions are not valid, allowance for these errors increases the required buffer zone.

There is also the problem of vertical positioning, which must be resolved for a mobile robot to avoid collision with unexpected obstacles. Unlike objects in the horizontal plane, short objects in the vertical plane could pass out of the beam's range or be completely beneath the sonar cone yet still be in the pathway. For extended objects, the closest point to this type object is always

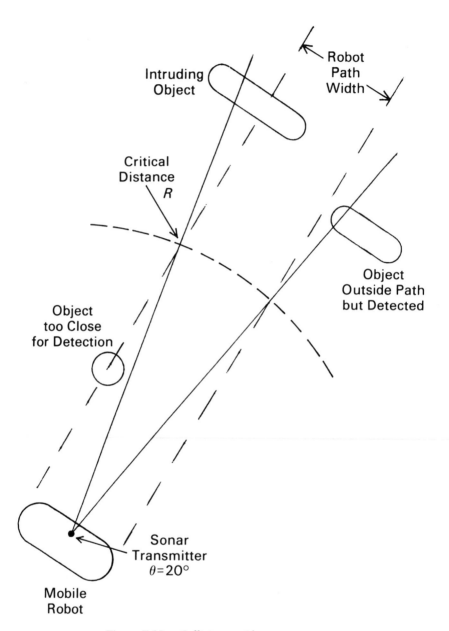

Figure 7-11. Collision-avoidance geometry.

on a line normal to the robot. Thus if a sonar system is given sufficient angular coverage to see the object along this normal, the minimum range to the robot can be measured directly.

To be absolutely sure that the robot can see anything that it is approaching would require angular coverage in the direction of travel of close to $\pm 90.°$ Furthermore, if the robot is approaching a wall at a shallow closing angle, it may not detect the potential collision until the last seconds.

One possible aid to collision detection for a stationary robot and a moving object is to analyze the return for a Doppler shift (is anything moving?) and then check the direction of the Doppler shift (is it moving away from the robot?).

Three types of algorithms have been suggested by Lozano-Pérez[12] for obstacle avoidance. Although developed for a fixed-base robot, they are easily adapted for a mobile robot. They are (1) hypothesize and test (the most obvious); (2) penalty functions (assigned to known objects); and (3) explicit free space (keeping track of available maneuvering space). The first approach estimates a path for the robot arm. It then tests this path for possible collisions. If a collision possibility occurs, it tries to modify the path. If no collision-free path modification is found, a new path is hypothesized and tested. The method requires geometric modeling of the environment. The second approach assigns a penalty for approaching an object, with the penalty rising as the object is approached and reaching infinity upon contact. The algorithm tries to find the shortest path with a reasonable penalty. The third approach stores free areas, perhaps after dividing the space into cubes of various sizes so that only the cube's corners and size must be stored. Then it is possible to ensure that motion can only occur through previously defined free space. This algorithm imposes the largest computational load.

A different approach, called the *configuration space method,* is often used with mobile robots. If you know the maximum diameter of the robot, you can consider that the robot is actually only a point source, but all objects then must be expanded by half this value. (The center of the robot can approach no closer than half its diameter without hitting an object.)

Mapping

Mobile robots need a map of their environment. Ultrasonic systems can be used to generate a map through a scan of the area. A map needs to have better resolution than is available through normal system beam width. Either a phased array method is needed to steer a small-diameter beam, or multiple, overlapping, wide-angle range measurements must be taken. If the sonar transducer is moved repeatedly over small steps (such as 2°), the resultant information can be used to plot obstacle locations.

Another mapping approach is to use multiple sensors around the robot to let it see in all directions at once, and then have the robot move around to cover the area from many view points. In one series of experiments, Elfes[13] divided an area into 42×79 cells and, using 24 transducers mounted around the robot's circumference, plotted whether each cell was occupied or unoccupied. Initially the 6-in^2 cells were classified as unknown, but as the robot moved around the room and kept collecting data it was able to determine with confidence the status of each cell.

The mapping task becomes much easier if the robot is given a map of the area to begin with and only required to keep track of its own position in the area as it moves around. In this case, ultrasonics can more readily handle the collision-avoidance and navigation update problems.

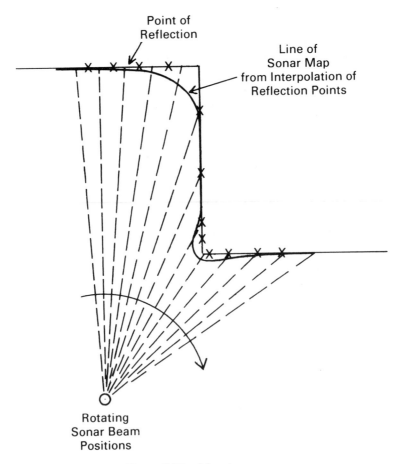

Figure 7-12. Mapping corners.

Because a sonar beam will be returned from either one wall or the other, never from the corner, mapping the position of inside corners in a room is notoriously hard to accomplish accurately. Depending on the beam width of the sonar and the position of the robot, walls can be plotted fairly accurately up to 12 in from the corner. The corner can then be located by recursive line fitting: a straight line is fitted to the wall data along each side, and the corner is projected from the intersection of the two sides.

Figure 7-12 shows a typical sonar map of an inside corner and an outside corner and demonstrates the results of line fitting. The locations of all sonar returns are shown by X's. The inside corner is cut short, and the outside corner is extended. Even with recursive line fitting, the corner location will still be off by a couple of inches unless the robot has an accurate map of the area in advance and can use it to correct its sonar-generated map. However, since the robot can never position itself in spaces near a corner, the slight residual error may not matter in practical applications.

7.7 APPLICATIONS

There are many potential applications for sonar ranging systems, but few existing commercially available ranging models. Few commercial mobile robots have been built, and even most of these have only limited range finding or mapping capability. One example of a sonar collision system is the mobile robot security guard being developed by Sanwa Seika of Japan and reported by Kajiwara.[14] This system uses 12 pairs (transmitter and receiver) of sonar sensors, in groups of three at four corners of the robot. They use sonars with a frequency of 40 kHz. To improve sonar accuracy, they use a technique that includes a least squares fit over the last 17 measurements.

In another example, an ultrasonic range finder that detects echoes from 8 in to 200 in was demonstrated on a mobile robot.[15] This sonar system had a range resolution of 0.4 in and an angular resolution of 10°. Even though neither figure is near a theoretical limit, the robot was still able to avoid two obstacles in its path (of which it had no prior knowledge). However, travel time was 100 s over a distance of only 16 ft (less than 0.1 mph). This delay was primarily due to the average time to prepare each new path plan (4 s).

REFERENCES

1. Kinsler, L. E., and A. R. Frey. *Fundamentals of Acoustics.* New York: Wiley, 1962, p. 232.
2. Frederick, Julian R. *Ultrasonic Engineering.* New York: Wiley, 1965, p. 26.

3. Kay, L. Airborne ultrasonic imaging of a robotic workspace. *Robot Sensors,* Vol. 2: *Tactile and Non Vision,* Alan Pugh, ed. Kempston, Bedford, U.K.: IFS Publications Ltd., 1986, p. 291.

4. Maslin, Gerald P. A Simple Ultrasonic Ranging System. Paper presented at the 102nd convention of the Audio Engineering Society, Cincinnati, Ohio, May 1983.

5. Biber, C., et al. The Polaroid Ultrasonic Ranging System. Paper presented at the 67th convention of the Audio Engineering Society, October 1980.

6. Masters, W. M., A. J. M. Moffat, and J. A. Simmons. Sonar tracking of horizontally moving targets by the big brown bat *Eptesicus fuscus. Science,* Vol. 288, June 1985, p. 1331.

7. Lawrence, B. D., and J. A. Simmons. Echolocation in bats: The external ear and perception of the vertical positions of targets. *Science,* Vol. 218, Oct. 1982, p. 481.

8. Fuzessery, Z. M., and G. D. Pollak. Neural mechanisms of sound localization in an echolocating bat. *Science,* Vol. 225, Aug. 1984, p. 275.

9. Simmons, J. A. Perception of echo phase information in bat sonar. *Science,* Vol. 204, June 1979, p. 1336.

10. Kuroda, S., A. Jitsumori, and T. Inari. Ultrasonic imaging system for robots using an electronic scanning method. *Robotica,* Jan. 1984, pp. 47–53.

11. Everett, LCDR B. *A Multi-Element Ultra-Sonic Ranging Array,* NAVSEA Technical Report 450–90G-TR-0001. Jan. 1985. Washington, D.C.: Government Printing Office.

12. Lozano-Pérez, T. Task Planning. *Robot Motion: Planning and Control,* M. Brady et al., eds. Cambridge, Mass.: MIT Press, 1982.

13. Elfes, Alberto. Sonar-based real-world mapping and navigation. *IEEE Journal of Robotics and Automation,* Vol. RA-3, No. 3, June 1987, pp. 249–265.

14. Kajiwara, I., et al. Development of a mobile robot for a security guard. *15th International Symposium on Industrial Robots.* Tokyo: Japanese Industrial Robot Association, Sept. 1985, pp. 271–278.

15. Everett, *A Multi-Element Ultra-Sonic Ranging Array.*

MOBILE ROBOTS

Most industrial robots are mounted on a fixed base and cannot move around. Their work envelope is thus limited to the volume reached by motion of the robot's joints. Hence, these robots must have their work brought to them. A mobile robot, however, can go where the work is and can transport items between fixed-base robots, thereby opening up many new application areas. Mobile robots can be used, for example, to move products, parts, inventory, and other lightweight items from one location to another in factories. They are quite valuable in hazardous environments, such as a nuclear installation, at the scene of a fire, or a bomb disposal site. In the home, most functions will require mobile robots, such as aiding the handicapped. They also may be used to monitor and troubleshoot production problems.

Mobility is becoming more important as robot technology continues to expand and enters the era of the mobile robot. This chapter examines many specialized requirements and techniques of mobile systems and supplies design trade-off information in important technical areas such as locomotion, steering, motive power, and navigation. In addition to the topics covered here, chapter 15 discusses a number of current applications of mobile robots while the advanced mobility technique of legged locomotion is discussed in chapter 16.

8.1 APPROACHES TO MOBILITY

Mobility in robots has been a desirable goal for years. Some of the earliest experimental robots were mobile, starting with Shakey, developed by Stanford

University in the 1960s. Today, many university laboratories are using experimental mobile robots to develop vision systems, artificial intelligence programs, specialized sensors, and navigation/mapping programs related to mobility.[1] Although these experimental models have contributed heavily to advancing the state of the art, no university mobile robot has made it into industry.

The first step in mobility for industrial applications was the gantry robot introduced by Olivetti in 1975. Gantry robots have improved since that early overhead robot and are now offered by about half of all robot manufacturers. In a gantry robot the base moves along overhead rails, allowing the unit to be maneuvered over manufacturing equipment in the workplace. When it arrives at its desired location, the gantry robot lowers its arm to reach objects. Since it is overhead, the gantry can clear most manufacturing equipment and thus is ideal in cluttered environments. Being on rails, the gantry can reach a much larger work area than conventional industrial robots can. For example, one system offers a 40 ft × 10 ft work area. Gantry robots, however, are still not mobile enough to handle applications requiring wide-ranging mobility.

The next improvement in mobility were wheeled vehicles, which can access an even greater area of the factory. The first wheeled vehicles developed were teleoperated robots that had a remotely located operator to indirectly steer and control the robot's motion. Commercially available, these robots have an onboard vision system through which the operator can view the scene. Teleoperated robots have performed very well in inaccessible or hazardous locations, but since they require a human operator they are too expensive for normal factory operations.

AGV Systems

From teleoperated robots it was a natural step to automated guided vehicles (AGVs). These models are like driverless forklift trucks. They can handle large loads from one area of the factory to another, but are generally constrained to follow a fixed path. The path is a wire buried in the floor or a paint stripe on the floor that the robot is programmed to follow. Some models operate on rails.

AGVs employ a radio frequency (rf) sensor (to follow a buried wire) or an optical sensor (to follow a visual path). In wire-guided systems, low-power rf signals are carried by wires imbedded in the floor. The AGV system interprets amplitude or phase variations of this signal to determine when it is going off course. A servo loop controlling the AGV steering system brings an off-course AGV back on center over the wire. Wire-guided systems have a higher installation expense and provide less flexibility for system changes, but generally need less maintenance than a photooptical system. They also have some advantages in complex systems,[2] since communication between vehicles and between one AGV and the system can be done through the existing rf link.

In a photooptical system, either paint or a reflective tape is used for path control and an onboard photoelectric tracking system detects the difference between the desired path and the floor. Maintenance for optical-path-based systems usually requires restriping about every six months in areas receiving moderate traffic.

Both optical stripe and wire-guided AGVs allow limited control over vehicle path and stopping points, but they are still tied to physical tracks along the floor, even though they do not need rails.

Figure 8-1 shows a wire-guided AGV from Jarvis P. Webb. The unit may operate on its own or be integrated into a complete automated material han-

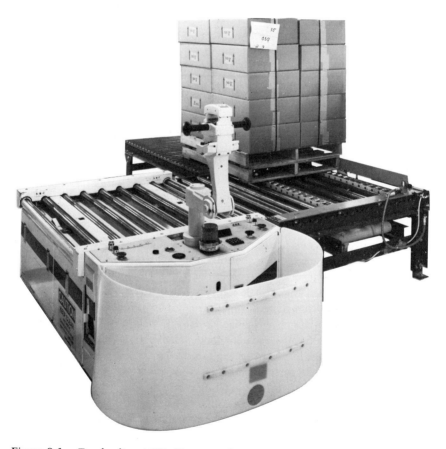

Figure 8-1. Top-loading AGV. (*Courtesy of Control Engineering Company, a Jervis B. Webb Company affiliate*)

dling system. It has its own microprocessor, but it can also report to a central computer through the same rf links used for vehicle control.

No operator is required: the unit automatically loads and unloads material at its stops, and its material-handling capabilities are varied. For example, for the unit in the figure three types of tops are available: a lifting top, a conveyor top, and a shuttle top. Each top can manipulate materials being transported and can transfer a load from 2,000 to 4,000 lb in just over 4 s. The vehicle can run for up to 16 hr on a battery charge.

The AGV does have some drawbacks. For example, if an AGV finds something blocking its path, it stops and waits for someone to remove the object. Even if the AGV could push the object aside or there was room to go around it, it still waits for a human to intervene. This limitation is safety related, but mostly it is due to the limited sensor and intelligence capabilities of current AGVs.

If a user wants a more autonomous AGV, he must wait for the development of complete collision-avoidance systems and better autonomous navigation and mapping systems. Researchers at Georgia Institute of Technology are developing off-wire capability for AGVs, using a Litton AGV. They are also using a combination of machine vision and ultrasonic ranging to perform automated loading and unloading. Martin Marietta is developing an autonomous vehicle for outdoor use, under a contract with the U.S. Army. This vehicle is designed to detect the edges of a highway and follow a road. It can also travel off the road.

Independent and system-controlled AGVs are both available. Independent AGVs have sufficient onboard computing power to operate independently of a remote computer system and are locally controlled. System-controlled AGVs operate under control of a remote, system-level computer and are less expensive but not as flexible. Some authors have referred to the locally controlled AGV as a "smart" AGV, but we reserve that designation for future AGVs that can leave their designated path when necessary or desirable.

All AGVs have local software to cover the path tracking function. Originally this function included only track following and automatic stop functions, because the earliest systems did not include complex layouts requiring routing control. Other examples of specialized software include vehicle operations, dispatch control, traffic management, and vehicle routing. These functions may be system- or locally controlled.

Coded information can be given to the AGV at various locations along its path to give the AGV some precise locating points so that it knows where it is, to control traffic and prevent vehicle collisions, and to alter its routing to meet changing needs in the workplace. In this last instance, the AGV may be directed to switch to an alternative branch or to skip a location at which it normally stops. Some coding has been provided by simple bar codes read by an optical system, but most require more complex IR or rf communication systems.

Sometimes the AGV must take a branch path instead of following the main loop. This switching arrangement can be handled in several ways. At the central system level, certain rf-carrying wires may be switched off and others switched on. The AGV will follow the one that is on. Another rf approach uses different frequencies for different paths, and the AGV is told to switch frequencies. This approach is often followed in a completely locally controlled system. In this case, the AGV may be manually switched to follow a different path. The locally controlled AGV will know what path to follow and will switch its receiving circuits appropriately, thus "seeing" only one path.

A different approach is used for central control of an optically based system. Path choice is also possible, usually through locking the steering system. If the AGV is to turn, it is allowed local control over its steering, and the tracking system will follow the turn. If the system wants the AGV to continue straight ahead, it sends an external command (via rf or IR) that locks the steering wheel and forces the AGV to continue in a straight line over the switch.

In systems with multiple AGVs, traffic control must prevent potential collisions between vehicles and route vehicles to ensure that all areas of the system contain sufficient AGVs for their current needs. Traffic control may also direct the vehicle to a recharge station. AGVs are powered by batteries, which must be recharged. Some systems recharge batteries overnight, and others do so during waiting (non-busy) periods.

Complex systems often require communications between AGVs and a central system. The imbedded wire or a separate radio or IR-based communication system can be used. Vehicles may be started manually as needed or controlled by the system according to overall needs. The AGVs can circulate indefinitely and check each of their way stations to determine whether a load is present, or they can be dispatched only when needed.

Autonomous Robots

The most desirable type of mobile robot is completely autonomous. That is, it can travel anywhere without an explicit path. A few autonomous, limited-capability, industrial-grade mobile robots are commercially available. One model is built by Cybermation (Fig. 8-2). This robot is well-designed for industrial use and experimentation and will be used as an example in some of our discussions.

Experimental and/or educational robots are also being developed. Although not immediately applicable to current production operations, these robots can help us understand the principles of mobility. Most of the pioneering compa-

Figure 8-2. Commercially available mobile robot. (*Courtesy of Cybermation Inc., Roanoke, Va.*)

nies in this area (RB Robots, Androbot, and Artec Systems) are no longer in business, but they helped develop mobile technology.

8.2 DESIGN CONSIDERATIONS

Design specifications of mobile robots include things that are unique to mobile robot development. For example, general mobility problems such as sloping floors, door jambs, travel between floors, and wheel marks on the floor must be resolved. The robot must not run into people, boxes, or even wastepaper baskets. It must watch out for items hanging from the ceiling or protruding from benches, and avoid electrical cords and air hoses. The robot must be designed to make turns, especially wide-angle turns, to remain horizontal, to not run off edges (such as loading docks). If the robot has reversing capabilities, then additional locomotion, steering, and sensor ramifications must be taken into account.

Completely autonomous robots are still in the experimental stage. Although the techniques being investigated seem to be within the state of the art, no mobile robots commercially available are fully autonomous in unknown, unstructured factory environments. Unfortunately, the design of some models imposes unacceptable constraints on where they can operate.

A mobile robot may have many of the same features (manipulators, vision systems, and specialized sensors) as their fixed-base cousins. But at least three factors make them unique, all of which are due to requirements imposed by the robot's mobility. The mobile robot needs (1) a locomotion system to allow it to move, (2) steering capability, and, most important, (3) a mapping function (to determine where it is), a goal planning function (to determine where it wants to go), and a collision-avoidance system. More than likely, it will also need an autonomous navigation capability.

Conflicting requirements impact all areas of robot design. For stability the robot needs a low center of gravity, but it must be high enough to clear expected obstacles, such as door jambs, piping, or cables. Steering implies that the robot should be flexible enough to maneuver in tight quarters, but, unless the design is kept simple, cost, weight, and reliability can easily be severely affected. Robots that can turn on their own axis, which some mobile robotics companies seem to consider important, are more unpredictable in the workplace and, therefore, more dangerous.

Another concern is motive power. The robot must have sufficient onboard battery power to operate for long periods without having its batteries recharged. But longer running times require more battery capacity, which adds weight to the robot. This added weight means more torque is needed from the motors, so their weight increases. All of this additional weight could reduce the robot's payload-carrying ability. The converse problem, often overlooked, is that the longer the robot goes without a recharge, the longer it will take to charge it when needed, increasing the robot's down time. These trade-offs (and others, such as the number of wheels) must be addressed in the design of mobile robots.

These considerations have led to a conceptual design for a mobile robot consisting of three sections: a base that provides mobility, a midsection that provides intelligence, controls, and sensors, and a working section that performs the robot's application-oriented functions. In most mobile robots, in order to keep the center of gravity low, the base usually holds the drive mechanism, the locomotion and steering systems, servo control and power amplification, and a source of mobile power.

The midsection has the bulk of the robot's electronics, sensors, and computing capability. Sensors supply necessary collision-avoidance and navigation data and handle special-purpose requirements imposed by the application. A few sensors, such as contact switches, may be mounted on the base, but most of them must be higher to obtain the necessary line-of-sight clearances.

The working section may be completely passive, or it may have a conveyor arrangement or even an arm that interacts with its environment. AGVs have had small robot manipulators installed to perform load and unload functions.

8.3 LOCOMOTION

Locomotion is the subsystem that gives the robot its mobility. Wheels, tracks, and legs can be used as locomotion mechanisms. In theory, the choice depends on trade-offs in power requirements, costs, desired locomotion characteristics, and the surfaces over which the robot must transverse. In practice, however, the choice is usually wheels. (Legged systems are still in the laboratory stage.) Wheels perform well if the terrain is level or has only a limited slope, is reasonably smooth, and has good traction. Otherwise, tracks can be used; they are well-suited for outdoor, off-road applications. They can even climb stairs.

Legged robots have the greatest capability of going over most types of terrain and climbing stairs. But they are difficult to design, the computations required for leg control and balance are numerous and difficult, and their power drain is so large that these robots are currently seen only as experimental models in university or R&D laboratories. These robots are further discussed in chapter 16.

Since wheeled vehicles are the primary type of mobile robot in use, and expected to be used in the future, we discuss them in detail. We also briefly discuss tracked vehicle designs.

Interrelated variables to be addressed in the selection and design of wheel configurations include the total number of wheels, the number of drive wheels, wheel size, steering method employed, torque requirements, and desired speed. Steering is covered in section 8.4; the other topics are discussed in the following paragraphs.

Wheeled Vehicles

Robots usually have three or four wheels, although six or more are possible. Some configurations even have sets of double wheels. Four wheels offer more stability, more traction (if all are driven), and slightly better clearance capabilities. Three wheels offer less complex steering, the elimination of shock absorption mounting, less weight for the wheel assembly, and some navigational improvements.

Larger wheels can roll over minor obstacles and allow the robot's undercarriage to clear small objects on the floor. But larger wheels raise the robot's center of gravity, which makes it less stable and contributes to the reduction of the robot's available dead-reckoning positioning accuracy. In other words, when a wheel goes over an object, the apparent path length increases, which affects the robot's internal position calculations. Because of stability and accuracy considerations, most systems have adopted small wheels (from 4 in to 1 ft). In one study conducted by the author, he determined that a wheel diameter of 10 in $\pm$ 2 in was optimum for a mobile robot 48 in high.

There are many ways to construct, mount and steer the wheels. Figure 8-3 illustrates five types of wheel configurations. Figure 8-3a shows the standard three-wheel approach, with two drive wheels and a single (front) steering wheel. Figure 8-3b shows all wheels providing drive and steering, which is Cybermation's approach. Carnegie Mellon's Pluto robot has two wheels on each of three support legs, as shown in Figure 8-3e.

Figure 8-3d is Weinstein's approach[3] in which three rotating wheels spaced 120° apart allow the robot to climb stairs. The robot is normally supported by two of the wheels. When it comes to stairs (or other obstacles of less than half the wheel's height), contact forces make the configuration rotate automatically and allow the robot to climb the stairs. (Because of the sharp climbing angle, this system has difficult stability and torque problems.) Weinstein also provides design information on the specialized drive mechanism required.

Another clever approach is the wheel chosen for TOPO, a personal robot developed by Androbot in 1983. This technique, shown in Figure 8-3c, has the wheels at an angle to the floor, making the robot easier to steer and more stable. Other designs include the University of California's rotating roller wheel. The rollers are mounted lengthwise around the circumference of the wheel, which lets the robot move in any direction. Wheel action permits forward movement, and the roller action permits sideways movement. Some

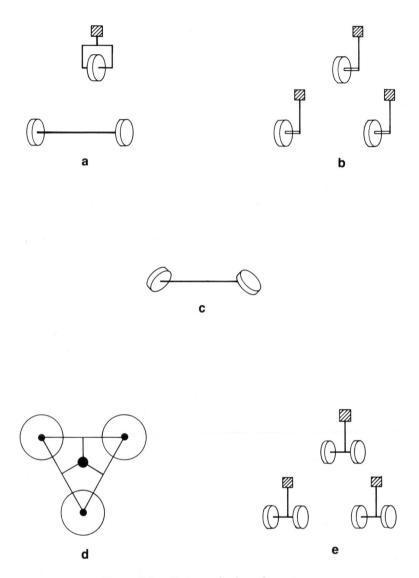

Figure 8-3. Various wheel configurations.

manufacturers use larger drive wheels with smaller idlers that are free to turn on their own axes. In this arrangement no steering mechanism is provided, and the robot steers by differential power to its two drive wheels.

Cybermation has a three-wheeled robot in which each wheel is attached to a separate leg assembly and is independently steerable. The wheels are 8-in pneumatic tires, which provide some shock absorption. The base uses a 1-hp centrally located main drive motor to drive all three wheels through an ultra-high efficiency (98%) 17-to-1 drive train. Speed ranges from 0.03 to 6.5 mph,* and full speed is available within 6 ft of travel. Through proportional-integral-differential (PID) algorithms, the robot's velocity can be precisely controlled, thus providing smooth start/stop action.

A three-wheel system has much to recommend it. They rest firmly on the ground without independent suspension even when the surface is uneven. (Four-wheel systems need a way to handle uneven floor surfaces.) If steering is based on turning one wheel rather than two, the steering system is not as complex. Dead-reckoning calculations are easier and more accurate. The weight saved by using one less wheel and a simpler steering mechanism means the robot needs less torque and battery capacity.

Torque Calculations

Torque is the production of rotational force around an axis and is equal to the vector product of the force and the distance to the point at which the force is applied (the wheel radius). Torque is an important factor in determining robot locomotion. How much torque is needed from the mobile robot's drive motors? Since force equals mass times acceleration, the answer depends on the robot's weight and the acceleration desired. Acceleration is needed to increase the robot's speed and to counter gravity when traveling up an incline.

How much acceleration is needed depends on how fast we want to increase speed (for example, it should not take an hour for the robot to reach operating speed) and on the worse-case ramp angle that the robot must climb. Accelerating up a ramp increases torque requirements. We will use a $10°$ ramp, which can be considered the minimum, although under some conditions it might be desirable to provide for steeper ramps. This angle is approximately a 1-ft rise per 12 ft of travel, a typical ramp value used for access for the handicapped.

We will compute the torque requirements independently for accelerating and for climbing a ramp. Acceleration depends on the starting and ending velocities (V_s and V_e) and on the time to change speeds:

$$a = \frac{d}{dt}(V_e - V_s) \qquad (8\text{-}1)$$

---

* The usual unit of robot speed is feet per second (ft/s). To convert to miles per hour (mph) or meters per second (m/s) use the following equations:

1 ft/s = 0.68 mph	1 mph = 1.467 ft/s
1 ft/s = 0.305 m/s	1 m/s = 3.28 ft/s

Typical human speeds are 3 ft/s (a slow walk) to 25 ft/s (a fast sprint).

If we assume the robot started from rest and reached a speed of 4 ft/s in 1 s, the required average acceleration is 4 ft/s². Climbing a ramp requires opposing the force of gravity g. This opposing force equals

$$f_g = mg \sin \theta \tag{8-2}$$

where m is mass and θ is the ramp angle. For a $10°$ ramp $\sin \theta \approx 0.172$, resulting in a minimum acceleration of 5.55 ft/s². To increase speed while climbing a ramp, a higher figure would be needed. To provide some safety factor, we might want a minimum acceleration capability of 7.5 ft/s².

With a $10°$ ramp, a 200-lb mobile robot needs 34.5 lbf to climb the ramp.* With a 10-in wheel, its torque on the wheel is 172.5 in-lb or 14.4 ft-lb. With a top speed of 4 ft/s and a 10-in wheel diameter, rim speed is approximately 1.5 rps (90 rpm). A typical drive motor can run at 1800 rpm, allowing a 20/1 speed reduction and a 20/1 increase in effective torque. The result is a minimum motor torque requirement of 69 oz-in for each of two drive motors, plus the torque that is necessary to overcome inertia, friction, and drive train inefficiencies.

Dead-Reckoning Considerations

The robot needs a method of dead reckoning to know approximately where it is at all times. Even if an accurate mapping system were available, dead reckoning is still desirable in order to handle conditions between periods of position determination, during periods when the robot may be out of touch with a navigational aid, or just to double-check the navigation system.

Dead reckoning involves measuring the direction (or angle) of travel and the distance traveled and adding these measurements to a known starting position. Since errors will occur in both measurements, dead-reckoning systems have increasingly large errors as the robot moves. They must be periodically updated by some other navigation system to keep the error acceptably small.

The easiest part of dead reckoning lies in distance traveled; if the robot continues in a straight line, no angle measurements are necessary. The distance traveled by the robot for each complete revolution of the drive wheels is

$$X = 2\pi R_w \tag{8-3}$$

where R_w is the radius of the wheel. The total distance traveled is

$$X = 2\pi R_w N \tag{8-4}$$

---

* We will use the term *pound* for pounds of force (weight, lbf) and pounds of mass (lbm). The two are essentially the same for conditions in which the acceleration due to gravity is earth standard. If we were designing a moon roving robot, we would have to differentiate between the two.

where N is the number of wheel revolutions. Although knowing the speed of the wheels and the time of travel lets us calculate travel distance, it is much more accurate to use wheel encoders to determine the number of revolutions and, from that, the distance traveled.

Equations are provided in section 8.4 to determine angular direction based on steering changes.

Tracked Vehicles

Tracked vehicle design has been around for many years, on tanks and bulldozers, for example. The two tracks can be bands or belts and have continuous or linked element arrangements. Driving wheels may be smooth or use sprockets. Some wheels are driven and others, called *idlers,* are allowed to freewheel. Some idlers support the vehicle weight as it runs along the ground, and are then called *rollers.* Figure 8-4 illustrates the arrangement between drive wheels, idlers, and rollers.

The distance between tracks affects vehicle maneuverability, as does track length. For good maneuverability, the width must be about 70% of the track ground contact length. (Ground contact length is that portion of the track in contact with the ground all at the same time.) For example, a tracked vehicle 5 ft long might have a ground contact length of 42 in, requiring a minimum width of 28 in for maneuverability.

For the tracked vehicle to climb obstacles (including stairs), the general track shape must be trapezoidal. Thus, the robot has its own built-in ramp. The height of the ramp must be greater than any obstacle the robot is expected to cross. If a robot needed to climb a stairway (most steps are 7 or 8 in high), a minimum vertical dimension for the track assembly should be, say, 10 in plus the radius of the top pulleys.

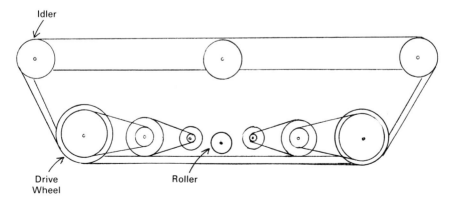

Figure 8-4. Tracked vehicle drive.

Other design considerations include the spring tensioning of idler wheels and the method of mounting the drive motors.

In addition to maneuverability considerations, the length of any tracked vehicle is a function of the width of any unsupported area that the vehicle must traverse. As a minimum, the track length must be at least twice the length between the supports, and proper design would use three times the unsupported area. For example, if the robot is to clear standard 30-in doorways, it should not be wider than 27 in (to leave some maneuvering room of 1½ in per side). According to the 70% width-to-tread-length ratio, the robot tread length (in contact with the ground) would be 38 in. This would allow an unsupported gap of 12–18 in to be crossed.

For a stair-climbing robot, track length should be long enough to contact at least three steps. Most conventional industrial stairs are 13–15 in between risers, requiring 39–45 in of track in contact with the ground. Note that some design compromises must be made if a tracked vehicle is to climb stairs, go through doors, and be highly maneuverable.

Being lower, tracked vehicles have more clearance problems than do wheeled vehicles. They can also cause greater floor damage. Although the tracks distribute loads over a greater area, they can still damage wires they may run over by pinching them. Total weight of the mechanism is much greater than for wheeled models and drive efficiency is poorer, so the vehicle uses much more power.

8.4 STEERING

Approaches

Locomotion and steering are related. In many configurations the same wheel provides locomotion and steering, often simultaneously. Other configurations use different steering methods: at least six are available.

1. Use one steerable front wheel with two rear drive wheels (separates drive from steering)
2. Use two steerable front wheels linked together or independently steerable.
3. Use steerable front wheels with drive wheel steering. Drive one rear wheel forward and a second wheel backward to make the robot rotate about the center of that wheel axle.
4. Use method 3 with no steerable front wheels, just idlers.
5. Provide independent steering of each drive wheel (offers maximum flexibility, requires maximum complexity).
6. Use front-wheel drive and a single rear wheel for steering.

Choosing a steering method depends on the intended use for the robot and the availability of other navigation methods. For example, slippage problems increase if a drive wheel also provides steering; thus, this approach is not suitable if dead reckoning is heavily used, but it gives the most capability for maneuvering in tight spaces. As another example, when a robot uses a separate steering wheel, the resulting path requires a larger turning radius, so clearance problems at the corner become more significant. However, calculating the turn and controlling the wheel for turning are much easier.

In any turn, time is important. The longer a turn is continued, the greater will be the length of the turn. Likewise, for the same turning time, the faster the robot is going, the greater the turn angle will be.

Calculations

Figure 8-5 shows the effect of a varying turning radius on the robot following a path. In case 1, a sharp curve, the actual path is shorter and the robot remains

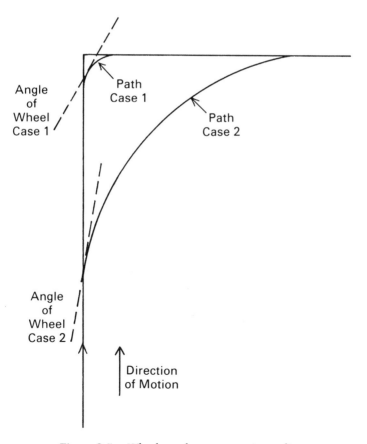

Figure 8-5. Wheele angle versus turning radius.

closer to it than in case 2. But turning errors (both steering and timing) are more critical, and sideways stability is poorest (the robot might slip sideways or even tip over).

In case 2, a wide turn, the actual turning path is much longer and its deviation from the desired path is greater than in case 1. This approach produces an error contribution if there is any error in timing the turn. The wider turning radius also provides better stability for a given forward speed, but chances of a collision are greater. Because the turn must start earlier, the data input and computations must be performed earlier and at a greater sensor range.

After the turn is made, the robot needs to know the actual angle of that turn. Figure 8-6 illustrates one method of determining the new coordinates (X_n, Y_n) of a turning robot and the new heading angle θ_n. This approach is based on the number of encoder counts (N_L and N_R) for the left and right wheels. The following steps illustrate this approach, assuming both wheels have the same radius R:

1. Compute the average number of counts:

$$N_A = \frac{N_L + N_R}{2} \tag{8-5}$$

2. Compute the difference in count between sides:

$$N_D = N_L - N_R \tag{8-6}$$

(if $N_D < 0$, then $\theta < 0$ and the robot is turning left).

3. Compute the change in heading:

$$\Delta\theta = \frac{2\pi R N_D}{N_T D} \quad \text{(in radians)} \tag{8-7}$$

$$\Delta\theta = \frac{360 R N_D}{N_T D} \quad \text{(in degrees)} \tag{8-8}$$

where D is the linear distance between the wheels and N_T is the total number of encoder counts per wheel revolution. Distance traveled along the arc depends on the number of counts that occur for each revolution of the wheel and on the radius of the wheel.

4. The new heading is thus

$$\theta_N = \theta_0 + \Delta\theta \tag{8-9}$$

5. Now

$$X_N \approx X_0 + \frac{2\pi R N_A}{N_T} \cos(\theta_0 + \Delta\theta/2) \tag{8-10}$$

where X_0 is the old value of X.

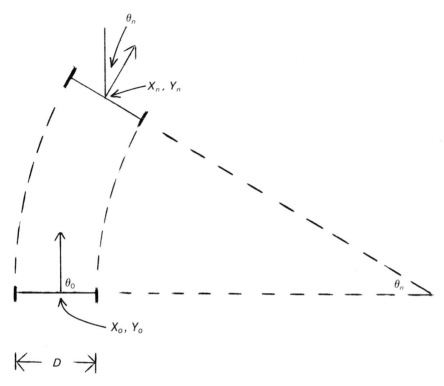

Figure 8-6. Turn calculations based on wheel encoders.

6. And

$$Y_N \approx Y_0 + \frac{2\pi R N_A}{N_T} \sin(\theta_0 + \Delta\theta/2) \qquad (8\text{-}11)$$

where Y_0 is the old value of Y.

The trajectory to be followed is usually a series of straight lines (Fig. 8-7), but the robot usually follows a curved path, so the added width required for clearance during turns must be considered when we compute the desired trajectory. In some tight spaces the robot might have to slow down and make a tighter turn.

The arc length traversed by each wheel is different during turns, which means the wheels might slip. However, if the wheels are driven independently, or if the wheel is an idler, there should be little forward wheel slippage for moderate turn angles since the changing load on each wheel acts to offset this type of slippage. Lateral slip is another matter, however, because the inner edge of both wheels (during the turn) must slip slightly outward due to the

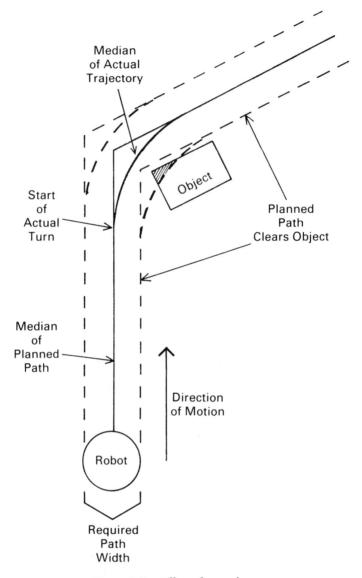

Figure 8-7. Effect of curved turns.

width of the wheel itself. The sharper the turn, the worse this slip is. Another consideration to keep in mind is that the turn puts the load on the edge of the wheel, and not the center, affecting turn calculations.

Differential steering refers to driving the wheel on one side with more torque than the wheel on the other side. The robot then turns in the direction of the wheel with less torque. This type of steering simplifies the mechanical design but requires more accurate control of wheel speed. It also has slip sensing problems.

To allow the robot to turn in even tighter quarters, one rear wheel can be driven forward and the other in reverse. The robot would then turn around the center point between these two wheels (assuming the other wheels are only idlers). Figure 8-8 illustrates this turn, with the front wheel turned parallel to the rear wheel axis by direct steering for a steered wheel or by idler action for a freewheeling wheel. Normally, this type of turn makes the robot follow a path along the center of its drive wheel axis. If a turn along the physical center of the robot is desired instead, the forward moving wheel must travel further than the wheel turning backward by an amount related to how far the physical center is from the wheel axle center.

Another approach, used for example on Cybermation's robot, is to turn all three wheels simultaneously to turn the robot. If all wheels are turned the same amount, the robot will change its direction of motion around its center

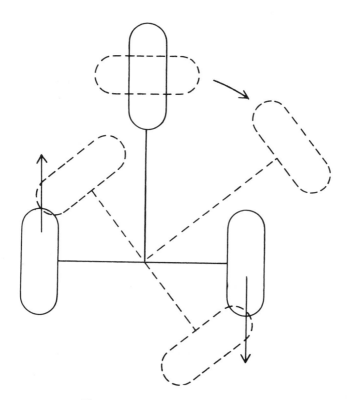

Figure 8-8. Turn about wheel axis.

point. The orientation of the base does not change—in effect, the robot will move on an angle instead of forward. This movement complicates the interpretation of information coming from sensors, and the robot may have to carry more sensors to obtain coverage in all turning directions. On the Cybermation model the sensors are mounted on a turret that turns along with the wheels. Thus, even though the base is moving sideways, the sensors are looking forward.

Turn Acceleration

Fast turns can cause stability problems. Even during a constant-speed turn, the robot is accelerating due to its changing direction of motion:

$$a = \frac{V^2}{R_T} \qquad (8\text{-}12)$$

where V is a scalar representing speed and R_T is the turn radius. Since $F = ma$, the centrifugal force is

$$F_C = \frac{mV^2}{R_T} \qquad (8\text{-}13)$$

from the center of the turn. Note that this force is proportional to the square of robot velocity. This force combines vectorially with the force due to gravity,*

$$F_G = (32.15 \text{ ft/s}^2)m \qquad (8\text{-}14)$$

to produce the resultant force on the robot (as shown in Fig. 8-9). This force has two effects: it tends to move the robot sideways, and is limited only by the coefficient of friction between the robot's wheels and the floor; it tends to make the robot want to tip over, if the resulting vector falls outside the stability triangle. The force holding the robot to the floor is proportional to both the wheel area in contact with the floor and to the coefficient of friction.

Assuming a 10-ft turning radius and a 1-ft/s speed, there would be negligible centrifugal force, equal to about 0.3% F_G. Cutting the turning radius to 1 ft and increasing the speed to 4 ft/s gives a sideways force of about 50% F_G, which almost certainly will cause the robot to slip or tip over.

---

* Acceleration due to gravity is usually approximated as 980 cm/s² or 32.15 ft/s². To be more accurate, it could be adjusted for latitude (over a range of -2 cm/s² at the equator and $+3.2$ cm/s² at the poles), as well as a minor decrease with altitude (less than 1 cm/s² at 5,000 ft).

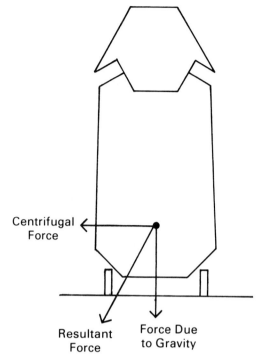

Centrifugal Force

Resultant Force

Force Due to Gravity

Figure 8-9. Centrifugal force due to turn.

8.5 POWER AND STABILITY

Battery Considerations

Batteries use different combinations of materials and electrolytes. Perhaps the best combination for a small-weight, high-capacity battery is the gelled cell. Gelled cells contain a gelatinous electrolyte instead of a liquid. They can be recharged many times, and they will not leak corrosive fluids or emit corrosive fumes.

To keep motor noise out of electric circuits, electronic/computer power and locomotion/steering power must be supplied from separate batteries. Locomotion/steering power is generally greater than electronic power. Computing the drain on the electronic battery is straightforward, since the drain is fairly constant.

Locomotion/steering battery capacity depends on the average current needed by the drive and steering motors, the amount of time the robot is

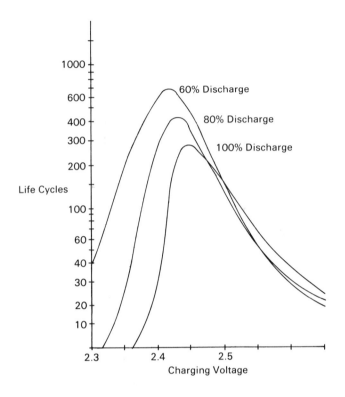

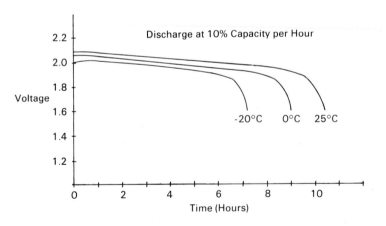

Figure 8-10. Typical battery curves. *(Based on data from Gates Energy Products, Inc.)*

turning versus the time it moves in a line, the percentage of time the robot is moving during its operating shift, and the approximate number of starts and stops that will occur in one shift. As a first approximation, average current = operating time percentage × the sum of average drive current and the product of average turn current × the percent of time turning.

After calculating average current drain, we can determine battery size by choosing the amount of time between charges. Battery discharge time is not a linear function of supply current. For example, increasing the drain more than proportionally reduces the discharge period. Available current is also a function of temperature; as temperature decreases, the battery's output capacity decreases. Therefore we must use the manufacturers' curves for the battery we want. Figure 8-10 shows two examples from typical batteries. The lower curve shows the effect of temperature, the second the effect of discharge depth on life cycles. Note that capacity is often tested at a 20-hr discharge rate.

If the robot is going to monitor its own voltage to determine when the battery needs recharging, the robot must monitor the battery under average load conditions because the voltage will always drop under heavy load. The effect of transient peak loads is not important.

Battery weight can be considerable. The weight of a typical 12-V gel-type storage battery with 28 A-hr capacity might be 21 lb. In fact, 20% of a mobile robot's weight can be from the many required batteries.

Stability Requirements

Mobile robot stability involves static and dynamic factors. We have already mentioned turning stability. We will now look at stability associated with ramps.

For purposes of stability, the robot can be considered to have all of its weight located at a single point, called the *center of gravity* (CG). The lower the CG, the more stable the robot, so heavy items such as batteries and motors should be mounted as low as possible.

Under static conditions, a mobile robot must always keep its weight vector inside the triangle defined by the CG and the robot's wheels. See Figure 8-11. A 48-in-high robot, for example, with three wheels spaced along a 24-in equilateral triangle, and an 18-in-high CG located directly above the center of the wheels could not climb a ramp greater than about 30° without falling over backward. If the robot tried to climb this ramp with insufficient torque, it could not make it. But in its static position it may still tip over. Therefore, if there is any possibility that the robot will attempt to climb a steep ramp, the robot should be equipped with some device to sense it. One device is a mercury switch set to cause a contact closure that alerts the robot to a potentially dangerous climb and, perhaps, automatically causes it to back up.

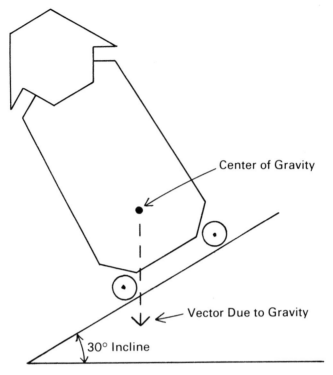

Center of Gravity

Vector Due to Gravity

30° Incline

Figure 8-11.   Effect of incline on stability.

An indication whether a robot is currently going up an incline can be obtained by comparing the robot's speed with its motor current. The current will be constant on level ground, but increases when the robot is climbing at an angle (assuming constant speed) since more torque is required.

A downward ramp is a bigger problem. By the time the robot sensed it was going down too steeply, such as if it was falling off a cliff, it would probably be too late. Two solutions have been proposed. The first uses a sonar system aimed downward ahead of the robot. With proper design it is possible to detect range to the floor; therefore if the floor drops off or if there is a steep downward ramp, the robot can be alerted. The second approach uses a small pilot wheel connected to the front of the robot. When this wheel goes over an edge, it could signal the danger in time for the robot to stop before its main wheels go over.

Another problem occurs on inclines and declines. Although the static situation may be stable, the dynamic situation may not be. For example, a robot that tried to accelerate up a ramp or tried to stop too quickly while going down a ramp might suddenly become unstable.

Reaction time is another important factor. Even if the robot does detect that it is reaching an unstable position, will it have sufficient time to react to the danger? Will its reaction be correct? Such safety procedures as cliff detection should be able to interrupt any other activity the computer was engaged in so that a maximum amount of time is available for reaction. Some consideration must also be given to what the robot does with the knowledge. For example, if the robot is moving forward rapidly and it detects a cliff directly ahead of it, slamming on the brakes could guarantee that the robot will fall over the edge because of its forward momentum. In this case, a slower deceleration could be safer.

8.6 INTELLIGENCE

Where normal programming leaves off and intelligence takes over is a fine distinction. Suppose the robot is going to retrieve a part in a supply room. It knows to go down a hallway, count doors, and enter the second door. But what happens if the robot misses a door, perhaps because of a different design on the door or poor lighting in the hall? The robot would pass the correct door and continue to the third door. But this doorway could be an entrance to a stairwell, which would put the robot in severe danger. Fortunately, the robot has safety programs to protect it. One detects the cliff ahead and stops the robot.

Nontechnical people would probably say that the robot planned to go through the door, but, when it noticed the stairs, it became frightened and refused to follow its previous instructions. They would credit the robot with intelligence, which is correct if you define intelligence by the resulting action rather than by the logic leading up to the action. The point is whether or not some of the programs qualify as intelligence, the robot must be able to handle problems as if it had intelligence.

Three areas of mobile robot programming are the most difficult and require some intelligence: collision avoidance, path planning, and mapping.

Collision Avoidance

Collision avoidance is the ability of the robot to not run into things. To the extent that obstacle locations are known in advance, they can be stored on a global map and be considered during path planning. Relying exclusively on this approach, however, does not allow the robot to enter new (i.e., unprogrammed) environments or cross known environments when there is the chance that some unexpected obstacle may appear. To be able to handle all potential collisions, the robot must be able to see horizontally and vertically with sufficient detail to alert the robot when it approaches an unknown object on a collision course. Present collision detection systems fail to detect some objects and err in determining clearance to detected objects.

Collision avoidance is divided into collision detection and avoidance path planning. Collision detection recognizes the existence of an object in the robot's path. Avoidance path planning provides a local collision-free path around the object.

To embrace potential collisions from objects approaching the robot, or when the robot makes a sharp turn or backs up, full 360° coverage is required. Several assumptions simplify the handling of these cases, however. First, we can assume that the responsibility for preventing collisions is not with the robot when another object is approaching it from behind (after all, we do not have eyes in the backs of our heads either). Second, we can assume that the robot has a map of any area that it may wish to turn into or back through, since the robot could easily have made such a map while approaching this area. This map can alert it to potential collisions in the new area. Third, we can assume that if the robot needs to back up, it is acceptable for a backwards collision detection system to require more time to operate, since this is a rare event, and the robot should take more time to decide what to do.

One significance of these assumptions is that different types of sensors and/ or different algorithms should be considered to solve the problem. Let the major sensors handle forward movement. Add an extra sensor or two for sharp turns or limited backward motion. It might even be possible to back up "blind," with only a contact sensor in case something is in the way, if the robot knew (from a map) that the space should be clear and had found nothing extra there when it went through the area previously. Under worst-case conditions, the robot might have to turn around so that it is once again moving forward, even if it is going in the opposite direction.

Other problems with any collision detection systems are the effects that the environment (temperature, darkness, and electrical/acoustical noise) has on the system, the effect that traveling on a ramp has on the sensors, and the effect that robot motion has on the sensors.

Collision detection systems must operate in real time while the robot is in motion. They must also contend with obstacles in the most unlikely places. Objects may hang from a ceiling (chandelier), protrude from a wall (window air conditioner), or rise from the floor (waste paper basket), and yet be so placed as to not be seen from a given collision-avoidance sensor, which has a limited angle of view.

Although it is possible to perform collision detection based on vision (which is the way humans do it), it is very difficult to recognize objects and determine the range to them fast enough with current vision systems. Probably the best results to date using vision have been reported by Waxman,[4] who describes a mobile robotic vehicle navigating on roadways. Most current collision detection systems are based on ultrasonic ranging, although proximity sensing may serve as a backup. Specific considerations regarding the use of ultrasonics in collision detection systems were covered in chapter 7. Further discussions on collision avoidance, including current status of some existing systems, is given in chapter 16.

Path Planning

Global

Two types of path planning operations are necessary for mobile robots: global and local. In global path planning, the primary purpose is to plan a collision-free path from where the robot is to some distant goal point. Global path planning then is strategic and can be done in advance. In local path planning, the primary purpose is to find the best way around unexpected obstacles. Local path planning is tactical, that is, carried out as a reaction to an unexpected event.

In global planning, the robot must determine where it is and where its goal is. This determination is normally done in relation to a global map of the area that the robot has in its memory. Once its current position is updated on the map, the robot can plot a path between itself and its goal. If no obstacles are present, plotting the path is straightforward. With obstacles, the robot must use a path planning algorithm to determine the shortest route to its goal.

Once the path has been made (or updated), the robot can move forward. During this motion, the robot typically follows three steps. (1) It turns to correct the heading between itself and its goal. (2) It moves toward its goal a small distance, depending on the accuracy of its map, its sensors, and the turn. (3) After this set distance is traversed, the robot again computes where it is and where its goal is and plots a new path to the goal, correcting any small errors that occurred in the first move. Since the calculations take time, there could be a series of starts and stops.

Local

Local path planning adjusts robot motion as necessary to handle unexpected obstructions that the robot finds blocking its path, or handles local docking to charging stations, material pickup points, and similar tasks. Local path planning is usually more accurate than global path planning because the available sensor resolution and map resolution cover a much smaller area. It is usually necessary to be more accurate because precise movements must be made.

In general, the task is not to find the shortest route to a goal or charging station, but to come up with *any* solution for clearing the unexpected object or for docking properly. Since the resulting local path will always be reasonably short, the first solution can always be used and the robot need not optimize it.

If the robot can find no short-range solution to its problem (someone has shut a door, for example), the local path planner turns the problem over to the global path planner to see if it can determine an alternative route to the desired goal (perhaps the global map shows a second door).

Navigation Example

Figure 8-12 illustrates the interaction of global path planning, collision avoidance, and local path planning. The robot needs to go from point A to point B as part of its overall movements. At the global path planning level, it is necessary to plan a path that will traverse these points while clearing known obstacles in the room. In this example, assume that the robot plans and executes the path shown as a dashed line. While moving, the robot must approach object 1, without coming too close to it.

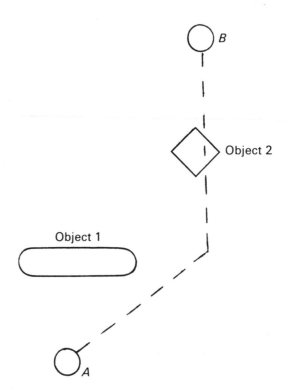

Figure 8-12. Global mapping.

The robot can approach an object closely, even though it was unplanned. This situation might occur due to errors in the path planning algorithms, errors in the actual robot motion, or movement of the object itself from the time that the robot plans its path to the time it arrives in the vicinity of the object. Referring back to Figure 8-12, we see that the collision-avoidance system assures the robot that it is safe to proceed around object 1.

The robot gets safely past this object. However, it finds an unexpected obstruction (object 2) that was hidden from it when initial global path planning was performed. The collision-avoidance system alerts the robot to a potential collision, and the robot must use another program (local path planning) to get around this object.

Mapping

The function of mapping is to place all known obstacles on a two-dimensional map of the area and to update the robot's position on this map as it moves. This update can be based on dead reckoning, onboard inertial navigation (gyroscope), active mapping systems (ultrasonic or vision based), or external devices. One popular external device is a beacon, usually infrared. The robot is able to triangulate on known positions of several beacons and to position itself on its map. Another external device is a corner reflector, which onboard active sensors (such as a laser) can locate.

Onboard mapping, whether based on visual data or ultrasonics, has several problems. In a vision system shadows will change what the object looks like. In fact, an area in darkness may appear to be open and the robot may head for it. Or if the range to the object was not calculated properly, then phantom objects will be placed in the map.

Ultrasonic systems have problems as they move about in matching their maps with maps of the area. Because their resolution is poor, maps taken from various positions do not always match. Because of edge effects, openings (such as doorways) appear less wide than they really are, and corners often appear in the wrong place. As long as there is room to spare, ultrasonic systems can be used for navigation. When the available space becomes limited, more accurate systems are needed.

Robot systems can easily update their map when they are at a known location, such as a pickup station or recharging point. The needed accuracy during periods of travel is then based on the available clear space, and often a dead-reckoning approach may be adequate for this function. Cybermation's robot, with an advanced dead-reckoning system, can hold positional errors to within a half inch after extensive travel.

Some companies have used IR beacons to allow the robot to determine where it is. One very good use for beacons is to assist the robot in locating a recharger or a workstation. Although very accurate, beacons have a few problems. Several must be placed wherever the robot may wish to go, thus increasing total system cost. Provision must be made for dead reckoning to cover periods when the robot cannot receive one or more of the beacons (a person may be in the way). Finally, other sources of energy (such as lights) must be accounted for to ensure that the robot does not triangulate on the wrong object.

8.7 ERROR CONSIDERATIONS

A fixed-base robot has a fixed global coordinate system, and, if necessary, the robot can realign its sensors by moving the arm to its "home" location. Mobile robots can only do this when they reach a fixed location, such as a workstation. During travel between fixed locations, errors occur that can affect the actual position of the robot. Everett[5] has an excellent discussion of ultrasonic and other error sources related to mobile robots.

Many sensor systems are mounted on a movable turret for positioning at any angle. The use of a movable head for the sensor is a further source of inaccuracy, because the exact position of the moving head to the body of the robot and to the direction of travel is never known precisely. Even a provision for shock absorption between the body and the wheels can cause oscillations in data arriving from the sensors. In fact, some path planning algorithms require the sensor platform to be motionless and absolutely level; these requirements are difficult to guarantee, and they put undesirable constraints on robot capabilities.

Since the robot has height, horizontal and vertical sensing must be used. Most sensors provide a cone of coverage that decreases in size close to the robot. Near the robot, then, some areas will not be seen, and a collision might occur. This problem is particularly troublesome immediately after a sharp turn. If the robot had no prior sensor data on this area, it could collide after the turn with something that it never saw.

Position errors occur regardless of whether beacon systems, comparisons of visual or ultrasonic maps with a stored area map, or dead reckoning are used for position calculations. When the robot uses dead reckoning, the resolution of the wheel encoders is quite important, as are slippage problems due to surface variations or ramp motion. For example, one problem in using a difference in wheel rotation to indicate the amount of turn is that going over a bump or obstacle will increase the travel of that wheel and therefore can indicate a turn when there is none.

Most experts would agree on the desirability of both an accurate dead-reckoning system and a method of periodically updating this information through some type of accurate global navigation. This arrangement can be through onboard sensors such as vision or sonar, through onboard self-contained systems such as a gyro-based, inertial navigation system, or offboard tracking or beacon systems. There is always the possibility that if two systems are used they may disagree sharply. Although this situation implies an error of some type, it is not always obvious where the error is and which position is accurate.

Errors may also occur with the collision-avoidance algorithm used. In order to calculate whether the robot will clear an obstacle, many systems shrink the robot to a point and expand the object by an equivalent amount. The computation of whether the robot will clear the object may now appear straightforward. However, many additional factors must be included that can indicate a larger clearance than is necessary. These include the errors and tolerances of the guidance, locomotion, and sensor systems, and the effects of nonsymmetrical robots.

8.8 CURRENT APPLICATIONS

Although autonomous mobile robots are not used extensively in factories, some models are being used. A major department store in Tokyo uses robots for its customers and for inventory. On entering, each customer takes a robot shopping cart. The cart automatically follows the customer around the store as she shops so that she can place her purchases in it. This same store uses other types of mobile robots to unload trucks and store the received inventory. The Japanese also have a concrete-finishing robot that travels over wet concrete and trowels the surface to make it level and smooth. With the exception of experimental/research mobile robots, only a few mobile robots are commercially available in the United States (not including AGVs), including Cybermation's robot and several educational robots.

One research mobile robot can serve as a robot lawn mower. For the past several years, Dr. Ernest Hall[6] and some graduate students at the University of Cincinnati have been working in this area. The robot they developed has a novel omnidirectional vision system to detect the line between cut and uncut grass and uses ultrasonics for obstacle detection. For safety purposes, a human operator checks the field for people, drives the mower to the work area, and starts the engine. Once started, the robot cuts the grass by following the uncut grass line. Teach mode techniques were used first to lay out the general shape of the area. Automatic operation includes the necessary high-level path planning, path positioning, and collision-avoidance capabilities.

In the teleoperated area, there are several interesting projects. Analytical Instruments, Ltd. of England has built a remotely operated vehicle called Ro-Veh designed to rescue people. It has a thermal imaging camera to locate a human in a smoke-filled building. It can climb stairs and go over rubble, using tracked locomotion. It can also be used in police work. The New York Police

Department has purchased several teleoperated robots, which can be used for bomb disposal or in hostage situations.

SURBOT, a surveillance robot for use in U.S. nuclear power plants is being tested in Brown's Ferry nuclear plant. A three-wheeled, teleoperated robot, it has a vision system (color camera and spotlights), temperature and humidity sensors, an air sampler, and radiation detectors. A robot arm can obtain contamination smears. The camera is on an extendible boom tower that can extend 5 m straight up, to see over things or to see onto catwalks.

A final example comes from the U.S. Navy, which has developed a remote-controlled robot fire engine that can go close to a fire, carrying vision sensors and high-pressure hose, and can lay a stream of water on the fire. This robot is being developed at Aberdeen to fight shipboard fires, particularly on flight decks. The Japanese have also developed a firefighting robot.

REFERENCES

1. Julliere, M., L. Marce, and H. Place. A guidance system for a mobile robot. *Proc. 13th International Symposium on Industrial Robots.* Chicago. April 1983.
2. Larcombe, M. H. E. Tracking stability of wire guided vehicles. *Proc. 1st International Conference on Automated Guided Vehicle Systems.* Stratford upon Avon. June 1981, pp. 137–144.
3. Weinstein, Martin Bradley. *Android Design: Practical Approaches for Robot Builders.* Rochelle Park, N.J.: Hayden Book Company, 1981, p. 35.
4. Waxman, Allen M., et al. A visual navigation system for autonomous land vehicles. *IEEE Journal of Robotics and Automation,* Vol. RA-3, No. 2, April 1987, pp. 124–141.
5. Everett, H. R. A multi-element ultrasonic ranging array. *Robotics Age,* Vol. 7, No. 7, July 1985.
6. Hall, E. L., et al. Experience with a robot lawn mower. *Proc. Robots 10 Conference.* Dearborn, Mich.: Robotics International of SME, April 1986.

PART
III

COMPUTER HARDWARE AND SOFTWARE

According to Isaac Asimov, a computer is an immobile robot, and a robot is a mobile computer. Portia Isaacson has said that today's computers are merely incomplete robots. Unfortunately, current systems have not quite progressed so far. There is still a vast difference between the stand-alone power of a computer and that portion of a computer's power that is applied to robotics control, but we are making progress in utilizing this power in the industrial robot.

It has been this power of the computer applied to the robotics field that has allowed the field to make its most rapid advancements. Software techniques (especially in visual processing and artificial intelligence), improved languages, and faster, smaller, less expensive hardware have improved the robot's capabilities in performing current jobs and have opened up new application areas also. Chapters 9, 10, and 11 examine how computers are used with robots, and discuss robot computer hardware and software, robot languages, and robot intelligence.

COMPUTERS FOR ROBOTS

9.1 HISTORY

The earliest automatons, such as the Writer (see chap. 1), used a cam-and-lever system to control motion (Fig. 9-1). Since the cam system was adjustable, these robot forerunners could even be considered programmable.

The earliest robot systems did not use computers or programmable controllers to control the robot. Instead they relied on plugboard connections, relay logic, limit switches, and position stops. (The Planobot, the first industrial robot, used plugboards to give it a programmable capability of 25 steps.) Machine sequence control used a relay logic system based on a ladder-type programming environment in which each step implied the next. Unexpected jumps in the program, such as might be necessary to cope with input/output (I/O) from sensors, could not be handled. Because the robots were operations oriented, they could not deal with the complex decisions required with sensors.

Control mechanisms developed over time, paralleling the improvement in other robot technologies. Control technology has continued to evolve and now uses the latest microprocessors and intelligent sensors to control robot action. Perhaps the first improvement occurred in the early 1970s, when programmable controllers were introduced. These early controllers emulated relay logic for controlling the robot, similar to machine sequencer control. Programmable controllers have improved significantly in the past 15 years, and now offer many advanced functions with higher memory capability, operating speeds, and I/O capability.

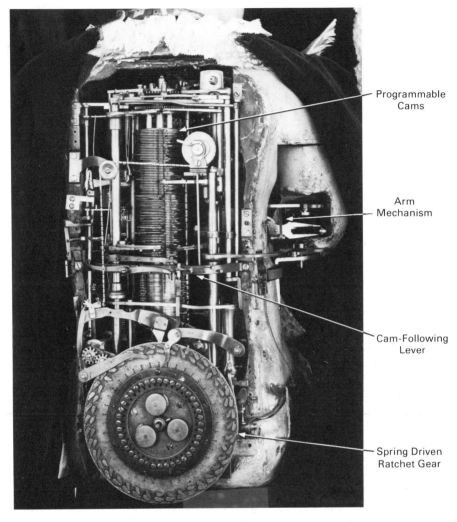

Figure 9-1. Writer mechanism. *(Courtesy of Musée d'Art et d'Histoire, Neuchâtel, Switzerland)*

Some sources refer to programmable controllers as PCs, which also refers to a personal computer, and personal computers are being used in many areas of robotics. Therefore we reserve the acronym PC for a personal computer.

Computers were first used in the factory to prepare paper tapes off-line for numerical control machines. As the technology evolved, controllers took on more of the characteristics of a computer, and computers began to be used both for off-line programming of robots and for central control of several robots. The result is that the distinction between controllers and computers has become blurred. Some people still think of dedicated computer chips within a robot as part of the robot controller, and they use the term *computer* only for off-line systems. A more accurate description for most modern robots is that they incorporate a complete computer system that serves several functions, one of which is robot control.

9.2 FUNCTIONS

The development of functions that can be performed within a robot has matched the rise of robot technology. Early robots were only designed for position control. Since they were not based on closed servo-loop techniques, the robot could only approximate the desired position. As robot use expanded, the need for precise position and velocity control became paramount and servo-controlled robots became standard. Now that external sensor inputs are being used more often, such as vision sensing and tactile or pressure sensing, a degree of robot intelligence is also required. Each of these developments improved robot capability by increasing the use of computer technology.

Basic moves of the robot are still under the control of circuitry that may more readily be identified as a controller than as a computer. But many other functions are now performed in robots and their associated intelligent sensors, and it is more accurate, for most robots, to talk of them as computer functions, even if separate robot joint controllers are present.

Robot computer functions include at least seven areas.

1. *Overall system supervisory control.* This control may be exercised over one robot or several robots in a work cell.

2. *Supporting I/O activity.* The robot often has a special computer chip handling this I/O, especially if the robot is performing extensive I/O operations. In associated robot sensors, a similar function relates to a computer accepting raw data and serving as a preprocessor.

3. *Data processing.* Within the robot, computers combine data from various sensors and perform necessary joint coordinate calculations and conversions. In vision sensors, a computer performs image segmentation and feature extraction.

4. *Decision making.* In the robot, computers handle safety considerations and determine appropriate responses that the robot should make to changes in its environment. In vision sensors, the computer makes decisions when the processor performs pattern recognition.

5. *Motion control.* At this lowest level, a computer chip will accept the real-time commands from higher-level programs and send them to the appro-

priate robot servo system (which may have its own computer chip for digital filters or PID calculations). They accept internal sensor feedback to determine when and if the robot reached the designated position. They can also communicate with simple external devices such as a gripper.

6. *Off-line development of computer programs.* This earliest use is still a major contributor to robot operations.

7. *Off-line simulation and test of computer programs.* By allowing programs to be completely tested off-line, down time on operating robots is reduced.

In future robotics applications, the computer can be expected to provide more functions as newer technology is applied. It will probably contain a knowledge base, operate under some type of expert system, and learn from its experiences. There has been at least one prototype robot design based on a second-generation expert system. Mina[1] reports using reasoning techniques and interfacing with a knowledge base as part of the design for an aerospace robot application.

Any of the functions we have described could be carried out by a single computer system, and some by a single computer board. However, when added together these functions exceed the capacity of any current computer, even stand-alone minicomputers. Thus, actual robot systems usually have several computers supporting these functions. For example GMF's Karel series contains three computer chips within the robot, an approach discussed in more detail in section 9.4. Even in this modern system, additional external computer systems are needed for vision system processing or robot simulation.

9.3 PROGRAM ENTRY

Before a robot can do any work, it must learn the necessary sequence of steps. These steps include all of the robot commands necessary for manipulator positioning, speed control, gripper activity, interpreting sensor data, and synchronizing the programs with conveyor movement. Entering these steps is, in fact, programming the robot, so the robot needs at least one method of program entry. In this section we discuss three ways to enter programs in the robot: lead thru, teach mode, and off-line programming.

Lead Thru

Lead thru techniques allow the operator to lead the robot through the desired motions by physically moving the robot arm. The series of locations through which the arm is moved are stored in the robot's memory and used as coordinates when the robot later follows that path.

The earliest lead thru programming did not allow the robot to recalculate the path to provide motion smoothing or to improve cycle time. It also did not accept sensor data to control the robot, nor did it allow more than limited end-effector/gripper control. These capabilities have been added, but since an operator is still making the fundamental moves, lead thru can never offer the full capabilities of newer techniques of program entry.

Nevertheless, lead thru is an excellent method for entering certain types of programs into some robots and can be easily learned by factory workers. It is a manual method, and no separate computer terminal or teach box is required. The operator physically grasps the robot arm and moves it through the desired positions. If the robot is too heavy or too awkward to move easily, a smaller version of the robot arm could be maneuvered instead. The large arm then follows the motion shown to the smaller arm (a master-slave relationship), an approach necessary with many current robots.

An entry button lets the operator select motions and positions to be stored. The operator activates the button during motions to be stored. When the desired points are stored in the robot's memory, the robot can follow them at slow speed (to check the programming) or at normal operating speeds. Lead thru is particularly good for complex motions often found in applications requiring continuous and regular motion patterns (such as spray painting). The technique can even be used with an external timing source. For example, the robot can be started from a conveyor signal and remain synchronized with it.

Lead thru is easy with hydraulic robots because they are easy to move with the hydraulic power off. Since most early robots were hydraulic, the lead thru technique became popular. On the other hand, electrically powered robots are difficult to move, especially if their motors use gears. They must generally be directed via a teach box.

Other lead thru disadvantages are that

1. It is difficult to incorporate sensor inputs when programming the robot.
2. It is difficult to correct an entry, and the complete sequence often must be reentered.
3. It is often not always possible to transfer a program to another robot, and the same type motion often has to be separately entered into a second robot that is to perform the same function.
4. The production line must generally be stopped as each robot in turn has its instructions entered.

Teach Mode Operations

Teach box capabilities were added to control robots more effectively than lead thru. Teach mode requires a separate control box, such as in Figure 9-2. In this

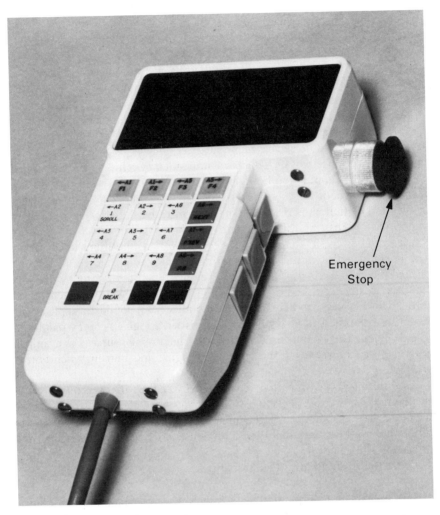

Figure 9-2. Teach pendant. *(Courtesy of Thermaflex Corporation)*

approach, the robot is moved by a sequence of commands until its arm reaches various desired locations. When each desired point is reached, the operator directs the robot to store it. After all key positions have been entered, as well as appropriate gripper commands, the robot is able to reproduce the desired motion by computing the path between stored points. The teach box also allows robot speed to be entered for each path segment.

Some teach mode systems use a standard program to describe robot motion. Trajectories are described by different variables for arm start, intermediate and end points, and end-effector orientation. These specific move parameters are then collected from the robot position as it is led through the steps by the operator.

The teach box can control each axis of the robot independently. It also controls robot speed, gripper action, and other tool functions. Teach mode entry has also been called *teach pendant operation,* because the teach box hangs like a pendant from a cord attached to the robot controller.

Through velocity smoothing, conditional program branching, and other features, teach mode operation has improved robot performance and reduced the required teach time. The total number of programmed steps that can be stored has also been increased from 25 to, for example, 1750 programmed points in each of seven programs for T3 language, and 10,000 steps in a single program for VAL II.

Today, the most advanced teach methods combine the advantage of operator movement demonstration with high-level program functions. In these robots, teach mode allows the operator to specify auxiliary commands, such as I/O handling and gripper control, and to provide capability for the robot to smooth out or speed up the motion as appropriate.

Even the latest systems (such as Karel) still support teach pendant robot control, through single-function keys and menu-driven soft keys on the pendant. For example, minor position changes, program startup, and error recovery can still be routinely done with the pendant. To support more complex applications, Karel has a much more complex pendant than many earlier systems. Its teach pendant even includes a light-emitting diode (LED) display with 16 lines of characters, 40 columns wide.

Teach mode is still perhaps the most widely used method of entering programs. For simple operations, such as many pick-and-place tasks, it is easier for an operator to direct robot motion than it is to write a program, test it, download it to the robot, and then test it a second time on the robot. There is also the added advantage that no programming skills are required, making this method of program entry the preferred one for individuals without computer training. Teach pendants are also used for robot testing and for manual positioning of the robot, even in systems using off-line programming.

Although both teach mode and lead thru techniques have been used on all types of robot applications, they have some drawbacks. They must be performed on the robot, thus removing the robot from production activities during teaching, even to shutting down the complete production line. Data available in other factory data bases, such as a computer-aided design (CAD) system, cannot be used. There is often no provision for recalling an earlier taught program, so the complete sequence has to be stored again. There is no guarantee that the steps taught by the technician are the optimum ones for the process. Each robot performing the same task must often be taught independently, which takes valuable technician time. Off-line programming gets around these problems and supports additional functions necessary in the work cell.

Off-Line Programming

In off-line operation, a program to control robot motion is designed and entered into a computer that is *separate* from the robot. Hence, robot operations on the factory floor are not interrupted. A single off-line system can support several robots, and programs written for one robot can be easily modified to serve a second robot. The term *off-line* relates to the fact that it is not connected to the robot. After programs are entered into a separate computer, the system can be used to test these programs. Originally, testing could only find obvious program errors in the code but could not determine some run-time errors, such as arm-workplace collisions or application-specific timing errors. Now complete robot simulation systems (discussed later) have been developed that allow more complete testing and help in the program design.

The evolution of robot applications from simple pick and place to complicated vision-controlled operations has profoundly affected the associated robot programs. Originally arm motion was the primary part of a program. Today, internal logic, sensor I/O, program computations, and communicating with other robots requires many more program steps than motion commands. These steps are difficult or impossible to support in lead thru and teach mode program entry. Because of its ability to provide all of these necessary steps, off-line programming is now used for all complex applications.

For example, in welding the robot needs to know: Is the welding wire stuck to the weld? Which way does the seam turn? How much current does the arc need? This change in program emphasis has affected program languages, and they became much more sophisticated. It also changed the way programs could be entered into a robot, since off-line programming came to be the only practical way of handling many types of complex programs.

Some estimates indicate that half of all computer-controlled robots will be programmed off-line by 1995; others estimate even more. Some robot applications will still be better served by lead thru (such as painting), and test operations will still need teach modes.

Off-line programming has other impacts on robot design. It requires a remote computer to be connected to the robot(s) and a provision for downloading the robot with the entered programs. The robot computer must also be designed to accept these externally entered programs. The off-line system often provides utility programs to assist in entering and editing the application programs. It also implies some type of network communications, since the remote computer prepares programs for more than one robot and then communicates these programs to the appropriate robot. Although this communication is possible without a full network (the earliest numerical machine tools used paper tape), it is much easier when all the robots are tied into a local network with the off-line system.

9.4 COMPUTER HARDWARE

Approaches

Various levels of computer hardware are supplied with robots. Companies can still buy inexpensive pick-and-place robots that operate under pneumatic power and are controlled by limit switches and commands from a low-priced controller. Another level of robot uses encoders to monitor the robot's position, relegating limit switches to backup functions, and provides much more accurate motion control. However, most current robots are more sophisticated and need elaborate hardware.

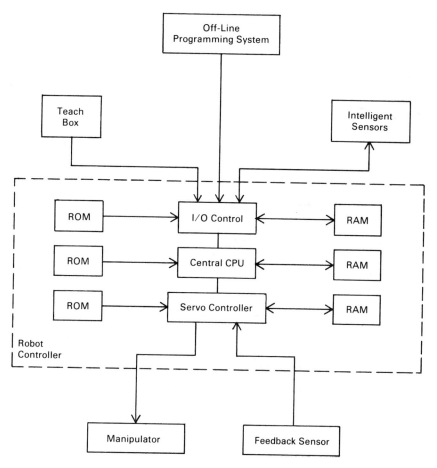

Figure 9-3. Basic robot computer system.

Robot computer systems generally have servomotor joint controllers, a central processor, I/O interfaces, program memory (ROM), and working memory (RAM). Figure 9-3 is a basic system block diagram, and Figure 9-4 illustrates typical hardware.

What makes computer hardware for robots different from hardware used for typical data processing tasks is the amount of I/O activity, the number and type of peripherals, and the real-time nature of the tasks. In many complex robot systems, a separate CPU handles just the I/O subsystem functions. In addition, since more and more of the peripherals are intelligent (contain their own processing capability), communication and coordination between the various computer systems are becoming very important. Even robots with limited or no external sensor capability may have internal microprocessors at each DOF joint to handle feedback signals, sensors, and fine-motion control.

The real-time nature needed for many responses (such as force feedback to prevent damage to parts) requires sophisticated and rapid interrupt handling and more sophisticated languages and programs. The older approach of polling each sensor during program operation is no longer adequate.

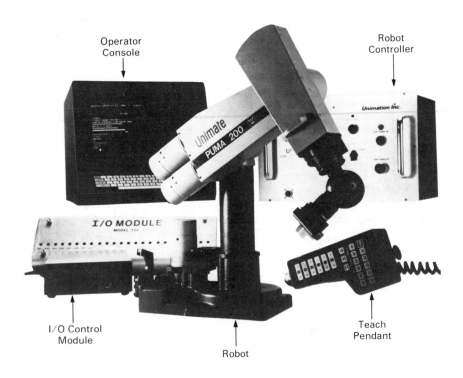

Figure 9-4. Typical robot computer hardware. (*Courtesy of Unimation*)

At the local level, adaptive control hardware uses sensory feedback to direct the servo system positioning and to make sure that each step was performed without error. Some provision for error recovery is often present. At this level, where only limited intelligence is required, the necessary computer power often resides in a microcontroller—a controller on a chip. The power of some of these chips gives robots needed computational ability and saves space as well. Microcontrollers can be located anywhere on the robot arm and handle many of the functions previously performed by large external computers.

Available microcontroller chips include the Intel MCS-96 and MCS-52. The MCS-52 is an 8-bit processor and can only handle limited computational tasks, such as are found with open-loop control. When closed-loop control and adaptive algorithms are necessary, the more powerful processing power of the 16-bit MCS-96 is used. Other programmable controllers, such as the AMD 2900 family, use bit-slice technology.

In 1985, about 75% of robots had integrated computers. In 1990 that percentage should rise to 95%. Integrated systems better support the incorporation of such complex functions as work-cell control, sensor interfacing, and vision systems. These computers are often placed in separate stand-alone control cabinets containing the computer, a system control panel, and an attached teach pendant. Separate control cabinets ensure that the operator does not come too close to an operating robot, and they offer the potential for controlling several robots. Figure 9-5 shows a typical robot computer system cabinet.

Current Systems

Recent improvements in intelligent controllers allow them to be used in work-cell configurations. Some of these systems can handle virtually any processing task (including vision, if necessary) while effectively controlling multiple robot operation within the cell.

One example of what a programmable controller can accomplish is in a heavy-duty gantry robot built by Clay-Mill Technical Systems for the automotive industry. This unit has five arms and a laser-based vision system that allows the robot to pierce holes in the automobile body to very tight tolerances. In this system, one programmable system controller serves as master controller and integrates the five individual arm controllers and the vision system.

An example of a controller with all of the attributes of a computer is GMF's Karel line of robots. These systems include a controller cabinet, manipulator arm, control console/teach pendant, the Karel command language, the Karel programming language, and various options. The heart of the system is three Motorola 68000 chips, which are 16-bit powerful microprocessors. The *main*

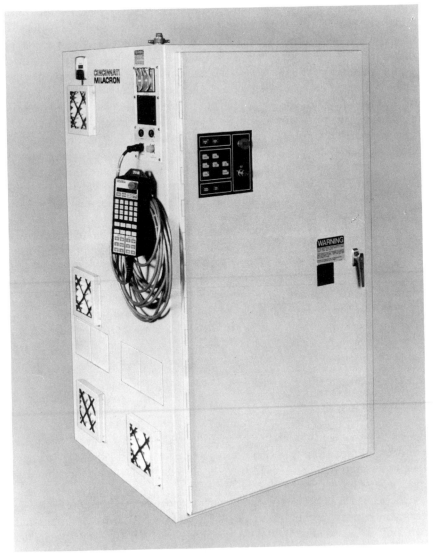

Figure 9-5. Computer cabinet. *(Courtesy of Cincinnati Milacron)*

central processor unit (CPU) controls all I/O handling and peripheral devices. The *sub* CPU provides for the interpolation of motion control statements, and the *axis* CPU drives six axes through two printed circuit boards (PCBs) (optionally, nine axes through three PCBs).

In the I/O area, the system can support 64 discrete inputs and 64 discrete outputs, 5 parallel digital words as I/O groups (16 bits maximum per word), and 5 sets of analog inputs and 5 sets of analog outputs (12 significant bits), allowing the user extensive control over I/O capabilities. The system also supports three serial RS-232 ports (one of which is also available as an RS-422 port) with communication rates of 300 to 9600 baud.

Optionally, higher speed I/O is available through their network command processor (NCP), which supports both DDCMP and MAP, two network standard interconnection protocols (see section 12.6). The NCP also supports various high-speed sensor interfaces. Other options include a line tracking system that matches the speed of the conveyor line to robot motion and a weld control option that detects if a weld wire has become welded to a workpiece.

Controller Limitations

Most controllers have only a limited ability to accept, interpret, and integrate sensor data. It is often difficult to handle six DOF and path planning calculations in real time. The controllers may have only a limited capability to work as part of an integrated robot team under central computer control and remote communications. Most systems cannot respond to unusual events, except by shutting down.

For better inverse kinematic solutions that approach real time, higher instruction speeds are necessary. In one approach, a floating-point processor with a 13.4-Mflop (mega [million] floating point operations per second) rate was used on a six DOF system, and inverse kinematic equations could then be solved in real time within a 1-kHz sampling period.

During the design of a spray-finishing robot controller, Nesse[2] found three significant limitations with robot controllers:

1. Not enough programming memory is included to support total path storage requirements.
2. Special options, such as a vision system, are very difficult to integrate into a system.
3. Adding modules is complicated by the nonstandard interfaces and low bandwidths in use.

Even robots with the latest computers are limited. Since most robots are fixed in position, work must be brought to them. They have few sensors, and more-sophisticated sensors, such as a vision system, must be separately mounted and controlled. Most robots have only one arm and often only one tool or end-effector. They have relatively little intelligence, and only the newest robots can be downloaded with data and programs from an off-line programming system. No robots are able to speak or to understand spoken commands.

Each of these factors impose limitations to the capability and applicability of robots. However, these limitations should not continue much longer. Computer techniques exist to solve all of these problems.

Accuracy Considerations

Accuracy in positioning the end-effector is quite important in many applications. But there is more to system accuracy than the response of the servo systems. Numerous potential sources of inaccuracies and limitations must be considered. For example, if a vision sensor is used to determine object offset position, the robot is expected to have sufficient accuracy to move to this new position and grasp the object. Sources of error include the servo system itself, the robot's global reference frame accuracy, sensor coordinate frame accuracy, and vision system distortions. Less obvious error sources are temperature effects on the length of the manipulator links, errors in initialization, and resolution limits of the vision system.

Day[3] has done a study of robot error sources and reports that robot accuracy is affected by

1. Environmental factors, including heat and electrical noise.
2. Parametric factors, including hysteresis, compliance, and link lengths.
3. Measurement factors, related to sensor resolution and accuracy.
4. Computational factors due to computer roundoff and truncation errors, inaccuracies in computer parameters (such as the value used for the sine of an angle), and various control errors. If the robot is mobile, there will also be a computational delay to determine the robot's position.
5. Application factors, such as installation errors, alignment errors, and coordinate frame errors.

An overlooked point about accuracy is that the system cannot perform better than its inherent errors and inaccuracies will allow it to, yet finding the limitations of any robot is not easy. Errors and error testing are further covered in chapter 13.

9.5 PROGRAM TASKS

Software for robots may be off-line, supporting program development or robot simulation. It may be located in the intelligent sensor (using templates to identify parts in a robot vision system, for example). However, the program areas discussed in this section are key to robot operation and they run under the robot's operating system. The next section discusses robot simulation.

To perform the functions listed in section 9.2, robot software modules are typically classified as

1. Supervision
2. Task creation
3. Task execution
4. Motion control
5. I/O handling
6. Remote communications
7. Error handling
8. Operator interface
9. Management reporting

Not all systems have all features, and a few have additional capabilities. These modules are all system level, not applications programs.

Supervision

The highest level module is supervision, and includes task selection and control, task timing, and handling task interruptions. Task interruption may occur because of operator intervention, error conditions, changes in input conditions, or even because another task needs servicing. System initialization is also handled at this level. Initialization includes setting motion and load limits, defining task parameters, and providing global system coordinates, particularly if the robot must work with others in a work cell.

When a robot is first turned on, diagnostic tests and alignment functions are performed. However, most error detection and handling routines are part of a separate set of software, since they must both react to the robot's external environment and interpret signals arriving from many sensors. For example, a hopper empty signal can occur at any time, not just when the robot is performing system diagnostics.

Task Creation

Task creation allows the generation, testing, and update of the application programs, either off-line or via a teach pendant. In off-line programming, task creation includes the robot control language, text editors, simulators, or other tools to help a programmer create application programs. For teach pendant operations, task creation includes the instructions to capture and store endpoint information and operator input commands (such as open gripper) as well as the capability of smoothing data, computing continuous paths, and accepting changes in the data.

Task Execution

Task execution supports the conversion of high-level application programs to the actual robot machine code. In some cases, a separate interpreter or compiler must be used to allow the program to be executed. Task execution also provides the necessary run time system to allow the program to be executed. Supporting the task execution functions are specialized debugging aids, such as single-step, breakpoint setting, and some task initialization or handshaking functions.

Motion Control

Motion control includes all of the algorithms provided to support robot movement: coordinate conversion, solving the inverse kinematic problem, computing required speeds, and calculating accuracies. Other functions performed as a part of motion control are interpreting when a desired location has been reached, interpolating between points (using curved or straight-line algorithms), and providing motion and velocity smoothing functions.

I/O Handling

When separate sensors collect and transmit data to the robot, it must have a method of processing and interpreting this information. The data to be processed include the determination of joint position from encoder inputs and the handling of visual and tactile inputs. This processing also involves the associated interrupt handling and the output of sensor system commands.

Remote Communications

There has always been a need to accept remote data. Most robots have some method of handling these communications, even if it is only support for an RS-232C interface. For more complex applications robots must accept data from sophisticated sensors, other robots, work cell controllers, and remote factory-level computer systems. Many types of interfaces and communications protocols must be supported. Just the necessary timing, handshaking, and parity/CRC checking on high-speed data lines can keep an I/O chip busy. The requirement to accept remote commands has also complicated this area.

Error Handling

Error handling is one area where a big difference can be found between types of robots and between languages. Many robots have such limited control languages that they do not offer sophisticated methods of handling environmental problems, task monitoring, or related error handling functions. Others, such as AML, provide extensive capabilities. A programmer must still write the error handling programs, which can be a difficult chore because the programmer must try to predict almost everything that can go wrong and determine how each potential error should be handled by the system.

Operator Interface

The principle operator interface can be found in lead thru and teach pendant controls. The operator uses a control panel to turn the system on and off, change operating mode, and perform maintenance. The robot has the ability to signal an operator under a wide variety of conditions, such as part jam, bin empty, emergency stop activation, out-of-tolerance conditions, and other error or abnormal situations. The signal can be as simple as lighting a few indicators on the robot control panel, or it may involve sending detailed information to a central computer to relay to plant operational personnel.

Even such a simple concept as stop has many ramifications to the operator. Is the stop due to activation of an emergency stop button? Is it an abnormal stop due to some out-of-tolerance condition (such as overrun)? Is it a commanded temporary stop (or hold)? Is it a conditional stop where the robot is waiting for some action (such as arrival of a workpiece)? Is it an apparent stop in the middle of a cycle while the robot is waiting for conditions to stabilize? In each case, restarting the robot is done through different actions. Any nearby worker certainly needs to know why the robot has stopped and under what conditions it might start again. Because of the importance of knowing when a robot is safely stopped, emergency stop activation usually lights a large indicator lamp on the robot. This assures the worker that the robot cannot restart itself.

Management Reporting

Most robots have some form of record-keeping ability. It may be as simple as keeping track of the number of hours of operation, which are needed for uptime calculations, or the robot may actually count the number of pieces of each type handled, the defects found in parts, the times it needed help, times when the hopper was empty, among other things. Information of this latter type is usually sent via a communications link to a plant computer as part of a management reporting system. For more limited data collection, the only information available may come directly from the robot.

9.6 ROBOT SIMULATION

The development of off-line programming techniques have freed the robot from having its current operations interrupted in order to develop a new program. However, the operator still must test programs written for the robot. Having a computer simulate the robot allows complete testing before down loading programs into the robot.

One problem with any off-line programming method is the difficulty of visualizing the exact action of the robot in response to its commands, and knowing in advance whether certain motions are possible (i.e., not blocked by part of the robot, the object being worked on, another nearby robot, etc.). This problem was solved by the development of complete robot simulation packages that can be run on a computer and will graphically simulate the robot's responses to program commands. Some packages are designed to simulate many types of robots in operation, and others are limited to simulating a specific manufacturer's robots. In addition to showing the robot's motions in pseudo three dimensions, many of these programs show a color change in areas of potential contact, to alert the operator. Errors, oversights, and collisions can thus be seen in simulation before the program is loaded into the robot. Simulation can also be used to determine whether the robot can perform the required task, based on speed, accuracy, or envelope constraints. Most experts view simulation as either desirable or absolutely necessary for at least half of all off-line programs.

Robot simulation is done on graphics terminals, since they must provide a realistic representation of actual robot motion. Three types of graphical images are available, and each presents a three-dimensional representation of the robot. The most common type, since it requires less computation time to generate and is less expensive, does not eliminate hidden lines. Illustrated in Figure 9-6, this approach provides line drawings of the robot, including links, joints, and base, and usually includes a line drawing of the object and workplace.

Calculating the placement of the multitude of lines required to provide a realistic robot simulation is a difficult task for current computers for three reasons: (1) The drawing must present a three-dimensional image on a two-dimensional screen, so all lengths must be adjusted for the viewing angle. (2) Robots are complex objects, so many lines must be computed. (3) If the robot is to be shown moving in real time, the calculations must be very rapid. For example, a typical simulation might require about 300,000 floating point operations per second (flops). Updating this at a 20-Hz rate (to show motion),

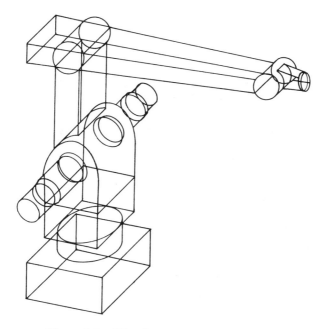

Figure 9-6. Wire-frame simulation example.

requires about 6 million flops (Mflops). The current state of the art for floating-point-based signal processing chips is 8 to 16 Mflops, and thus it is possible to provide wire-frame robot simulation with real-time motion representation.

Calculations are more difficult in the second representation, when more complicated simulations are attempted, such as removing hidden lines for more accurate robot representation and less operator confusion. Hidden lines are lines that represent parts of the robot that cannot be seen in its present attitude. Since the determination of which lines are hidden depends upon robot position and orientation, this feature must be constantly recalculated and appropriate lines blanked out during any robot motion.

The third type of representation gives a more realistic-appearing robot (Fig. 9-7), a solid-type image with texture and shading added. Hence, the image appears like an animated picture rather than a simple stick figure. Esthetically this picture is best, but the added computation time limits the showing of real-time robot motion. Figure 9-6 takes four times the floating-point computations required by the wire-frame approach.

An early computer graphics system was PLACE (position layout and cell evaluator) from McDonnell Douglas. It offered wire-frame representation of robots and simulation of the robot performing actual tasks. PLACE worked

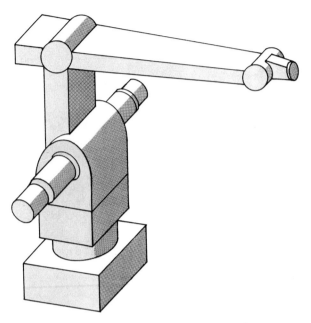

Figure 9-7. Solid simulation example.

with T3 programs and was loaded by a tape cassette. Although a good first step, it was limited in real-time generation and acceptance of sensor inputs.

Many available robot simulation systems provide graphical robot representation on a CRT screen. Some systems that show the robot, its motion, and, most importantly, its environment (object location, other robots, sensor locations, etc.) are RoboTeach (GM), Robographix (Computer Vision), SmartWare (GMF), Robot-SIM (Calma Division of GE), IGRIP (Deneb), and McAuto (McDonnell Douglas). IGRIP (interactive graphics robot instruction program) allows graphical simulation of robot programs before use to check for safety, interaction between the robot and its workspace, and interaction between robots.[4] IGRIP can represent solid objects with real-time animation. Written in C to improve its portability, IGRIP is specifically designed for the IBM PC/AT and the Silicon Graphics IRIS workstations. Objects may be presented as fully shaded, wire frame, or wire frame with hidden lines removed. The wire-frame version executes the fastest. The program allows automatic collision detection and real-time motion clocking.

Once the simulated work cell has been designed, programmed, simulated, and tested, the actual program for the real robot controller can be generated from IGRIP in several languages, including Karel, VAL II, and T3. There are also plans to accept programs in these and other languages. The simulation program gives the programmer control over various operating conditions, including initial conditions (e.g., initial position of the arm, object size and location) and program execution rules. Program execution rules include designating the operating speed, the placement of hold points, and the methods for handling errors. A hold point is a spot in the program where the robot image freezes in position, allowing the programmer to examine the situation. A related and important capability is the program's ability to replay commands so that the operator can back up and reexamine a portion of the program.

Robot cycle time analysis techniques are also available. One approach is robot time and motion (RTM), described by Lechtman and Nof.[5] RTM supplies prediction information (how long it will take for the robot to perform specific tasks). The system looks at 10 specific elements (such as grasping an object or moving the unloaded arm) and determines the time it takes the robot to perform each one. It then uses these times to check programs that will be run on the robot. The programmer can thus explore other robots or other approaches on the same robot to improve speed.

Some programs can emulate or simulate complete manufacturing systems, such as the hierarchical control system emulator (HCSE) developed by the National Bureau of Standards to emulate many types of automated manufacturing systems controlled by hierarchical control systems. These programs run on a DEC VAX computer under the VMS operating system.

9.7 WORK CELL CONSIDERATIONS

So far we have discussed only individual robots, but in many applications two or more robots must work together, either side by side on an assembly line or as part of a complete work cell. Each situation has an impact on which tasks, timing, and position information should be shared among the robots.

There are significant differences in programming requirements between a robot that stands alone and one that is controlled as part of a team; such as timing, communications, and collision avoidance.

Timing

A stand-alone robot has more flexibility in the length of time it takes to do different steps and in the exact time that a given step will start. When two (or more) robots are working together, they must be synchronized to report their progress to each other (or to the cell controller). One robot must often wait for the other to finish a particular step so that the two robots can remain in synchronization.

Communications

Stand-alone systems can get along with no data link if they rely on teach mode program entry. Even with off-line programming, the data is only periodically downloaded and the robot is normally in the halt mode at this time. In work cell operations, real-time data communication between robots, between a robot and the cell controller, and between the cell controller and a plant-level computer is necessary. Thus, the system must handle real-time, high-speed data communications, particularly of the evolving standard network protocol MAP.

Collision Avoidance

Although the robot program was probably tested under simulated operating conditions to ensure that the two robots will not collide, there are always unplanned circumstances when a collision might occur. To prevent collisions,

1. The simulation system must simulate both robots operating at the same time (an added burden which some systems cannot handle).
2. The programs themselves may include collision-avoidance calculations to determine whether the computed path is collision-free.
3. Some workcell robots have collision-avoidance systems (often based on sonar) as a run-time backup to prevent collisions.

Other Tasks

In addition to controlling the other robots in the work cell, the cell computer may have other system-level tasks to perform. For example, in a typical work cell, the cell computer may be made responsible for controlling all robots in the cell, collecting necessary manufacturing statistics, and providing various error detection and fault correction functions. In this latter category, if a "good" cycle has been defined for the system, the controller can constantly compare its current operations against those of the standard cycle and supply an alert when out-of-tolerance conditions occur. When the controller monitors robot operation in this manner, completely unattended work cell operation is possible.

Nagamatsu[6] listed the following functions that a computer system must be able to perform to operate as a cell controller:

1. Lock out individual robots for safety
2. Send operating commands to a robot through a local network
3. Receive error messages, production data, and current status from the robots
4. Respond to a host (factory-level) computer
5. Communicate with a central computer to upload and download programs

These and related tasks are shown in Figure 9-8 for a typical cell controller.

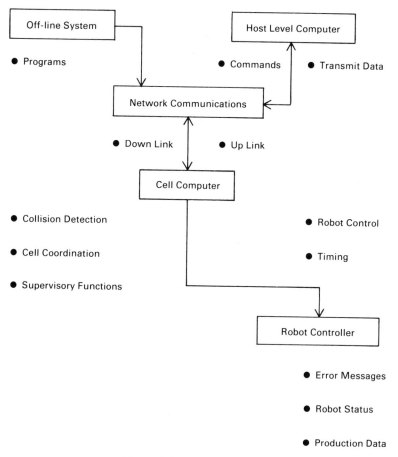

Figure 9-8. Program levels.

9.8 OTHER HARDWARE CONSIDERATIONS

Modularity

Standardization is highly desirable, especially for users and third-party software developers to produce programs that are transportable between different robots. Chapter 12 covers standards in general and the MAP communication standard in particular; chapter 10 reports on the status of the Japanese effort to standardize robot programming. But a type of standardization may also be found at the user level, developing application programs in blocks or modules. One of the trends covered in chapter 15 is modular design, which use is increasing in robot control systems. Modular design improves maintainability and reduces system integration time. Several countries are developing languages to support modular design.

One promising language is Sharp, which was developed at LIFIA/IMAG near Grenoble, France, to automate robot programming. Sharp[7] is based on three planning modules (grasp planner, motion planner, part-mating planner), and is a task-level language. It accepts as input a description of the objects, a structural model of the robot and environment, and an ordered description of the assembly tasks. Sharp then generates the robot control program to complete the assembly.

Operational Factors

Robots have several modes of operation, including run (or automatic) for normal operations, teach (for teach pendant operation), manual for lead thru operation (or direct axis control), halt (to suspend operations) and download (to accept program information from other computers). (The actual names may vary among manufacturers.)

The time to finish a complete operation obviously depends on the tasks performed. The following factors all improve (that is, decrease) required cycle time:

1. Increase arm motion speed
2. Reduce arm travel distance
3. Compute continuously while arm is in motion
4. Reduce required accuracy
5. Eliminate or reduce delays caused by peripheral devices
6. Reduce the number of on-line inverse kinematic calculations
7. Increase computer clock speed (rarely adjustable by the user)
8. Improve control algorithms (few users have the technical skills to do this)

Maintenance and Testing

Robot maintenance and repair require specialized skills, and most users should have the manufacturer, the subcontractor that installed the robot, or a firm specializing in robot maintenance provide periodic preventive maintenance and repair functions. However, users should be familiar with the following maintenance-related functions: (1) power-on diagnostics, (2) self-test, (3) teach-mode tests, and (4) calibrations.

All robots go through some type of internal diagnostic testing programs when first turned on, and indicate whether it passed these diagnostics. If not, the robot often indicates the general source of the trouble. If a user has any doubts about a robot during operation, the user should halt the robot and have it perform these diagnostic routines again. (Note that some problems will not show up in diagnostics.)

Some robots offer self-testing capabilities or have an expert system included in the utilities provided that can help to diagnose the problem. Chapter 11 looks at one such system.

The robot can be directed by a teach pendant. If the robot is unable to follow all of the commands given to it by a teach box, there is obviously a problem. Note, however, that a teach box has two testing limitations: (1) It usually directs the robot at slow speeds, which reduces the possibility of catching speed-dependent errors. (2) Problems with external sensors or interfaces often do not show up.

All robots need periodic recalibration. When a robot is initially set up, calibrations are performed, such as defining global coordinate reference points, and these should be periodically rechecked.

REFERENCES

1. Mina, Irene. A distributed knowledge-based system for robotics. *Proc. Robots 11 Conference.* Dearborn, Mich.: Robotics International of SME, April 1987, pp. 6-31–6-54.
2. Nesse, T. A., Kang G. Shin, and M. C. Leu. High-level design of a spray finishing robot controller. *Proc. Robots 11 Conference.* Dearborn, Mich.: Robotics International of SME, April 1987, pp. 6-15–6-29.
3. Day, Chia P. Robot accuracy issues and methods of improvements. *Proc. Robots 11 Conference.* Dearborn, Mich.: Robotics International of SME, April 1987, pp. 5-1–5-23.
4. Model, J. S. IGRIP—A graphics simulation program for workcell layout and off-line programming. *Proc. Robots 10 Conference.* Dearborn, Mich.: Robotics International of SME, April 1986.

5. Lechtman, H., and S. Y. Nof. *A Users Guide to the RTM Analyzer.* Research Program on Advanced Robot Control. Vincennes, Ind.: Purdue University, 1981.
6. Nagamatsu, Ikuo. An approach to software organization for control of industrial robots. *Journal of Robotic Systems,* Vol. 2, No. 3, 1985.
7. Laugier, C., and J. Pertin-Troccer. SHARP: A system for automatic programming of manipulation robots. *Proc. 3rd International Symposium on Robotics Research.* Gouvieux, France, Oct. 7–11, 1985.

CHAPTER

10

ROBOT LANGUAGES

This chapter starts with a historical review of robot languages and then summarizes 28 available languages. It gives examples of robot commands and an overview of robot language approaches and limitations. Because so many languages are described, we make no attempt to supply enough information to actually use the language. (This information can only be obtained by attending special courses offered by the manufacturer and/or studying the relevant programming manuals.) The languages are not equally important—some never left the research laboratory. However, taken in their entirety, they give us an understanding of the needs of robotic systems and the approaches for satisfying those needs.

Early robot capabilities were limited; hence language requirements were minimal. They were usually incapable of specifying operations at a system level or interfacing with complex external sensors; they were often designed to support or complement teach pendant operations and were difficult to modify. And few languages could be used on more than a few robot models.

As technology and applications changed in the workplace, higher-level languages were developed to allow the robot program to be written and tested off-line and downloaded to the robot. Newer languages also allowed limited changes to be easily made in the program to satisfy minor production changes. One drawback, however, is that these languages can only be used by trained programmers skilled in the specific language. Hence factory personnel can no longer handle robot commands, as they did with teach box entry.

It would be ideal if robot languages were so user friendly that almost anyone could enter commands to a robot. Unfortunately, no such robot language or robot applications programs have been developed, as they have in

249

other environments (such as office use). Many languages that are called high-level are really not. Finally, programs still need considerable modifications with associated system-level effort to provide the capabilities needed to interface robots effectively with many intelligent sensors.

10.1 EARLY LANGUAGES

Early in the development of robots it was realized that standard computer languages (such as FORTRAN and COBOL) would not serve the real-time and specialized I/O requirements of robots, so new languages were developed. Most of these languages are primarily of historical interest, since they either never became commercially available or were superseded by later languages. By 1982 only 8 of 22 early languages were in use and commercially available: T3, Anorad, JARS, SIGLA, VAL, HELP, AML and MCL. APT (an NC language), MHI, MINI and WAVE had been superseded. Seven others were operational but not commercially available: Funky, RAPT, Emily, RCL, RPL, AL and PAL. Autopass, Maple, and LAMA were not completely operational, although some modules had been demonstrated.

An excellent and detailed comparison of the features of 14 of these languages was published by Bonner and Shin,[1] expanding on earlier work by Gruver.[2] By 1987, numerous new languages had been added, and even the eight in use in 1982 were, with few exceptions, being superseded.

In the summaries that follow, dates are given for most of the languages. Since the development of a language takes several years, especially if offered commercially, we have arbitrarily picked a specific year for each, usually the year that the first paper on the language was published. Table 10-1 lists many of these languages and shows the high-level languages they are based on or similar to.

Table 10-1. Robot Language Source

Assembly Language	FORTRAN	Algol	BASIC	Pascal	PL/1	Lisp	Numerical Control
T^3	RCL	AL	VAL	HELP	Maple	MINI	Anorad
Emily	RPL^a			JARS	Autopass	RPL^a	APT
SIGLA	MCL			AML^a		AML^a	
ML	RAPT			RAIL			
				Karel			

^{a} Has features of two languages.

APT

Computer-aided manufacturing (CAM) was first applied to numerical control (NC) systems in which tapes were prepared off-line and used to control lathes, milling machines, and drills. APT was the major numerical control language used before robots were adopted. APT thus served as a base for several early robot languages and was implemented on many computers, including the IBM 370.[3]

MHI (1960)

Perhaps the earliest robot-level language was developed at MIT for their robot hand MH-1. MHI stands for mechanical hand interpreter, and its features were limited. MHI was built around the concept of guarded moves—that is, a move would continue until a sensor input was detected. It could also accept pressure sensor inputs from fingers on the hand. However, MHI had no capability for arithmetic functions, and had few language control structures.

WAVE (1970)

Perhaps the first true robot language was WAVE, developed in 1970 at Stanford University[4] as an experimental language for programming robots. It ran on the DEC PDP-8 computer. Since the early algorithms were so complex and the PDP-8 so much slower than current computers, real-time control could not be performed.

WAVE pioneered the concept of describing the position of the end-effector through Cartesian coordinates and the three Euler angles as well as the coordination of motion from various joints. The language required preplanned moves, and only small deviations during execution were allowed. It could accept contact sensor data to stop motion on sensing contact. WAVE was also interfaced with a vision system to demonstrate eye-hand coordination. WAVE served as a guide for some later languages, but has given way to later languages, notably AL.

MINI (1972)

MINI was not actually a new robot language but an extension of Lisp applied to robots. It was developed at MIT to add robot control functions to Lisp. Robot motion was handled in Cartesian coordinates with joint calculations independent of each other. MINI allowed the robot to sense force at the ounce level, whereas other robots (even current ones) can only sense forces at the pound level.

AL (1974)

AL (arm language) was a successor to WAVE. Also developed by Stanford,[5] it was one of the original languages for the Stanford arm. Based on the language ALGOL, with some features of Pascal, AL was originally compiled on a PDP 10 and executed on the DEC 11/45. The commercial version is based completely on the 11/45. AL was the first language that was both a robot control language and an advanced computer language. Improved by many individuals, it has served as a base for other later languages.

AL is particularly good for controlling manipulating tasks, such as in assembly applications. It was designed to control four arms at once, two Stanford arms and two PUMA robots. Force sensing and compliance can be implemented through condition statements and appropriate subroutines. Both SIGNAL and WAIT commands are provided to allow assembly steps to be synchronized. These same commands, along with COBEGIN and COEND, allow multiple arm motion to be controlled.

VAL (1975)

VAL was originally developed by Vicarm in 1975 for their robot manipulator. This small robotics firm was purchased by Unimation in 1977, and the language was further developed for the PUMA robots.[6] VAL was released in 1979 as the first commercially available robot language. It was revised completely in 1982 as VAL-II.

VAL was an extension of BASIC designed to run on the DEC LSI 11 microcomputer. It is used to control Motorola 6502's, the microprocessors that control the six motions of the arm. VAL has always been able to handle multiple arms and included some of the features developed for the AL language, such as SIGNAL and WAIT. Although the original version did not accept sensor information and had no provisions for changeable tools or vision interaction capability, later versions of VAL have these features.

VAL consists of a full set of English language instructions to write and edit robot programs. The robot program can be originally generated by a teach box routine and then edited under VAL. Taught positions may be stored as coordinate transformations or as joint angles. VAL is an interactive language operating in an interpretive mode for monitor commands. In addition to monitor commands, it has robot specific commands, such as LEft, ABove, ALIGN, and MOVE.

Various alternatives are provided for end-effector control, such as CLOSE (close gripper during next command), CLOSEI (close gripper immediately), and GRASP (close servo-controlled hand). VAL has a full set of conditional commands (IF THEN, THEN REACT, GOSUB) and trajectory accuracy control statements (CO—coarse control, FI—fine control). VAL even added a PICTURE command to cause a vision system to scan an area, analyze the scene, and prepare a table from memory of recognized objects.

TEACH (1975)

Developed at Bendix Corporation for their PACS system, TEACH[7] addressed two specific robot issues: parallel processing and robot-independent programming. In handling these challenging topics, TEACH introduced three key ideas:

1. Program segments can be executed in series or in parallel.
2. Comprehensive mapping is provided between the output devices supported in the program and the physical devices (robots, peripheral units) found in the workplace.
3. Movement specification is simplified because it is based on a local coordinate system.

TEACH was ahead of its time in many of these features, especially in trying to handle different robots through various driver routines.

ML, Emily, Maple (1975)

ML, Emily, and Maple are related languages developed at IBM. ML is a low-level robot language, similar to an assembler language, capable of controlling joint motion in Cartesian coordinates. ML can control two or more robot arms and offers touch and proximity sensors. It has no vision interaction.

Emily[8] was developed as an off-line assembler for the ML language. It ran on the 370/145 and system 7. Maple[9] has a PL/1-type base and is an interpretive language. It used ML to handle low-level manipulation functions. Maple received only limited use.

Autopass (1977)

Autopass is a high-level robot language developed by IBM to run on mainframes.[10] Based on PL/1, the name is a contraction of *auto*mated *parts assem-*

bly system. Autopass used a different approach to language. Most early languages are based on explicit programming (that is, each movement and intermediate point must be separately described), so many instructions are required for any reasonable task. Autopass, on the other hand, is an implicit language, describing the task directly, which allows fewer and simpler instructions.

Although all of its features were never fully implemented, Autopass served as a base for later languages, including AML. It can handle more than one arm, allows many different tools to be used, has sensor feedback capability, and allows interfacing to a vision system for verification and modeling. Autopass is an object-oriented language, supporting such statements as "rivet object 1 at place 2." It was also one of the first languages with tool-related commands.

Funky (1977)

Funky was developed by IBM to run on an IBM system 7. It can handle a single IBM arm, but has no vision interaction, and its sensor capabilities are limited to that of touch. Funky was superseded by AML.

AML (1977)

AML (a manufacturing language)[11] was designed as a structured language to run on the series/1 computer to drive the IBM line of robots. AML combines some features from APL, Pascal and Lisp. Highly transparent, it allows function calls to be made to AML from the user programs as if they were built-in routines. Commercially available in 1982, AML is an interactive language that simplifies debugging.

AML was designed around a series of modules. One module provides system interaction, another handles manipulator control, and a third handles real-time tasks assigned by the AML interpreter. AML can interface with AML/V, an industrial robot vision processing system. This interface is supported via subroutines to handle and control the camera and images.

One unique feature of AML is its ability to handle data aggregates so that vectors, different coordinate frames, and coordinate rotation can be handled better than with other languages. Data objects can be scalar or aggregate. With this complex capability, the language is too complicated for many robot operators with little data processing training. The language portion is quite similar to Pascal, and those familiar with Pascal would be at home with AML.

AML is an open-ended language structure allowing further extensions to be developed by users or robot system developers. It is possible to construct a top-level set of subroutines that handle many of its complicated functions as defaults.

Several versions of AML have been developed. AML/ENTRY is designed to allow specific robot commands to be entered into the system. For example, it includes such commands as PMOVE(PT(11.00,200.00,50.00)); DELAY(0.3); DOWN and GRASP.

One recent version of AML is AML/2, which runs on the IBM 7532-310 industrial computer and controls the IBM 7575 and 7576 robot arms. Its major point of interest is that it can be developed on an IBM PC equipped with DOS 3.1. A subset runs on the PC for the IBM 7535 arm. A development package available includes program editors, debuggers, downloaders, and uploaders and offers teach mode, teach simulation, and syntax checking.

LAMA (1977)

LAMA (language for automated mechanical assembly), developed at MIT,[12] is similar to Autopass and is a task-level language. LAMA was only partially implemented and never commercially available.

SIGLA (1977)

Ollivetti developed SIGLA for their Sigma assembly robot.[13] SIGLA has some similarities to ML and supports a set of joint motion instructions. It can handle limited parallel task processing, can accept inputs from a teach box, and has the necessary utilities for instruction execution and data storage. SIGLA was one of the few early languages that could handle multiple arms.

RAPT (1978)

RAPT (robot APT), was developed at the University of Edinburg as an extension of APT[14] primarily for assembly applications. RAPT has been adapted and refined by GEC, Britain's largest robot manufacturer. Written in FORTRAN, it uses actions and situations to describe its tasks. Body descriptions are in terms of their features (plane faces, cylinders, shafts, etc.).

PAL (1978)

Richard Paul, one of the initial developers of WAVE and AL, decided to extend the capability of Pascal to robots at Purdue. Commands such as MOV (move),

TOL (a reference to the tool in use), and ARM (a reference to the arm) were provided in PAL.[15] One significant advancement was that the data structure of PAL was set up to support matrix transforms describing the relationship between the arm and an object. Thus, PAL programs included the capability of solving robot motion through homogeneous transform equations. It could also define end position as a combination of other, known, positions.

MCL (1978)

MCL is a structured language based upon APT.[16] It picks up many of the geometric definitions and features found in APT (widely known at the time).

The Air Force awarded McDonnell Douglas a contract to develop MCL as part of the ICAM (integrated computer aid to manufacturing) project. Written in FORTRAN, MCL has postprocessing to allow it to be used with different robots. Unlike most other early languages, MCL can be used for off-line robot programming; however, it has few debugging aids. MCL was offered commercially in 1981.

Anorad (1979)

Developed by Anorad, the Anorad language ran on Motorola's 6809 and controlled a single arm. It does not have a vision interface nor any sensor interaction. Anorad does not even allow changes to the robot's tools, but, it does provide a powerful set of numerical control features and is one of the first languages to allow automatic configuring.

JARS (1979)

NASA financed the development of JARS (JPL autonomous robot system) at the Jet Propulsion Laboratory to support the teleoperation of robot manipulators. Like many other languages, it is based on Pascal and runs on a DEC computer (the PDP 11/34). It can control a number of arms, including the Stanford arm and the PUMA 600.[17]

RPL (1979)

RPL was developed by SRI for in-house research and to distribute to members of the SRI Industrial Affiliate Program. It runs on the LSI-11 under the RT 11 operating system, and can be used on PUMA robots from Unimation as well as on the SRI vision system. Unlike more complex languages, RPL was designed specifically for operators who know little programming. Arranged around a FORTRAN-type syntax, RPL borrows ideas from Lisp to handle commands. It is implemented as subroutine calls and is thus both modular and flexible.

T3 (1979)

T3 was developed by Cincinnati Milacron to handle its line of T3 robots. It runs on a controller based on the AMD2900, a bit-slice chip. The language is popular and offers many features, but it is not a true textual language. T3 is capable of specifying tool dimensions, has reasonable debug facilities, and can be interfaced to several simple external devices. However, it can handle only one arm and has no vision interaction or parallel task capabilities.

One clever approach supported by T3 is that it can track points on a moving conveyor belt by accepting resolver output from the conveyor as input to the robot. If the linear axis of the robot is set parallel to the axis of the conveyor, the resolver output can then be directly subtracted (or added) from the computed Y-axis value and the new value can be used to control the robot motion. The net result is that the robot can more readily find and pick up an object from the conveyor.

10.2 CURRENT LANGUAGES

A few of the languages covered in the last section are still in use; however, many new languages have been developed since 1981 to meet the needs of today's robots. This section discusses many of these.

In addition, other languages have been developed. Some are based on BASIC, such as those offered by Seiko, Machine Intelligence, American Robot (now CIMFLEX), and Intelledex. Many European car factories (especially in France) use ACMA manipulators with a language developed by Renault. Other languages are HAL (Hitachi), CIMPLER (GCA), MELFA (Mitsubishi), and PARL (Matsushita).

LM (1981)

LM is a high-level robot language developed at LIFIA (previously IMAG) in Grenoble, France. LM allows multiple arms to be controlled simultaneously. Latombe[18] has several good descriptions of this language. LM supports a world (as opposed to local) coordinate system, allows sensory interactions to be coordinated with arm motion and tool action, and enables interfacing with the outside world.[19] LM is also a powerful tool in robot vision. Scemi, a French robotics manufacturer has adopted LM for some of their robots.

RAIL (1981)

Developed at Automatix to control visual inspection and arc-welding-type applications, RAIL[20] contains a large subset of Pascal. Available commands include LEARN (to define points or paths), LEARN FRAME (to enter a reference frame), and SINGLESTEP (to execute one command at a time).

RAIL runs on a Motorola 68000 and is supplied in three different variations. Since RAIL is an interpretive language, it has the capability of interactive debugging. A more sophisticated language than most, RAIL is comparable to AML and LM, without the capability to support multiprocessing.

HELP (1982)

HELP is an interpreter, with some similarities to Pascal and to RAIL. It was developed originally by the Italian firm DEA and was later offered by General Electric under license for their Allegro robot. HELP offers continuous path motion through its SMOVE command, supports concurrent processes, and can define and activate multiple tasks simultaneously. Thus, it can control multiple arms for assembly applications.

VAL-II (1982)

VAL-II was conceived as a new language rather than an improvement of the highly successful VAL. Released in 1984[21] by Unimation, VAL-II had many improvements:

1. The computer it is based on operates 10 to 12 times faster than earlier machines.
2. It can be interrupted for necessary critical action within 28 ms.
3. It can modify the arm path through data provided by external sensors.
4. It has more powerful interface commands, real-time data communication, and cell control features.
5. It could probably be considered the first of the second-generation robot languages.
6. Mathematical functions such as ABS, SIN, SQRT, and INT were added.
7. It provides floating-point arithmetic with higher accuracy and less error buildup.

One VAL-II robot can now control another in a work cell; alternatively, the language allows supervisory control of the robot from a remote computer. Structured programming statements have also been added (CAS for case, WH for while, EL for else, etc.).

Karel (1985)

Developed at GMF to fully support network communications and various I/O devices, Karel[22,23] has proven to be a step forward in languages. It offers teach mode and off-line programming and can be fully integrated into work cell control systems. Based on Pascal, it builds on first-generation robot languages by adding more powerful and unique extensions. For example, Karel has four types of simple data structures, including string and Boolean.

In addition, Karel handles complex data, such as path data, vectors, and three-dimensional positions. Both location and orientation are represented in Cartesian coordinates rather than spherical coordinates, an important point for programmers. Karel can keep track of several auxiliary coordinate axes, such as are often needed for arc welding applications. Special motion control and monitoring commands extend motion control to relative, absolute, and directional moves under both wait and no-wait conditions. The system also features numerous operational, status, prompt, and error messages to improve operator understanding.

Karel allows data for one program to be accessed by another, thus facilitating robot interaction and sharing of calibration and initialization data. It includes a line editor to aid in program creation, modification, and debugging.

Most important, perhaps, is that Karel facilitates robot communications through RS-232, DDCMP (DEC's DECNET type of network communication), and MAP. A real-time sensor option allows sophisticated sensors, such as vision systems, to be synchronized with robot action.

ROPS (1985)

Developed by Cincinnati Milacron, ROPS (robot off-line programming system) allows off-line programming via an IBM PC/AT (under MS DOS) or a DEC VAX (under VMS 4.0). Written programs can be downloaded to the robot via a communications module. ROPS can accept data from CAD/CAM systems and upload existing robot programs for modifications.

ROPS is compatible with T3, Cincinnati Milacron's earlier successful language. It is a modular, file-based system that produces three types of output files: the actual robot program, a program listing, and a complete diagnostic system. ROPS also contains robot data files, alignment files, calibration data, and coordinate system files as input sources for the programming module.

10.3 LANGUAGE COMMAND REVIEW

Although there is no standardization within robot languages, some commands between systems are similar. This section reviews some of these commands.

Motion Control Commands

MOVE PICKUP—move the end-effector to a previously defined point (PICKUP). Some languages offer several types of moves. For example, VAL-II provides a MOVES command that requires straight-line, rather than slew, interpolation during the move for more accurate tracking. Karel specifies the type of move through setting a TERMTYPE (termination type) variable. Others offer move via commands, to make sure the arm goes through a specific point in space. Other commands are MOVE SLEW (move at maximum speed), MOVE CIRCLE (move end-effector along an arc), and DMOVE, in which the arm is given a certain distance to move, rather than going to a predefined location.

Related are the approach commands, where the robot moves close to the final position but remains a short distance away. An approach command could be used before a move to make sure a robot end-effector always made the final contact motion from a fixed direction.

Speed commands allow the robot arm velocity to be controlled, which permits lower-speed operation during training, setup, or program test, and allows slower, more accurate moves to be made during parts of the program. Speed can be defined as an absolute value (SPEED 50 ips) or on a relative basis (SPEED 75, which refers to 75% of normal speed or last set speed, depending on language).

To support teach pendant programming, a HERE command is usually available. When the operator has manually directed the robot arm to a desired location, he or she can type in a command such as HERE PICKUP to define to the robot the current location with the name PICKUP. Points may also be defined with coordinate values directly by using a DEFINE command. Connecting points in space to ensure that the robot follows a desired path can be done by the PATH command. For example, PATH_1 could be defined as a series of points, and then MOVE PATH_1 would tell the robot to follow this path.

If parts are to be designed around a local coordinate system rather than in the robot's general coordinate system, a new coordinate system can be specified by FRAME. For example, if the robot is to be directed to specific points on a pallet, the pallet could be defined as a frame; then even if the robot subsequently moved the pallet, it would still be able to find the desired location.

End-Effector Commands

All languages have various end-effector commands, such as OPEN and CLOSE. Some also offer OPENI (for open immediately) and CLOSEI; others accomplish a similar function with a NOWAIT command. Many languages will accept parameters with these statements, such as CLOSE HAND 3, CLOSE(1.5 inch), and CLOSE(2.5 lbs). Other commands available in a few languages include RELAX, CENTER, and OPERATE(tool).

Sensor Commands

Standard I/O signals include combinations such as SIGNAL (output), SIGNAL 3 OFF, WAIT SENSOR 3 OFF, DEFINE VOLT, REACT (interrupt), and REACTI (safety interrupt). Karel uses AIN and AOUT for analog input and output commands, with DIN and DOUT for digital I/O.

Some commands handle multiple data. For example, SENS IO reads in data from several sensors, such as strain gauges on the gripper. SENSOR could be used to determine if a part was in position in the hand (is LED path being blocked by part?).

VAL offers a special vision system command, PICTURE, which tells the vision system to accept a frame of data, analyze it, and report the recognized objects.

Special Control Commands

MONITOR causes an output when the force changes over preset limits. RAIL has a WELDSCHED command to allow setting feed-wire rate, adjust voltage, fill delay, preflow, and weave parameters. Other special control commands are available for spray painting, part detection, fail inspection, and conveyor control.

Configuration control alters the robot's approach to a position. RIGHT and LEFT can be used to approach from the right and left sides, UP and DOWN indicate whether the elbow is to be up or down, and FLIP NOFLIP refers to wrist position. These commands let the robot go to locations that it could not have reached with the original configuration.

10.4 PROGRAM EXAMPLE

We give an example (Fig. 10-1) of a limited program for removing chips from a chip feeder and placing them in a pile. I/O signals for gripper force, gripper

```
PROGRAM Move_chip

VAR
Chip_pos:    POSITION        --Location of Chip
Assem_pos_1: POSITION        --Destination of Chip
Assem_pos_2: POSITION        --Alternate Destination


CONST
Grip_open    = 1             --Assign Gripper Open I/O Signal
Force_pickup = 2             --Assign Gripper Force I/O Signal
Last_part    = 3             --Assign Last Part I/O Switch
F_Max        = 100           --Define Maximum Force Desired


NAME:  Move_part

   BEGIN
```

```
Move_part:
WITH $TERMTYPE = FINE          --Use fine motion control
MOVE to Chip_pos               --Move arm to location of chip
WHILE DIN[Grip_open]
CLOSE HAND 1 NOWAIT            --Grasp the chip
IF AIN[Force_pickup] >F_max    --Check force during closing
     THEN CANCEL               --Stop if force exceeds maximum
                               --Chip grasp
WEND
IF DIN[Last_part]              --determine if last part taken
MOVE to Assem_pos_2           --If so move to last part
                                 location
ELSE MOVE to Assem_pos_1      --Else move to regular position
OPEN HAND 1                    --Release Chip
UNTIL DIN[Last_part]           --Get next part unless end
     GOTO Move_part

END
```

Figure 10-1. Sample program for removing chips from a feeder and putting them in a pile. *(Adapted from KAREL Language Reference Manual, Version 1.3. Troy, Mich.: GMF Robotics, 1986; used with permission.)*

263

open, and last part detection are provided as input. The first I/O signal is defined as analog and the latter two as digital.

10.5 LANGUAGE APPROACHES AND LIMITATIONS

High-level languages should be hardware-independent; that is, they should not require the programmer to handle low-level interface commands. Current robot languages are too low level because they provide direct access to the hardware components of a robotics system rather than being used at the conceptual level. They are too hardware specific, so it is difficult to port programs from one robot to another.

How important is the language? According to the general rule of one line of code an hour, regardless of the type of language, shorter programs will generally cost less than longer programs. However, when an attempt was made to use several languages to code a robot pallatizing task, it was found that Lisp (an artificial intelligence language) produced the shortest program. Since Lisp is a difficult language to learn, it is unlikely many application programs will use it.

The basic aim of any high-level language is to allow programs to be written in a language closer to our own rather than a machine's. They can be based on specific robot commands, such as UP and CLOSE, or they can be oriented to a specific application. In this case the programmer can deal with the task to be performed rather than the method of doing it.

One breakdown of robot languages[24] separates them into several levels (Table 10-2). The lowest language level (that of a driver) uses assembly language. The next level (point-to-point travel), the first robot language level, are Funky and T3. At the next level, handling primitive types of motion, Anorad, Emily, RCL, RPL, SIGLA and VAL reside, and so on.

In first-generation robot languages, motion control was the primary function. Necessary commands were implemented to move the robot arm and to control the end-effector. Only limited I/O or timing commands (WAIT, for example) were provided. Primary drawbacks are their inability to handle complex sensor data, limitations on the type of complex arithmetic functions included, and an inability to work closely with other computers or robots.

Second-generation languages can be defined as those with complete control structures, full arithmetic capability, full I/O handling (including interrupts), and process control support. They can handle advanced sensor devices and modify sequences of instructions based upon sensory data. They also can interface with other computers and keep track of their own activity, thereby allowing reports and statistics to be prepared. Some error recovery techniques have been added to preprogram the robot to handle certain types of error conditions. Although some authors have classed this ability as a form of

Table 10-2. Intelligence Levels

Human intelligence	
.	(many levels between humans and robots)
.	
Artificial intelligence	
Adaptive programs	
Task-oriented	Autopass
Structured programming	AL, HELP Maple MCL, PAL
Primitive motions	Anorad, VAL Emily, RCL RPL, SIGLA
Point-to-point specifications	Funky T3
Peripheral drivers	Assembly

limited intelligence, the way that these procedures were implemented defines them as strictly preprogrammed error routines.

As with standard computer languages, robot languages must be able to handle arithmetic calculations (especially trig functions), fixed values (constants), and variables. They also need various program control and branching commands. In addition, robot languages must have special motion commands, which include location, speed, end-point control, and via (or pass-through) commands. The ability to operate in real time, to interface with a wide range of sensors and control devices, and to communicate with both the operator and other computers during programmed activity are also important.

Peck[25] suggested a four-step approach to programming robots:

1. List all of the actions to be performed by the robot, assuming that no problems develop.
2. Produce a top-down design based on these actions.
3. Generate a detailed design, adding WHAT-IF's to handle unexpected events (i.e., what if the object is not present?).

4. Identify common functions that have been used before, and include appropriate modules from other programs. Identify common functions that are used in more than one place in this program and place them in separate program modules.

Concerning step 3, problems can develop with the manufacturing process, the part, the tools, the environment, or the robot itself. Table 10-3 lists a few common problems that can develop in a typical installation. For more information see chapter 13.

An ideal language would support handling any type of unexpected event. Although no company has developed the ideal robot language, several attempts have been made to define an ideal robot language. According to Bonner,[26] the most desirable features are

- Powerful software development environment

- Extensibility as new features are needed

- Sufficient and clear language constructs

- Efficiency

- I/O interaction capability

- Easy access to the hardware

- Concurrency

- Clarity and simplicity

- Decision-making ability

- Interaction with the robot world model

Table 10-3. Examples of Unexpected Events

Failure	Result
Part out of tolerance	Will not fit
External sensor failure	Robot blind
Supply breakdown	Lack of material
Damaged tools	Product out of spec
Missing tools	Cannot continue
Part not in position	Cannot continue
Robot motion blocked	Wait for clearance
Coolant not present	Temperature rising
Part damaged	Faulty product is produced
Robot failure	Cannot continue
Process out of synchronization	Wait for assistance

Lewis Pinson added the requirement for a convenient mechanism for transition between off-line development, simulation, and test and on-line control of the robot. We add a requirement for easy transportability between different types of robots and computers.

The trend in the latest robot languages seems to be toward a block structure approach, as evidenced by Karel and JARS. Unfortunately, these languages are still limited in the robots they can be used with. Thus, the manufacturer who has purchased different types of robots needs programmers trained in many languages, and they must completely redo their original work to transfer a program between two different types of robots. It is also difficult for third-party firms to develop application-specific software, since there is so little standardization.

The Japanese are taking steps to support language standardization. Arai[27] reports on activity at three levels:

1. Communication level: protocols, data definitions, and electrical interface signals
2. Function level: fundamental command meanings, such as I/O control and movement commands
3. Application level: application-specific standards

They intend to standardize communications first and applications last. At each level is a separate surface language layer for third-party application software and an intermediate code layer for machine-level and peripheral driver code. These standards will only apply to robots that work with other robots or that are under central control. They will not apply to stand-alone robots. The proposed standard is ISO TC184/SC2 WG4, proposed by a committee under JIRA.

Further indication of the importance of standardization can also be seen. The Japanese Electrical Machinery Association has sponsored a study defining standards for a universal, next-generation robot language.

Methods of integrating user-generated programs, image processing programs, and other sensory-based programs written for specific devices must also be defined in any standard. The language needs to be open-ended to accept new robot improvements and techniques as they are developed.

The robot operating system usually depends on the system on which it will run. It communicates between the robot and the outside world, including accepting operator commands and sensor data. It coordinates and controls all action taken by the robot. Within the system are manipulator-specific modules to perform coordinate conversion, generate trajectories, handle feedback information, adjust for error or out-of-tolerance conditions, and supply data to the operating system.

The robot language interpreter/compiler is the interface between the programmer and the machine. This interface along with various utilities provides editing features, macro commands, and communications handling facilities.

Soroka[28] listed the following limitations he found in most current robot languages:

1. The language is developed for a specific robot arm and controller and cannot be integrated into the CAD/CAM systems of the factory.
2. Higher-level and less restricted commands are needed to interact with other parts of the factory.
3. The languages are neither flexible nor intelligent enough to handle unusual situations.
4. Robot languages are too hardware-dependent. Hence, programming must be redone for a different model robot, and programmers must learn several languages.
5. Languages are too narrow in application. The languages should match a wide spectrum of applications and a wide spectrum of user abilities.
6. Robot languages are developed principally for marketing needs rather than toward a standardized language.

Besides the effort in Japan, the French, in 1985, also tried to design a programming system that separates the programs into various layers,[29] thereby providing a much better interface with the programmer, the operator, the robot, and a central control computer. This unique approach to language development is being used in the European Cooperative Project as part of their ESPRIT program. In this method, an operator can use an implicit language (e.g., Karel) to provide the knowledge representation, task planning, and sensor monitoring controls of the robot (layer 1). This information is then used as the inputs to an explicit programming language (e.g., VAL), which develops the specific move statements, data integration, program branching, and related control tasks.

The output of this second layer is a code that can be put directly into a robot control language that handles program flow control, trajectory calculation, servo-control functions, and interrupt handling for a specific robot. This layered approach is completely independent at each layer. Because the interface between layers is controlled, it is much easier to develop standardization.

NATO has also recognized the importance of standardization, and it has organized a working group on robot planning languages. Their first report is now available,[30] and, among other recommendations, the group believes strongly that:

1. Robot programming languages are still in their infancy.
2. They should be built in a layered fashion.
3. Software support, ignored to date, is very important.

REFERENCES

1. Bonner, Susan, and Kang G. Shin. A comparative study of robot language. *IEEE Computer Magazine*, Vol. 15, No. 12, Dec. 1982, pp. 82–96.
2. Gruver, W. A., et al. Evaluation of commercially available robot programming languages. *Proc. 13th International Symposium on Industrial Robots and Robots 7.* Chicago. April 1983, pp. 12.58–12.68.
3. *System/370 APT-AC Numerical Control Program Reference Manual,* IBM manual SH20-1414. White Plains, N.Y. n.d.
4. Paul, R. P. Wave: A model-based language for manipulator control. *The Industrial Robot,* Vol. 4, March 1977, pp. 10–17.
5. Mujtaba, S., and R. Goldman. AL, a Programming System for Automation. Stanford Artificial Intelligence Lab Memo AIM-323. Stanford, Nov. 1974.
6. *User's Guide to VAL.* Danbury, Conn.: Unimation, June 1980.
7. Ruoff, C. F. TEACH—A concurrent robot control language. *Proc. IEEE COMPSAC.* Chicago, Ill., 1979.
8. Will, P. M. and P. D. Grossman. An experimental system for computer controlled mechanical assembly. *IEEE Transactions on Computers,* Vol. C24, No. 9, 1975.
9. Darringer, J. A., and M. W. Blasgen. *MAPLE: A High Level Language for Research in Mechanical Assembly,* IBM Research Report RC 5606. Yorktown Heights, N.Y.: T. J. Watson Research Center, 1975.
10. Liebermann, L. I., and M. A. Wesley. Autopass: An automatic programming system for computer controlled mechanical assembly. *IBM Journal of Research & Development,* July 1977, pp. 321–333.
11. Taylor, R. H., P. D. Sumners, and J. M. Meyer. AML: A manufacturing language. *The International Journal of Robotics Research,* Vol. 1, No. 3, Fall 1982, pp. 19–41.
12. Lozano-Pérez, T., and P. W. Winston. LAMA: A language for automatic mechanical assembly. *Proc. 5th International Joint Conference on Artificial Intelligence.* Cambridge, Mass.: MIT, 1977.
13. Salmon, M. SIGLA: The Olivetti SIGMA robot programming language. *Proc. 8th International Symposium on Industrial Robots.* Stuttgart, West Germany, 1978.
14. Popplestone, R. J., A. P. Ambler, and I. Belos. RAPT: A language for describing assemblies. *The Industrial Robot,* Sept. 1978.
15. Takase, Kunikatsu, Richard P. Paul, and Eilert J. Berg. A structured approach to robot programming and teaching. *IEEE Transactions on Systems, Man and Cybernetics,* Vol SMC-11, No. 4, 1981, pp. 274–289.
16. Oldroyd, A. MCL, An APT approach to robotic manufacturing. *SHARE 56,* 1981.
17. Craig, J. J. *JARS: JPL Autonomous Robot System.* Pasadena, Calif.: Jet Propulsion Laboratory, 1980.
18. Latombe, J. C., and E. Mazer. LM 1981: A high level programming language for controlling assembly robots. *11th International Symposium on Industrial Robots.* Tokyo: Japanese Industrial Robot Association, 1981.
19. Latombe J. C., et al. The LM robot programming system. *Proc. 2nd International Symposium on Robotics Research.* 1984.
20. Franklin, J. W., and G. J. Vanderburg. Programming vision and robotics systems with RAIL. *Proc. Robots 6 Conference.* Detroit: Society of Manufacturing Engineers, March 1982, pp. 392–406.

21. Shimano, B. E., C. C. Geschke, and C. H. Spaulding. VAL-II: A new robot control system for automatic manufacturing. *Proc. IEEE International Conference on Robotics,* March 1984.
22. Ward, M. R., K. A. Stoddard, and T. Mizuno, A robot programming and control system which facilitates application development by the use of high-level language. *Proc. 15th International Symposium on Industrial Robots.* Tokyo: Japanese Industrial Robot Association, Sept. 1985, pp. 145–154.
23. Ward, M. R., and K. A. Stoddard. Karel: A programming language for the factory floor. *Robotics Age,* Vol. 7, No. 9, Sept. 1985, pp. 10–14.
24. Bonner and Shin. A comparative study of robot language.
25. Peck, F. W. Advantages of modularity in robot application software. *Proc. Robots 8 Conference.* Detroit: Society of Manufacturing Engineers, 1984, pp. 20-77–20-93.
26. Bonner and Shin. A comparative study of robot language.
27. Arai, Tamio, et al. Standardization of robot software in Japan. *Proc. 15th International Symposium on Industrial Robots.* Tokyo: Japanese Industrial Robot Association, Sept. 1985, pp. 995–1002.
28. Soroka, B. I. What can't robot languages do? *Proc. 13th International Symposium on Industrial Robots and Robots 7.* Chicago. April 1983, pp. 12.1–12.8.
29. Latombe, J. C. Systemes de Programmation pour la Robotique. MICAD-85, Feb. 1985 (in French).
30. Volz, Richard A. Report of the Robot Programming Language Working Group: NATO workshop on robot programming languages. *IEEE Journal of Robotics and Automation,* Vol. 4, No. 1, Feb. 1988, pp. 86–90.

ROBOT INTELLIGENCE

The next generation of robots will have a much higher level of intelligence. They will accept all types of information (sensor inputs), process it, make judgments on what should be done, generate appropriate output commands, and effect actions. To accomplish these functions, robots will require extensive information about their environment. They will also need a different type of programming logic, providing an intelligent, rather than a rote, response to new situations. Therefore, the next generation may be called intelligent robots. This chapter reviews the current status of a wide range of techniques that improve a robot's intelligence level.

11.1 APPLICATION OF AI

Robot intelligence is related to the computer speciality called *artificial intelligence* (AI). AI is a branch of computer theory that deals with adding a level of intelligence to computer applications, such as the ability to learn, plan, handle unknown situations, or even understand human language. This last activity, understanding human language from speech or typed in words, was one of the first areas of AI research. Unfortunately, language understanding has proven to be much more difficult than originally expected.

Some of the earliest AI work was done at MIT and Stanford, and one of the earliest areas to receive intensive investigation, which often produced more frustration than progress, was language translation. More progress was made in machine vision and eye-hand coordination (the bin-picking problem).

Although work started in the AI field in the 1950s, to date few applications have become practical, and they are primarily in vision, education, and medicine. However, there is every indication that AI will offer major improvements to robotics in addition to its contribution to machine vision.

Concepts were tested through the development of specialized hardware, such as advanced hands and mobile path finding robots. Many companies then adapted some of the algorithms developed by Stanford for visual processing. Although quite effective for many simple applications, these early algorithms converted gray scale data to binary images. The resulting matching techniques were not able to distinguish complex objects, such as objects lying atop each other. Later techniques in image processing (chap. 6) expanded machine vision to complex objects and overlapping objects.

As techniques were applied to robots, different levels of capability became apparent. Calling all current robots "dumb," Dr. Ernest L. Hall[1] forecast higher levels of robot intelligent capability. In his classification:

1. The dumb robot is a completely preprogrammed robot that does exactly what it is told to do, with no capability of determining from its environment when that action might be inappropriate.
2. The expert system robot makes decisions based on hierarchical decision rules, with the aid of some sensor input from the outside.
3. The trainable robot (self-trained) has the added advantage of remembering actions from the past and thus can learn from its earlier experiences.
4. The intelligent robot can predict things based on past experience. Its sophisticated sensor inputs further support these abilities. It has a near intuitive level of artificial intelligence. This future level of robot intelligence may almost be compared to a minimum level of human intelligence.

In the same presentation, Dr. Hall made a distinction between three types of human intelligence. What we consider as normal intelligence is based on logic from the brain. When a person applies wisdom, he also considers the heart. When we talk of taking the smart approach, it often means considering the pocketbook. The current level of robot intelligence has enough difficulties when the robot attempts to solve problems by logic alone, but in the future we can expect a robot's intelligence to be expanded to consider wisdom and economics in its decisions.

Although this chapter emphasizes robot intelligence, not all robots need intelligence. Many applications quite rightly consider other features to be more desirable. For example, the major impetus for installing robots may be reducing costs or improving personnel safety. A robot can often serve either need with little or no intelligence. However, many applications will not open up to robotics until robots can demonstrate a higher level of intelligence.

What features of an intelligent robot are desirable? According to Dr. Hall,[2] the ideal intelligent robot should be able to

- Understand voice input
- Accept natural language instruction
- Have visual discrimination capability
- Be capable of intelligent motion
- Be capable of learning
- Be self-maintaining
- Have adaptive CAI capabilities

Substantial progress is being made in each of these areas. In a major overview of the AI field,[3] Barr and Feigenbaum listed nine areas of specialization within AI:

- Problem solving
- Logical reasoning
- Language understanding
- Automatic programming
- Expert systems
- Learning
- Robotics programs
- Machine vision
- Knowledge representation

These areas are applicable to a wide range of problems in business and industry. In addition, they will be all directly applicable to robotics. Enough progress has been made only in machine vision for direct use in present-day robotics, although principles from all areas are being applied. This use will continue to expand. Not only will future robotics applications gain from current AI research, but AI itself will be driven and encouraged by the special requirements of robotics.

In addition to these research activities, many companies apply AI directly to specific problems. Areas currently receiving extensive knowledge transfusions include expert systems, natural languages, voice recognition, machine vision, symbolic processing, and AI programming languages.

11.2 TECHNIQUES

One comparison between human intelligence and computer intelligence is based on which tasks are the easiest. Humans find pattern processes (normal vision) quite easy, symbolic processes (reading) slightly more difficult, and arithmetic processes (mathematics) the most difficult. Machine intelligence, on the other hand, handles arithmetic processes (number crunching) quite readily, has more difficulty with symbolic processes (list processing), and finds pattern recognition the most difficult of all. (For example, is that the edge of the road, or is it a tree?)

It has been theorized that this difference in problem solving is in the way we store and retrieve data (across a vast interconnecting network of neural cells). Compare this to the single-item storage found in robots (and other computers). Humans thus have parallel access to pattern information, rather than the sequential and mechanical execution of precise instructions that characterizes most current computer design. Parallel access is one reason that neural network computers offer such promise (see chap. 16).

To improve pattern recognition processes, researchers have made changes in their programming approach. Instead of being mathematically based (like standard programming), AI programming is paradigm-based (i.e., it is based on patterns). For example, many AI languages have a common ability to handle node and link relationships, thus allowing data to be represented differently than in conventional computers. Nodes represent all valid relationships between objects, and links represent all valid connections between nodes. In AI, the term *valid* means "following expressed rules," so AI languages can be characterized as rule-based.

Data is important to almost all computer applications, but in AI data is based on a complex knowledge base. This knowledge base includes information about objects, processes, goals, time, actions, and priorities. This type of data may be considered as background information, and humans take it for granted when solving problems. In AI, the data must be separately identified, defined, and placed in the knowledge base.

Knowledge Representation

Knowledge-based systems can "reason" from a set of facts and a set of rules, either by forward chaining (going from facts and rules to conclusions) or backward chaining (from a tentative conclusion to uncover the required supporting facts).

Knowledge representation is a step above data storage. It involves an interconnected data structure, where the connection points (links) represent various concepts. Equally important is the use of inference rules for obtaining data from the network. The links define the relationship between various levels of knowledge. For example, a dog is a mammal, and a dog is a pet. But not all mammals are pets, nor are all pets mammals. Thus the link connecting dog with both pet and mammal is a one-way link read dog *is a* pet and dog *is a* mammal. Deductive reasoning, also called first-order logic, can then be applied to the knowledge base to infer information not specifically spelled out.

When the knowledge base is large and complex, it is often organized into groups called *frames*. Although frames are used in many tasks, they are particularly applicable to machine vision. Knowledge bases are also very useful in developing production systems (discussed later). Knowledge management allows new knowledge to be represented, manipulated, and learned.

One way that knowledge-based techniques can be used in a factory, for example, is handing data about parts. Most inventory systems contain basic information about a part (part number, part description, quantity on hand, list of suppliers, and, possibly, identity of assembly in which it is used), but they do not describe its interaction with other parts or give details on its use, such as which parts mate with it, its function, environmental limits on its use, alternative methods for its use, and substitute parts that could be used. An AI knowledge base could contain this information.

Although in theory all of this data (and more) could be kept in a standard data base, in practice it would be impossible to implement in most cases. Every part change would affect not only that part's file but the files of all parts that interact with it and perhaps other files that are not as obvious. It would therefore take too long to identify all data needing correction whenever a standard file of this type required updating. In addition, some areas of the file needing update would undoubtedly be missed. On the other hand, AI-type knowledge bases are designed on a link-and-node basis, and changes, entered once, will update all appropriate information.

AI Logic

AI systems are based on logic, and a fundamental axiom in logic is that every statement must be true or false. There are no half truths in most computer systems (except for fuzzy logic, which is discussed later). These logical statements are then combined into a propositional calculus. Propositions, which serve as the basis for most mathematics, could be expressed as "if. . .then" or "*A or B*." For example, "If it is spring, then the leaves are green," or "The car is in the driveway, or it is in the garage." In AI a slightly different approach is usually taken—predicate calculus. Predicate calculus applies a function to an object to determine its relationship to others. For example, the function *location of* when applied to *car* might return the value *garage*.

Problem solving (one of AI's branches) provided some of the initial successes, such as in chess. Solutions to problems were often based on a divide-and-conquer strategy.

Once you have a computer system that can form a logical conclusion by reasoning from known or assumed facts, you have an *inference engine*. Inference engines are often used to develop expert systems.

Production Systems

An area of AI systems that has received a lot of attention is production systems. Production systems were developed initially at Carnegie Mellon University as a model of how humans think. They represent knowledge as a set of conditional action rules. A production system can be used for many purposes, but is best used in applications in which the information is very diffuse (many facts), actions are primarily independent of each other, and combinations of facts are not common. The systems have three parts: a rule base (or set of production rules), a data structure, and an interpreter. Production rules might include the following examples:

1. If the robot arm is moving, then indicate it is functioning.
2. If there is no part to work on, then indicate the robot is waiting.
3. If the robot is not moving and if the conveyor line is moving, then indicate the robot needs attention.
4. If the robot is not moving and its cover is off, then indicate it is being repaired.

The interpreter must decide what actions to take with the foregoing information and, for information in conflict, what rule to use first. The interpreter could make the choice on the basis of which is first in the list, which had occurred first in time, which had a predefined priority, which statement matches the most terms, or it could simply select a rule at random. Each approach has its advantages.

If the interpreter acted on the basis of the order listed, the production system would never get to statement 4 as long as the conveyor was moving, since it would always stop and perform statement 3. (After a rule is executed, the interpreter goes back to statement 1.) One way around this problem is to have the interpreter skip any rule that has already been acted on once. Determining priority is more difficult than might be expected. For example, if statement 3 is given priority, then action may be taken when it is not called for, that is, when the robot is simply waiting for a part.

Production rules can be readily implemented to suggest repair strategies, and some experts feel this implementation is how production rules can best be used in robotics systems. For example, a set of production rules to repair a robot might start off this way:

1. If you turn on the control switch and
   the ready light does not go on
   Then check to ensure that the interlock is not open.
2. If the interlock is OK
   Then check that the robot
   is getting electricity

One production system language is OPS, developed at CMU 12 years ago.[4] The latest version is OPS5,[5] which can be used in the development of expert systems. OPS5 productions are not handled sequentially, but are executed when their condition elements (also called LHS, left-hand side) are best satisfied by the data. Execution order thus depends on the best match of knowledge with condition elements.

Expert Systems

Expert systems were developed as one of the first practical AI applications. Many of these systems are diagnostic troubleshooting systems, which is probably why they are of most immediate interest in robotics. However, other (nonrobotic) applications have also been developed. One system developed by Texas Instruments (TI) has been used in many types of applications, from car assembly to cable connections.

A different approach is used in the Kurzweil reading machine. It converts written text to speech and includes a scanning system plus a speech synthesizer. Other expert systems are

NOAH and ABSTRIPS, developed by SRI for factory design.

XCON, an expert system developed by DEC to configure computer systems. This system is 98% accurate compared to a 75% human accuracy.

EXCABL, an orbiter payload cabling system, was developed under OPS5, a production system language.

All expert systems are composed of a knowledge data base, a set of logical rules (or inference engine), and an interface to the user. Since nonprogrammers are the primary users of expert systems, the interface is quite important.

The knowledge base in an expert system comes from the combined experience of many human experts. The inference engine then operates on this knowledge base to solve problems.

The way a computer program represents knowledge affects how it applies and manipulates that knowledge. If the program is to demonstrate intelligence, it must be able to manipulate this knowledge easily. Typical programming methods interleave knowledge (data) and program control (logic). Expert systems separate knowledge from program control and provide knowledge-based programming.

Most expert systems are still limited. They increase a computer's speed and accuracy, but they still use only a preprogrammed hierarchical search procedure. However, they offer great potential, and are even available on microcomputers.[6]

Fuzzy Logic

Normal programs deal with statements that are true or false. If *all* the facts in the statement are not true, the statement is false. In real life, however, we must consider imprecise situations where true and false alone are not sufficient. For example, relative concepts (young, old, pretty, and strong) are not so definitive. Any decision based on these concepts will be relative, and formal rules will not apply. (Is a person old at 55, 65, or 75?)

In fuzzy set theory, a graded membership to a set is used. Some AI systems deal explicitly with this type of information (it is then considered a certainty factor). For example, we may consider a 6-ft 6-in male as tall with a 100% certainty, whereas a 5-ft 11-in male may be considered tall with a much lower certainty. In a fuzzy language this type of uncertainty is handled by the logic, usually through a *membership function.*

A fuzzy logic structure changes the normal two-valued parameters to multilevel parameters. Fuzzy values can still be true or false, but many values will be *true-like or false-like.* (There are also unknown values.) Thus, fuzzy arithmetic will assign a certainty value to any proposition. Kaufmann[7] offers a more complete discussion of fuzzy arithmetic.

Fuzzy logic is directly applicable to robotics. For example, when a robot has completed a program step and is ready to start the next steps, the first step is actually "completed" with some level of uncertainty. Robot motion is not perfect, and the end point will really only approximate the desired position. (Is the gripper near enough to grasp the object?) Fuzzy logic properly implemented will allow for the *probability,* not the *certainty,* that the gripper is near enough.

11.3 VISION SYSTEM RESEARCH

Needle Diagram and Generalized Cylinder

Two techniques that have made important contributions to intelligent vision are the *needle diagram* and the *generalized cylinder* (Fig. 11-1). The needle diagram helps give a three-dimensional look to a two-dimensional image. It is an important step in preparing 2 1/2 D sketches and provides needed facts about surface construction.

In needle diagrams arrows are added as surface normal vectors by "sticking" the arrows into the center of an object. Since the resultant arrows are always

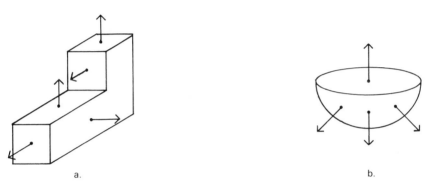

a.   b.

Needle Diagram

c.   d.

Generalized Cylinder Technique

Figure 11-1. Three-dimensional representations.

perpendicular to the surface, they effectively represent all types of shapes. For example, by showing arrows pointing in slowly changing directions along a surface, with no sharp edges between them, the needle diagram can easily show that the surface is curved (as in Fig. 11-1b).

The generalized cylinder approach takes a two-dimensional object (such as a square) and fills three-dimensional space with it through motion. The two-dimensional object can stay the same size when moving, perhaps forming a cube (Fig. 11-1c), or it can change in size, perhaps forming a pyramid (Fig. 11-1d).

Both of these approaches allow complex objects to be described to a computer. With the description in its data base, the computer is much better equipped to recognize complex objects, one of the most difficult vision tasks computers can be asked to perform.

For example, assume that a robot is asked to pick up a telephone out of a collection of complex, three-dimensional objects. How do we tell the computer what a telephone is? How does the computer recognize a telephone when it sees one? Both the needle diagram and the generalized cylinder techniques provide us with methods to describe complex objects to a computer. Other vision approaches (such as the DOG filter), aid the computer in answering the second question.

DOG Filters

A mathematical approach that seems to approximate human vision uses image differencing. In this technique, a two-dimensional Gaussian filter is first applied to an image, followed by two-dimensional image differencing. This process is often called a DOG filter, for difference of Gaussian. If two steps of differencing are applied, the result is equivalent to a single-function filter having the shape of a Mexican hat (hence the name sombrero filtering). The resultant signal from a sombrero (or modified sombrero) filter applied to an image matches almost exactly the signal observed from primate eyes when optic nerve signals are monitored. In fact, if you plot values for this type of DOG operator, the graph turns out to be the same as that found in human vision, using the center surround operator.

Computing DOG values is not the type of task you give to an overloaded processor. A typical filter may require 32×32 operators for each image point. Since the image resolution may be $1,000 \times 1,000$ pixels, taking its convolution operator requires about 1 billion operations on each image. Fortunately, there are shortcuts that can be used. For example, the computations can be done in one direction at a time and the results multiplied to reduce the total computation.

Motion Detection

A novel technique being investigated by researchers to detect object motion is optical flow. Optical flow identifies which parts of an image are in motion (i.e., have a velocity vector). Most current vision systems take snapshots and cannot easily handle objects in motion, so adding this capability will increase vision system utilization. Optical flow has also been combined with stereo imaging to produce more accurate range information.

Motion has been addressed at MIT by computing the velocity field along known contours in the image. By detecting discontinuities in this velocity field, the robot can distinguish moving objects from the ground and from the background. Optical flow is an interesting concept and quite different from how we normally view motion.

Hough Transform

Another approach to vision processing is the powerful, general-purpose vision algorithm, the Hough transform.[8] The Hough transform, as originally proposed, was a means for detecting only straight lines. It was subsequently modified to detect objects of a few other shapes and then generalized to handle any shape. Unfortunately, the more general the shape, the more preprocessing that must be done, the more memory that is required, and the more complex the computations become. Because of this, it is more applicable to parallel hardware computation, although successful implementations on conventional serial machines have been done.

The Hough transform has been applied to the general template correlation problem, where the template may be distorted in size or shape and be translated or rotated with respect to the image. When these variations are present, object recognition is difficult or impossible for most image recognition algorithms. The Hough transform predicts the center of an image from all its edge points and compares it to the center of the stored template. It then makes adjustments for size and rotation and comes up with a probability of match.

The Hough transform has also been used to detect motion in an image. Casasent[9] extends this transform to the detection of trajectories (predicting where the moving object within the image will go). Both straight-line and various curved path motions can be predicted.

Three-Dimensional Images

Different approaches are used to handle three-dimensional images. Stereo images obviously contain this data, but extracting it is quite difficult. The problem in any stereo imaging system is matching the many lines in each picture to determine which ones represent the same objects. It is difficult enough to recognize an object in a single-image frame. Each camera does not see exactly the same scene, reflectivity may alter the contrast, and certain parts of an object are hidden in one scene but not in the other, so matching objects between stereo frames can be quite difficult.

One approach is to start with a coarse resolution image. This type of image has much less total information in it, so it is easier to match the larger objects. If this picture is subsequently looked at with finer and finer resolutions, all parts of the picture can be matched accurately, and the range to each object can be found from these matches.

Ideas from computer-aided design and differential geometry are also used to determine curvatures. Figure 11-1b illustrated a line drawing in a single plane that gives a strong impression of being a curved surface.

A different approach converts shading and light patterns to contours. The contours can then be converted to a local parallel process and used to find skewed symmetric surfaces. Surface normals, such as provided by needle diagrams, can be quite useful in this area.

Extensive work is being done on applying three-dimensional data to the problems of pattern recognition. Information on local surface orientation, surface shading, and other highly structured data can be used to come up with a conclusion (i.e., this is a bottle). These values can be converted to develop fairly accurate three-dimensional maps (within about 2°).

There are other ways to handle advanced three-dimensional objects. Objects can be represented by basic shapes, such as are found in most commercial vision systems, or by geometric shapes, such as cubes, cones, and so on. You start with these basic objects and then put them together to build higher-level models.

Some of the work being done is aimed at how to present three-dimensional images in a manner that allows key features to be determined by the computer. For example, taking a cross section of an object and moving it produces an area that expands and contracts along the object.

Parallel Processing

Parallel processing is one method of processing complex images. A technique often used in handling low-level vision problems is distributed memory parallel processing, in which numerous limited-capability processors act in parallel, with each processor only being able to access a small part of the total memory.

Many parallel processing machines have been released commercially, including systems from BBN (Butterfly), Loral (LDE), Segment (Balance), and Thinking Machines (the Connection Machine). This last system has 64,000 single-bit processors and has been available since early 1986. Efforts are now underway to improve I/O and programming techniques. The warp systolic array, developed by Carnegie Mellon, processes data on an assembly line approach, with each processor adding its own unique operation to the data. Both GE and Intel are developing hardware versions of this machine.

11.4 AI LANGUAGES

Overview

Today's robots are only suitable for applications in which the goals are economy, safety, productivity, or quality control. To move into areas that require even the most limited amount of intelligence, the robot needs different types of programming languages than are currently used. In fact, it is estimated that half the needed commands have never been added to conventional languages.

Programs with limited intelligence can be written in numerous languages, even in assembly language, a step above the computer's internal language. However, certain languages are better adapted than others to the certainty factor–data handling methods needed in AI. Furthermore, there are different levels of difficulty in writing in different languages. To free the programmer from many difficult and tedious tasks, languages specifically designed for AI and related applications have been developed.

One of the many differences between AI programming and conventional programming is the design method employed. In standard application programming one must understand in general the entire program flow before beginning actual coding, but not in AI. There is even a saying in AI that if the problem is well understood at the beginning, it does not belong in AI. AI has six areas of significant differences from conventional processing. The first three include a requirement for symbolic processing (low level and high level), large amounts of data (requiring unique methods of knowledge management), and dynamic execution (the need for changes during problem solving). The other three are the potential for parallel processing, especially for problems that belong to the OR parallelism class, the nondeterministic nature of the computations (i.e., it is usually impossible to plan in advance what procedures are required), and open systems design, to allow the acquisition of new knowledge and continuous refinement of the operating environment.

AI programs all include some uncertainty. This uncertainty may come from the complexity of the task (no one is able to see the total interactions of the data base) or from changing requirements (data is constantly changing) or even from unpredictable human behavior. The best approach for AI seems to be designing and testing the software on an experimental basis and thus learning how to handle the problem during the design.

Because of these factors, most AI languages have been designed to put as few constraints as possible on the programmer. For example, the amount of storage used in AI programs is dynamic. When different parts of the program need more storage, they can request it. When they no longer need the extra storage, the system reclaims it for later use. Other important differences to the programmer are the ability to dynamically change the variable type in the program and the ability to dynamically change various subprocedures when invoked by the calling code.

What makes this type of programming possible is that the language itself is able to keep track of and understand to a sufficient level exactly what the programmer is doing. It is just as if you had an assistant handling many of the mundane but important tasks for you. In fact, an expert system is available to do this, called The Programmer's Apprentice.

Also of major importance to the AI programmer are the available tools that he is given, such as a powerful graphics workstation that allows a program to be described or understood in many different ways. An example of a powerful graphics techniques is the use of tree layouts to show levels of dependence between data.

Current Languages

There is a debate about which current AI language is the best. Each language is described here without entering into the debate, as each has its own strong points.

AI programming languages may be object oriented, logic oriented, or procedural oriented; some might consider production systems a fourth style. Lisp is a procedural-oriented language, Prolog a logic-oriented language, and Smalltalk an object-oriented language. Most AI languages are examples of rule-based software, an approach that allows the handling of uncertainty. Lisp and Prolog are two languages most often associated with AI, but many new languages, such as Smalltalk, are also becoming quite popular. In addition, progress has also been made in other languages. One example is Sharp, a system for automatic programming of manipulation robots, developed in the IMAG laboratory of LIFIA.[10] Another is Karma, a knowledge-based system developed at the University of California to program robots.[11] Forth[12] and LOGO are also proposed as advantageous. Many expert systems are being developed around a UNIX shell written in C, Pascal, or some other language, rather than the more conventional AI languages. Perhaps the newest AI language is Q'Nial, an APL-like language developed at Queen's University in Ontario and offered commercially by a New York firm for under $100.

There are numerous language-based AI computers, especially in the Lisp area. Examples include: the TI Explorer (a Lisp machine), the Tamura Machine (a Prolog machine developed at Kobe University), and ALICE, developed at Imperial College, London, which supports other languages such as Hope[13] and Parlog.

Lisp

Lisp (*list processing*) is one of the major AI languages. It has the ability to process symbols and lists of objects rather than conventional arithmetic handling commands. It is most popular in the United States. Lisp was developed by John McCarthy at MIT in 1957 to apply special processing techniques to symbols. Since there was originally no Lisp standard, several versions have developed. Lisp is now available in different dialects and two principle forms, MacLisp and Interlisp. In an attempt to standardize the language, Common Lisp was proposed. A standard subset of most modern Lisp languages, Common Lisp contains most of the features found in MacLisp and Interlisp, in addition to user-controllable features. Limited versions of Lisp have also been designed to run on various microcomputers, such as the IBM-PC AT.

In addition to these "standard versions," one implementation, LOGLISP, was developed at Syracuse University to provide Lisp users with some features of Prolog. However, unlike Prolog, a backtracking process is not used to explore alternatives. Instead, alternatives are examined in parallel. LOGLISP contains two distinct portions, a Lisp portion and a logic portion (or set of logic primitives). What makes this combination strong is the ability of each portion to call the other.

Lisp can run much faster on a computer that has been specially designed to handle some of its processes. One Lisp system, built by Lisp Machine, Inc., provides about 10,000 compiled functions in a system using 4 million words of storage.

Lisp deals with two kinds of objects: atoms and lists. *Atoms* are symbols used to name objects, and *lists* are a series of atoms or a conjunction of lists enclosed in parenthesis. Lists are operated on by the rules of predicate logic, which yield only *true* or *false* when the rules operate on data.

For example, one could define a class of objects as INVENTORY as follows:

```
(SETQ INVENTORY '(HOES RAKES SHOVELS))
```

In this case hoe, rake, and shovel are atoms; inventory is a list. The construct CONS can be used to add a new object to the inventory:

```
(CONS 'LAWNMOWER (INVENTORY))
```

Although there are some differences between dialects of Lisp, the following points generally hold:

1. All requests to Lisp are enclosed in parentheses.
2. The function to be performed is always first.
3. A single quote is used to specify all following elements as atoms, that is, part of a list, not functions or actions to take on a list.
4. Even in a list processing language, about half of the common functions are arithmetical, such as ADD1 (add 1 or increment), SQRT (square root), TIMES (multiply), and MAX (maximum value).
5. Many of the functions do not use easily understood mnemonics, making the language difficult to learn.

As an example of language difficulties, consider that words such as CONS do not readily explain their function (JOIN would be better).

Lisp is a function evaluator. Once the function and its arguments are entered, Lisp will output the result. For example, enter (PLUS '6 2), and the computer will answer 8. Entry of (GET 'ARM LOCATION) could yield INITIALIZED. The power of Lisp starts to show up via its property values, which are special true (T) and false (NIL) functions that result from a set of tests applied by predicates or by the COND function. Lisp also provides true and false results. For example (by convention), if the argument of the predicate FCN is an atom, the result is true; if its argument is an expression, the result is false.

The biggest strength of Lisp may also be a weakness to many programmers. Lisp is a highly mathematical language, containing properties of set theory. It is also a very recursive language. Both of these concepts are very desirable for AI applications, but they are difficult for programmers who are not mathematically inclined. Programmers need extensive experience to become proficient in Lisp. Another drawback is that Lisp requires much more memory than do other AI languages. Thus add-on boards containing 24 Mbytes of memory have been developed for the IBM-PC to support Lisp.

A trajectory planner is now available in Common Lisp to run on a VAX/750. A Lisp interface between the VAX and the SCEMI robot controllers has also been developed. This interface also allows interactions with high-level vision functions.

Prolog

Prolog was developed in France by Alain Colmerauer at the University of Marseille in 1972, and is the most popular AI language in Europe and Japan. It can handle such problems as finding a collision-free path between two points for a mobile robot. Prolog[14] employs a clausal form of predicate logic and uses a depth-first search with backtracking for problem solving. To better understand backtracking for a robot, we will take an example of a robot operation in a chemistry lab, where the robot adds reagent to a mixture. If the task requires satisfying several conditions at once (move reagent to point A while keeping container upright), it may be that the first attempt to plan this move would require the robot arm to turn in such a position that would tilt the gripper and perhaps spill the reagent. Backtracking means that when a program finds the current path to its goal blocked, it will automatically back up to the last point where an alternative branch could have been taken. Backtracking is potentially very powerful, and offers an advantage to Prolog over Lisp.

To give some idea of the structure of Prolog, we provide a recursive loop that is used to solve a collision-free path between two points, start point A and end point B. In the example, *path* is the predicate with A and B as inputs. During initialization, the necessary data is created and entered into memory by assertion of facts. After initialization, the clauses defining the desired path ["path_bet(A,B)"] include

```
path_bet(B,B):-retract(move(_,_,_)),!,
path_bet(_,B):-smallest_v_open(Va,Ba,Fa),
expand(Va,Ba,B),
assertz(close(Va,Ba,Fa)),
path_bet(Va,B).
```

The second clause determines the smallest path open to the robot at any time. This set shows how Prolog can solve this problem through recursion. The first clause is used as the end of the recursion. The following four clauses serve as the main loop. The clause is called recursively until a complete path has been determined.

Prolog has a first-order logic. That is, it can use deduction based on its own knowledge base and can solve problems by deductive reasoning. It is also a resolution-based system. It is not complete, however, because there are logical consequences that it cannot handle. For example, the order in which functions are input to the system can affect whether Prolog is able to perform the calculation. It may also produce an error by claiming some things as valid that are invalid.

Prolog is particularly suited for natural language applications and as an implementation language for developing expert systems. Like Lisp, several versions of Prolog have been implemented, including Prolog-10, Poplog, and Micro-Prolog. Versions of Prolog are commercially available to run on PCs and IBM-mainframes, on the DEC VAX and on Sun workstations.

Smalltalk

Smalltalk was developed in the 1970s at Xerox Palo Alto Research Center. Five key words describe it: *object, message, class, instance,* and *method. Object* refers to conventional objects (such as numbers and character strings) for special-purpose classifications such as dictionaries, file directories, budgets, programs, and text editors. The dictionary, for example, associates a name (i.e., Roger) with a value given to it (i.e., 3). The object is stored in a contiguous memory called *heap.* It has a two-word header and a body. A *message* is a request to an object to carry out one of its operations. It can also report the result of the operation to the requesting function. The message concept also helps to ensure the portability of the language, because the method of operation is not specified.

Class is an object that describes a set of similar objects such as rectangles. Smalltalk's building blocks (or classes) define the object's structure and behavior. *Instance* is an example of an object in the class (i.e., square). It has memory and responds to messages. *Method* indicates how the object will perform its operation. It includes a message pattern, variable declarations, and a sequence of expressions.

Note that each of these key words refers to objects. Smalltalk has an object-oriented style and is particularly applicable to tasks with a clear hierarchical object classification. It has two unique features: data abstraction, which is a method of summarizing or encapsulating the data into a direct data representation, and the concept of inheritance, in which similar data does not have to be stored in multiple locations but can be passed on by inheritance to certain other groups.

Since the object's attributes are hidden, the same message (or command) will work on most objects. Like all AI languages, Smalltalk needs a lot of memory and thus requires that data no longer needed be removed (garbage collection). Garbage collection can be done by marking the "good" objects (ones accessible to other objects) and deleting the remainder, or by providing a reference counter on each item. The first approach is best for limited memories. The second keeps track of how many other objects are referencing the object. If the counter gets to zero, no one is using the object and it can be deleted.

Object behavior can be redefined or added to dynamically, an important factor in AI applications. The designation of problems is also different, with all communications occurring over the sending and receiving of messages.

As an example of Smalltalk syntax, the following line demonstrates the use of two pseudovariables (amount and reason) used to refer to objects in a class:

```
spend: amount   for: reason
```

These pseudovariables would then be replaced by *instances:*

```
RepairBudget spend: 300 for: 'servomotor'
```

Smalltalk 80, one commercial version, offers over 200 classes of objects and several thousand methods (similar to a function in a conventional language).[15] A total of 64,000 objects can be referenced. A 500K memory is the minimum required.

Smalltalk is inexpensive. It costs less than $100 for a system running on the IBM-PC. It is also available for the Sun workstation. Smalltalk was also the basis for the Macintosh interface, supplying such items as pulldown menus and mouse pointers.

Forth

Forth has been heavily used in the creation of expert systems such as Expert2, also known as MVP Forth/Expert Systems Toolkit. Like Lisp, it is an inherently extensible language; that is, it is designed to allow easy user definition of new functions within the language.

Forth's most unique feature is that it is a stack-driven language.[16] A program consists of pushing data on the stack and applying operators called "words" to that data. The words may be primitive arithmetic functions, stack manipulation operations, or user-defined functions. Arithmetic expressions are given in the post-fix notation. For example, the Forth command to calculate and print $(3 + 5) * 10$ is

```
10 3 5 + * . ⟨return⟩
```

First 10, then 3, and then 5 are pushed onto the stack. When the word "+" is executed, the last two numbers on the stack, 3 and 5, are popped off, summed, and the result of the summation is placed back on the stack. The stack therefore has 10 and 8 on it. Next the word "*" is executed. Again the top two numbers are popped off the stack. They are multiplied and the result, 80, is pushed onto the stack. The word "." prints the top item of the stack.

User definition of new words is cumulative; the second new word may use the first, the third may use the second, and so on. Each word thus defined goes into an internally kept dictionary, which is itself stored in a stack. The system keeps a pointer to the last defined word, but since each word points to the word previously defined, a dictionary search becomes simply a matter of a search through a backward-linked list.

The following program illustrates the extensibility of Forth. It defines a word, poly, which evaluates a fourth-order polynomial of the form $6x^4 + 3x^3 + 2x^2 + 4x + 5$, where the value of x is residing at the top of the stack. The line numbers are not part of the Forth language (unlike BASIC) and are given here only for convenience.

```
1 : sqr dup * ;
2 : cube dup sqr * ;
3 : frpw dup cube * ;
4 : poly dup dup dup
5        frpw 6 *
6        swap cube 3 * +
7        swap sqr 2 * +
8        swap 4 * +
9        5 + .
```

This program, which consists of the definition of four new Forth words (using the : form), uses the stack manipulators dup and swap. Dup duplicates the top of the stack, and swap interchanges the top two stack elements. This program, called poly, is executed by loading the block containing the word definition and keying in the command:

```
3 poly ⟨return⟩
```

The data element 3 is pushed onto the top of the stack, making it the value of x. Then the dictionary is searched for the word poly, which would then be executed.

Forth works almost at an assembler language level for program storage. Absolute block numbers are used to reference the location of a program on disk. Although this procedure seems awkward at first, it allows easy implementation of overlays for large programs. Simply load the first few blocks of the program, define and execute those blocks, and include at their end the instructions to clear unneeded memory and load the next blocks of the program.

Most versions of Forth have a full complement of control structures (IF . . . THEN . . . ELSE, DO . . . LOOP, BEGIN . . . WHILE . . . REPEAT, and BEGIN . . . UNTIL, as well as a CASE statement) and data structures, such as single- and multidimensioned arrays. Even allocation of space for arrays is stack-driven. The word "allot" functions by popping the top number from the stack and adding that many bytes of storage to the dictionary area of the last defined word.

The power of Forth lies in its extensibility and in the way it takes advantage of the stack architecture used in today's microchips. Thus Forth can be considered for some types of AI tasks. In addition, Forth's stack-driven nature allows rapid execution of instructions, which is so important to control robotic movement. Furthermore, Forth is ideally suited for use on the microchips used to build the current generation of robots.

Logo

Developed by Seymour Papert, Logo was initially designed for teaching children about mathematical relationships and computers. It provides a simplistic instruction set for manipulation of an object (called a *turtle*). The object can be real (a small robot) or simulated on the graphics screen. Either way, it has the properties of orientation and location in terms of Cartesian coordinates. The property of orientation allows spatial relationships such as "backward," "forward," "right," and "left" to have validity. The penup and pendown commands allow the turtle to draw geometric figures, thus providing an immediate visual feedback. This feature makes the logo/turtle combination a good teaching tool.

Logo is actually the first popular AI language designed specifically for robot manipulation, although the original robot (a turtle) was little more than a toy, and its sole ability was movement and drawing lines in various patterns. Since then, Logo has been used on personal robots, and it has been suggested for certain industrial robot applications.

Like all AI languages, Logo is extensible, allowing generation of new commands on the basis of the supplied command set and previously defined new commands. Like BASIC and Forth, Logo includes file manipulation commands within its structure, including CATALOG, ERASEFILE, and SAVE.

In Logo, variables and file names are called *words*. References to variable names are preceded by double quotes ("). To assign a value to a variable, use the MAKE command:

```
MAKE "TEST 4
```

References to the value of a variable precede the variable name with a colon. To print the value of the variable TEST, use the command

```
PRINT :TEST
```

Procedures are defined by the key word TO followed by the procedure name (i.e., name of new command) and the input parameters, if any.

Logo supports recursion and list manipulation commands, similar to CAR and CDR in Lisp, and is ideally suited for the exploration of the recursive graphics patterns of fractal geometry. The following procedure definition (adopted from Thornburg[17]) illustrates the use of recursion in Logo to draw a tree with fractal branches.

```
TO TREE :SIZE :LIMIT
IF :SIZE < :LIMIT [STOP]
LEFT 45
FORWARD :SIZE
TREE :SIZE * 0.61803 :LIMIT
BACK :SIZE
RIGHT 90
FORWARD :SIZE
TREE :SIZE * 0.61803 :LIMIT
BACK :SIZE
LEFT 45
END
```

11.5 APPLICATIONS

VLSI Layout

AI techniques are being explored to aid in the design of complex integrated circuit layouts, to allow the designer to specify a set of requirements for a chip and have a custom-made chip produced. Researchers, including the AI group at Rutgers, are looking into the design of the chip as well as analyzing the final design. After developing a limited tool (called Redesign) to help engineers redesign transistor-transistor logic (TTL) level parts, this group is now working on a much larger system, called Leap, to analyze and refine the design of a very large scale integration (VLSI) chip.

Another group active in this type of application is Carnegie Mellon, which has used programs written in Lisp to design specific microprocessor chips, including placing the complete IBM 370 instruction set on a chip. This design was then compared (favorably) to the design that IBM engineers had produced without this type of design aid.

A program written in Prolog at Schlumberger Palo Alto Research can verify that the design of complex VLSI chips is correct. In one case a VLSI chip containing 30,000 primitive parts was examined and verified in 10 min. Although this chip was only about one fifth as complex as current chips, it is a good start. A more ambitious project is to put Lisp itself on a chip, and TI is trying to accomplish just that under a Defense Department contract.

Robotic Related

One of the earliest AI programs (STRIPS) for a robot application was written by Richard Fikes and Nils Nilsson of SRI.[18] The robot was placed in an environment of a series of rooms, doors, and objects, and it was given the task of developing a plan for achieving a predefined goal (a typical global path planning problem). STRIPS developed the plan, and a different program was used to direct the robot to carry out the plan.

Dr. J. Kenneth Salisbury of MIT's Intelligence Lab has been working on using Lisp to control and interpret motions of a robot hand.[19] Grasp functions and object recognition algorithms are being implemented.

Chang and Wee describe an approach for applying knowledge-based systems to plan robot assembly steps.[20] They point out that to a human the order in which to assemble items is usually obvious. (One does not put the cover on a unit until all of the internal parts are in place, nor put a chip in a socket before the socket is mounted on the board.) By using the knowledge base of workpiece structure, general assembly rules, and robot limitations, they developed a goal-oriented assembly planner that selects the order of assembly steps with limited searching.

Digital Equipment Corporation uses an expert system for automated material handling in which two robots are controlled by two expert systems.[21]

ASEA has developed a Maintenance Assistant for its robots that helps to determine production problems. Figure 11-2 is one of a series of displays provided to the operator under this expert system. In this step, the operator is asked to indicate the plane in which the error would be most visible and his or her certainty of that fact. After a series of such questions, the Maintenance Assistant will make its recommendation as to the most likely source of the problem. Other manufacturers with troubleshooting systems include REIS (SERVICE Expert), KUKA (an expert data base), and GMF (Problem/Solution Database).

Potential Applications

Three AI areas of particular future interest are

1. Learning on the job (extension of the Chang and Wee approach)
2. Recognizing unusual events and the actions to take when they occur
3. Recognizing internal malfunctions and self-diagnose its problems (extension of current expert maintenance programs)

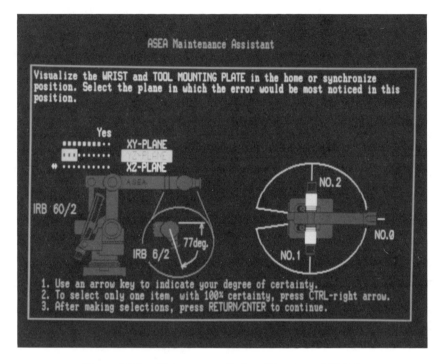

Figure 11-2. Maintenance display. *(Courtesy of ASEA)*

In a presentation at Robots 11 Ronald Meyer listed potential robotics applications for AI.[22] This list, summarized in Table 11-1, covers just about every area within Manufacturing. Although there is no doubt that AI is applicable to each of these areas, in robotics at least several steps remain to be solved. The intelligent robot will need the capability for visual discrimination, understanding voice inputs, learning, and self-maintenance, at least to the level of reporting what is wrong.

As a minimum, the following constraints must be met:

1. The robot must be able to detect and recover from errors.
2. Calculations must be fast enough to be performed in real time without slowing down the robot.
3. Programming capability must be flexible and capable enough to perform on a wide variety of functions.
4. The cost must be low enough to make the system cost-efficient.

Unfortunately, it will take years before the status of AI programs will allow these constraints to be met.

Table 11-1. AI Applications to Manufacturing

Expert systems	Business planning
	Design
	Process control
	Equipment diagnostics
	Scheduling
	Requirements planning
	Shop floor control
	Materials selection
Robotics	Material handling
	Welding
	Machine loading
	Assembly
	Spray painting
Natural language understanding	Data base retrieval
	Factory communications
Voice recognition	NC programming
	Voice data entry
	Robotics control
	Factory security
Speech synthesis	Human–machine interfaces
	Control systems
	Security systems
Machine vision	Inspection
	Recognition
	Measurement

REFERENCES

1. Hall, Ernest L., and Lewis J. Pinson. *Intelligent Robot Systems,* SPIE Symposium. Cambridge, Mass., 1984, p. 2.6.2.
2. Ibid., p. 2.6.4.
3. Barr, Avron, and Edward A. Feigenbaum, eds. *The Handbook of Artificial Intelligence,* Vol. 1. Los Altos, Calif.: William Kaufman, Inc., 1981.
4. Davis, R., and J. King. An overview of production systems. *Machine Intelligence 8,* E. W. Elcock and D. Michie, eds. New York: Wiley, 1977.
5. Brownstone, Lee, et al. Programming expert systems in OPS5. *An Introduction to Rule Based Programming.* Reading, Mass.: Addison-Wesley, 1985.
6. Webster, Robin. Expert systems on microcomputers. *Computers & Electronics,* March 1985, p. 69.
7. Kaufmann, Arnold, and Madan Gupta. *Introduction to Fuzzy Arithmetic.* New York: Van Nostrand Reinhold, 1986.

8. Hough, P. V. C. Method and Means for Recognizing Complex Patterns, U.S. Patent #3,069,654 (1962).

9. Casasent, David, and Raghuram Krishnapuram. Detection of target trajectories using the Hough transform. *Applied Optics,* Vol. 26, No. 2, Jan. 1987, pp. 247–251.

10. Laugier, C., and J. Pertin-Troccaz. SHARP: A system for automatic programming of manipulation robots. *3rd International Symposium on Robotics Research.* Gouvieux, France, Oct. 7–11, 1985.

11. Kirschbrown, R. H., and R. C. Dorf. Karma: A knowledge-based robot manipulation system. *Robotics,* Vol. 1, No. 1, May 1985, pp. 3–12.

12. Park, Jack. Forth and AI. *Computers & Electronics,* March 1985, p. 72.

13. Pountain, D. Parallel processing: A look at the ALICE hardware and Hope language. *Byte,* Vol. 10, No. 5, May 1985.

14. Warren, D. H. D. *Implementing Prolog—Compiling Predicate Logic Programs,* Technical Reports #39 and #40. University of Edinburgh, 1977.

15. Goldberg, Adele, and David Robson. *Smalltalk-80: The Language and Its Implementation.* Reading, Mass.: Addison-Wesley, 1983.

16. Harris, K. The FORTH philosophy. *Dr. Dobb's Journal,* No. 59, Sept. 1981.

17. Thornburg, David D. *Discovering Apple Logo: An Invitation to the Art and Pattern of Nature.* Reading, Mass.: Addison-Wesley, 1983.

18. Fikes, R. E., and N. J. Nilsson. STRIPS: A new approach to the application of theorem proving to problem solving. *Artificial Intelligence,* Vol. 2, 1971, pp. 189–208.

19. Salisbury, J. K., and J. J. Craig. Articulated hands: Force control and kinematic issues. *International Journal of Robotics Research,* Vol. 1, No. 1, 1982, pp. 4–17.

20. Meyer, Ronald J. Applications of Artificial Intelligence. Presentation notes from tutorial at Robots 11, Chicago, Ill., April 1987.

21. Stauffer, Robert N. Artificial intelligence moves into robotics. *Robotics Today,* Vol. 9. No. 5, 1987, pp. 11–13.

22. Meyer, Applications of Artificial Intelligence.

PART
IV

ROBOTIC
APPLICATIONS

Applying robotic systems to the wide range of tasks found in industrial environments requires the consideration of a number of factors. Perhaps the most important to be considered are in the area of standards and interfaces, until recently an area overlooked as each manufacturer tried to force its own ideas into the workplace. With the support of the Robotics Industry Association and various national organizations, progress is now being made in all areas of standards. Chapter 12 covers robotic standards, and its related area, robotic interfaces.

Chapter 13 covers applications engineering and includes the planning, justification, and installation of robotic systems. The chapter concludes with a discussion of safety-related techniques.

In chapter 14, we examine some of the design issues found in specific types of applications. Requirements imposed on robotic systems change as a function of the application. Knowing how the robot will be used is the first step toward knowing what design specifications are important.

ROBOT STANDARDS

Standards are important in designing any complex computer-based product. In the early days of computers, the standardization of languages allowed applications written for one machine to be transported to another. The standardization of data communications allowed machines to share information. On the other hand, the lack of standardization of power-line frequencies, voltages, and receptacles between countries has hurt sales, since it requires special adapters or country-specific models.

Although there was little standardization in the first 20 years of robot system design, standardization activity has increased significantly in the last five years. Safety standards were logically the first to be formally adopted, with Japan leading the way. Other standards have since been developed, especially in areas such as robot testing and communications, with many other standards in a draft and review cycle.

This chapter discusses the status of many current standards programs. Although the information is primarily applicable to industrial robots, related standards, such as communications, clean room, and AGV, are included.

12.1 JAPAN INDUSTRIAL ROBOT SAFETY STANDARDS

The primary purpose of safety standards is to eliminate or reduce industrial accidents. Few injuries or deaths from robots have occurred. In July 1982 the Japanese surveyed industrial robots installed in 190 workplaces. Accidents and near accidents reported are shown in Table 12-1. Eleven people were injured in these accidents; two of these people died, and two others were permanently disabled.

Table 12-1. Accidents in 190 Japanese Workplaces

Year	Number of Accidents	Number of Near Accidents
1978	2	2
1979	2	0
1980	0	2
1981	6	13
1982	1	20
Total	11	37

Using other information collected in this survey, the Japanese Ministry of Labor published the *Technical Guideline on Safety Standards in the Use, etc. of Industrial Robots* on September 1, 1983. These guidelines were expanded and explained in a document published in 1985.[1] They cover the construction, installation, and use of robots, joint man-machine operations, inspections, and training. The following paragraphs summarize part of this standard.

1. A most important requirement is the emergency stop button must have direct control over shutting down robot activity by disabling the primary power. In other words, emergency stop must not depend on the use of programming or interrupts. This precaution is wise, because if the robot malfunctions there is no way to ensure that an emergency stop program would be executed or, if executed, would halt the robot.

2. The holding capability of a robot must not be reduced by an emergency stop or a lack of power. There should be no danger of the robot dropping an object. Mechanical means or battery-operated power supplies can be used to maintain the robot's grip.

3. Activation of the emergency stop button must turn on an alarm light. Alarm lamp color recommendations are

Red—robot in run mode
Yellow—robot in emergency stop
Blue—robot in teach mode
White—robot waiting on supply of material
Green—stoker full, robot not in run mode

4. It is customary for the robot to operate at a slower speed in teach mode. This standard proposes a maximum value of 140 mm/s as the teaching speed (typical operating speeds are 1 m/s).

5. The major environmental provision is that the robot have an explosion-proof capacity if it is installed in an area containing inflammable vapors, combustible gases, or combustible dust.

6. There must be means to protect workers from accidentally approaching too close to an operating robot. Possible methods include stationing a watchman with authority to prevent entry and enclosing the robot in a railing.

7. If an enclosure is used, it must have only one entrance and be provided with a safety device or interlock to actuate an emergency stop if someone enters. Safety devices that include switches operated from safety mats, photoelectric sensors, or ultrasonic/infrared human sensors may be used in place of an interlock system.

8. When operations must be within the robot's movable range, other requirements for speed, existence of a dead-man switch, and proper training are specified. In local operation, plugging in a teach box must automatically disconnect the main operating panel (except for its emergency switch) so that another worker cannot start or control the robot if someone is near the robot. Any worker who approaches an operating robot should carry an emergency stop control, or else a second worker, with access to an emergency stop button (buddy system), must be stationed where he or she can watch the robot and the first worker.

9. Periodic inspections are required to check emergency stops, braking devices, and interlocks, damage to wires, abnormalities in operation, and unusual noises or vibrations.

10. The robot must be able to detect many abnormal conditions and shut itself down automatically. Abnormal conditions include memory parity errors, power supply overload conditions, power supply failures, overtravel or overlimit conditions, watchdog timer timeouts, excessive end-point errors, and overheating.

11. Operator training, both theoretical and practical, is required, with a minimum of 9 hours of theory and 4 hours of practical instruction.

12.2 RIA STANDARDS PROGRAM

The Robotics Industry Association (RIA) represents most manufacturers in the United States, and they have established subcommittees in six areas, under their R15 executive committee on robotic standards. RIA also serves as the administrator of the U.S. Technical Advisory Group to the International Standards Organization (ISO).*

---

* Information in this section was adapted from an article by Prange and Peyton[2] and updated through discussions with Jim Peyton.

R15.01 *Electrical interface standards:* This subcommittee submitted a standard entitled "Color Coding of Electrical Wiring for Industrial Robots" to the American National Standards Institute (ANSI), the U.S. organization responsible for issuing and revising standards. That standard is in the final stages of release approval. A second standard, "Flex Life of Cables and Cable Assemblies for Industrial Robots," is being drafted.

R15.02 *Human interface standard:* A standard entitled "Human Engineering Design Criteria for Hand Held Robot Control Pendants" has been released to ANSI for final approval.

R15.03 *Mechanical interface standard:* This group is working on a standard for interface flanges, located between the end-effector and robot wrist. The standard will improve part interchangeability by ensuring that parts manufactured by different companies still match. The proposed draft standard is entitled "Circular Mechanical Interface Standard for Industrial Robots."

Another end-effector standard, the ISO/TC184/SC5, is being used as a guideline.

R15.04 *Communications/information standard:* This standard, "Industrial Robot Messaging Standard," has been completed and submitted to ISO, the International Standards Organization. It will be used at the open systems interconnection level (application layer 7) of MAP (discussed later). It is intended to be a companion standard to the EIA proposed National Standard RS-511 ("Manufacturing Message Service for Bidirectional Transfer of Digitally Encoded Information"). This latter standard defines the services and protocol provided by the application layer and extends the scope of this layer to numerical controllers, robots, process control system, and vision systems.

R15.05 *Performance standard:* The objective of this standard is to determine true performance of an industrial robot in such areas as accuracy, compliance, cycle time, payload, repeatability, overshoot, and tolerances. It specifies a standard test path. A "Robot Performance Criteria" working paper has been prepared and approved as a draft standard. This performance standard has also been sent to ANSI for review and release.

R15.06 *Safety standard:* This standard is the first U.S. standard approved by ANSI (June 1986, under the title *ANSI/RIA R15.06 1986—American National Standard for Industrial Robots and Robot Systems—Safety Requirements*), and it is currently being revised. This standard covers 12 areas. In some areas, the standard is very good. In others it may not be sufficient to prevent injury, particularly in operations with an operator present. A summary follows.

1. The release of stored energy, such as hydraulic pressure release after a power failure, must be controlled.

2. There must be a means of indicating operating status. (The operator should not have to guess whether the robot is stopped temporarily or permanently.)
3. The emergency stop must be under complete hardware control, override all other controls, and remove drive power.
4. The robot must not be capable of being placed in automatic operation while the teach pendant is plugged in.
5. Perhaps one weak point is the coverage of slow-speed operations. The selected maximum slow speed is not only higher than the Japanese suggest (250 mm/s vs. 140 mm/s), but it also allows even higher speeds under teach control if the operator selects them. (In this situation the operator better not get careless.)
6. There must be a lockout/tagout means to shut off power to the robot, which must be located outside of the restricted work envelope.
7. The restricted work envelope must be conspicuously identified, such as by stripes on the floor.
8. Although devices that can sense a person's presence are discussed, unfortunately the standard requires only a simple device, such as a warning sign. This point is quite different from the Japanese requirement that there be an automatic shutdown when an operator approaches the unit.
9. During some types of maintenance or repair, the standard allows the bypass of safety devices.

RIA has also published a glossary and will keep it up to date. It will serve as the basis for a proposed American National Standard (ANS) robotic glossary.

The Automated Vision Association (AVA), a part of RIA, has established five standards subcommittees:

1. A15.01, Communication Standards and Protocols between Machine Vision Systems and Other Devices
2. A15.05, Performance (standard issued)
3. A15.07, Terminology (prepared draft glossary)
4. A15.08, Sensor Interfaces (prepared a draft connection standard and a draft interface standard)
5. A15.09, Marking and Labeling of Items to Aid Machine Vision (prepared draft specification)

The only vision standard currently released is ANSI/AVA A15.05/1—1989, entitled "Automated Vision Systems—Performance Test Measurement of Relative Position of Target Features in Two Dimension Space." It was released on December, 1988.

Most of the robot and vision standards are still in the discussion and planning stages. A few were published in draft form during 1987. One was officially issued in 1987, with two more issued in 1988.

12.3 TESTING STANDARDS

Some progress is being made in testing. Underwriters Laboratories has developed a new standard (UL 1740) that will evaluate robot design against fire, shock, casualty, and explosive hazards. It covers welding, paint spraying, assembly, transfer, and inspection robot systems. It also includes requirements in robot construction to reduce hazard levels in several areas. Tests are done on the robot under electrical surge conditions, power failures, and weight overloads. Evaluations are made on all electronic safety circuits and sensors.

The American Society for Testing and Materials (ASTM) is also looking into testing standards. Becker[3] identifies the following standard topics that ASTM recommends be tested:

Repeatability	Center of gravity
Payload deformation	Velocity
Payload capacity	Workspace footprint
Accuracy	Compliance
Power requirements	Overshoot and undershoot
Resolution	Articulation
Dither	Path errors
Slip in joints	

Two machine standards that have been available for several years are the National Machine Tool Builders Association (NMBTA) *Definition and Evaluation of Accuracy and Repeatability for Numerically Controlled Machine Tools* (2nd ed.), 1982, and the *European Machine Position Accuracy Standard* (VDI 3441). Both standards provide useful information but are inadequate for modern robots because of differences in work envelopes, weight, and movement.

12.4 OTHER STANDARDS ACTIVITY

1. *AGV standards.* The oldest AGV safety standard is FEM4.009C, published in 1979 by the Fédération Européenne de la Manutention (FEM), which covers driverless industrial trucks. Another AGV standard is ANSI B56.5.1978 *Electric Guided Industrial Tow Tractors.* Updates are being considered for both documents.

2. *IEEE (Institute of Electrical and Electronic Engineers).* The IEEE has supported many standards that are applicable to robotics systems, among them ANSI/IEEE 802.4-1985, *Token-Passing Bus Access Method and Physical Layer Specifications.* This standard has been adapted as part of the ISO Interna-

tional Standard 8802/4, and is directly applicable to MAP. Another IEEE standard of interest is the IEEE-488 instrument bus.

3. *ISO.* This organization is working on a number of standards. In addition to the end-effector draft standard mentioned earlier, the ISO is circulating a suggested standard proposed by West Germany. Entitled *Safety Requirements Relating to the Construction, Equipment and Operation of Industrial Robots,* ISO report number ISO/TC 97/SC 8/WG 2N32, 1983. A third ISO standard that is available as a working document covers performance and testing: *Manipulating Industrial Robots—Performance Criteria and Related Testing Methods,* ISO/TC184/SC 2/WG 2 N7, Rev. 1, 1986.

4. *NMTBA.* The NMTBA is developing language and definitions for safety requirements for automotive manufacturing systems and work cells.

12.5 DEVICE COMMUNICATION STANDARDS

Many robot sensors, especially those without associated logic circuitry, are interfaced with the robot's computer via discrete I/O lines. There is no standard here, and modules to support single I/O lines are available to accommodate voltage levels from 0 V (contact closures) through 5 VDC (TTL levels) up to 24 VDC and 120 VAC.

The next step above single lines is a parallel port interface, with eight or more data lines and several control lines. In theory, a parallel interface should be the best and quickest method of handling sensor I/O, but there is little standardization and little specialized peripheral driver software available.

The first level of truly standardized I/O is the RS-232 serial communications, long used by computer manufacturers to communicate with peripheral units (such as terminals) and other computers over a distance. RS-232C standards imply three levels of data standards, although only the first type is actually included in RS-232C:

1. *Hardware connection level.* Specific pins are assigned specific functions, and minimum and maximum voltage levels are established. Table 12-2 lists these standard connections for a data terminal equipment (DTE) device. The DTE standard is applicable to robot computers; however some manufacturers use a few other lines for nonstandard purposes.

2. *Data transmission level.* Start/stop timing, transmission speed, parity, and bit configuration may be adjusted as necessary for different applications. Figure 12-1 shows some available alternatives. Unlike the connection level, the system designer has multiple choices available, although both ends of the communication link must agree on the same selection.

Table 12-2. Standard RS-232C Signals

Pin	Name	Mnemonic
1	Protective ground	
2	Transmit Data (output)	TXD
3	Receive Data (input)	RXD
4	Request to Send (output)	RTS
5	Clear to Send (input)	CTS
6	Data Set Ready (input)	DSR
7	Signal ground	
8	Signal Detect/Carrier Detect (input)	CD
20	Data Terminal Ready (output)	DTR
23	Data Signal Rate	

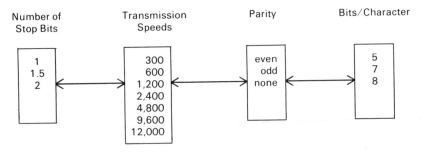

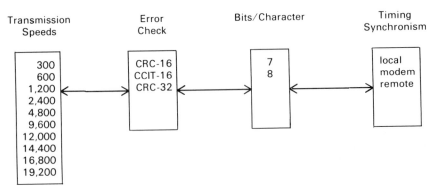

Figure 12-1. Data transmission alternatives.

3. *Software protocol level.* Methods of signaling are used to communicate between devices. Some examples used in asynchronous communications are ACK/NAK and XON, XOFF. In synchronous communications FLAG characters and PN signaling sequences may be used as part of HDLC.

There is also a standard for data formats on the lines. The American standard code for information interchange (ASCII) data formats are supported by most systems. However, only 7 bits are defined in ASCII, and most data is transmitted in 8-bit bytes. The format of the remaining characters (when the eighth bit is 1 instead of 0) is not standardized, and many variations are found. Some companies use these codes for graphics purposes, others for foreign alphabets, and still others for special control characters. Many systems also employ binary formats for transmitting data files. In this case, special provisions must be made to handle control character transmission.

Other communication standards include CCITT international standards, such as X.25 and V.24, and U.S. standards RS-422 and IEEE-488. Although they are applicable to communications in general, they are not used as often for local communications between robots or between a robot and the central computer. At the local level, the robot is often tied into a network, as discussed in the next section.

12.6 NETWORK STANDARDS

A significant part of the costs of installing a complete computer-integrated factory is communications. Standardization will lower these costs. Many vendors provide robots and related equipment to automate various functions in the manufacturing arena, thus creating islands of automation. However, when systems of one vendor cannot interface with systems of another vendor, it is difficult to control and coordinate any factory-wide system.

Early attempts to get around a lack of standardization included customized hardware interfaces and specialized gateways, both quite expensive. Some standards have gradually evolved, and we will discuss three: Ethernet, MAP, and SECS. There is some industry disagreement over the relative merits of Ethernet and MAP for robotic applications.

Ethernet

Ethernet is a local area network originally designed at Xerox's Palo Alto Research Center and later developed as a standard through the joint effort of Digital, Intel, and Xerox (often called the DIX standard). It is a broadcast bus-type system interconnected with coaxial cables.

In a broadcast approach, each transmitter that has a message first listens to see if the line is clear. When it detects no transmissions, it transmits its own message. Since another transmitter might also select this same time for its own broadcast, simultaneous transmissions (causing packet collisions) are possible.

Collision detection capability is provided to immediately stop competing transmissions, thus improving system efficiency and throughput. The Manchester data transmission technique is used to ensure that at least one signal transition occurs per bit time, thus improving error detection. Each transceiver taps into the network, loading it as little as possible. The transceiver performs carrier detection, collision detection, and collision consensus enforcement. After any collision, later transmissions are rescheduled randomly, thus reducing the possibility of another collision. The transceiver must have a watchdog circuit to disconnect the transceiver from the network in case of malfunction; otherwise the entire net could fail after a single local malfunction.

The maximum span of an Ethernet is 500 km (about a third of a mile), with transmission at 10 Mbps (megabits per second). Packet repeaters and packet filters can be used to expand the network. The format of the packets contains a 64-bit protocol, 48-bit destination address, 48-bit source address, 16-bit type field, 64- to 1500-byte data field, and 32-bit CRC (cyclic redundancy check) field.

Ethernet has been used in many robot installations. Figure 12-2 shows one way of tying together a series of robots on an assembly line using Ethernet.

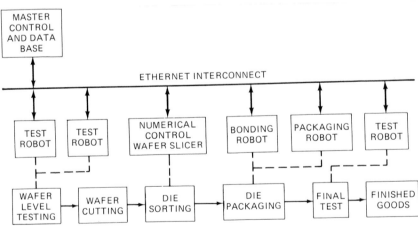

Figure 12-2. Ethernet system example. (*From W. H. Holzbock, Robotic Technology: Principles and Practice. New York: Van Nostrand Reinhold, 1986, p. 331; copyright © 1986 by Van Nostrand Reinhold*)

Manufacturing Automation Protocol (MAP)

An emerging standard getting a lot of support is MAP. General Motors led the drive to standardize robotics communications by creating a task force in 1980 to develop a factory communications specification. This effort culminated in a demonstration at the 1984 National Computer Conference, where seven vendors (AT&T, DEC, Gould, Hewlett-Packard, Honeywell, IBM, and Motorola) demonstrated that their equipment could be interconnected with this new protocol. The following year, 19 vendors offered MAP standard equipment. By 1986 there had been successful implementation in at least four GM plants, as well as at Kodak and Deere.

The MAP protocol has gone through several variations,[4] and MAP 3.0 is the current de facto standard. There is still a lack of a complete MAP specification, however, and many vendors use slightly different implementations. MAP has been designed to fit into the seven-layer ISO/OSI model (OSI = open systems interconnect). Table 12-3 illustrates this layered approach, fundamental to most network protocols.

Examples of available software include the network module MicroMAP, a full implementation of MAP 2.1 developed by Motorola. This software provides users with the important middle layers (data link, network, transport, and session) requiring the user to provide only the necessary application packages at the top end and the hardware interface at the base end. MicroMAP can be developed under such environments as SYSTEM V/68, an AT&T validated equivalent of UNIX for the M68000 microprocessors.

By early 1987, there were 12 MAP installations in the United States. Robot

Table 12-3. OSI Reference Layers (as used by MAP)

Level	Layer	Typical Standard	Function
7	Application	EIA-RS511	User programs and network utilities
6	Presentation	Unspecified	Translation of data
5	Session	ISO-IS8326/IS8327	Communications management/ synchronization at the highest network level
4	Transport	ISO-IS8072/IS8073	Data transfer between end points
3	Network	ISO-DIS8473	Message routing
2	Data link	IEEE-802.2 LLC Class 1	Framing, error control
1	Physical	IEEE-802.4 10-Mbps broadband	Isolation, encoding, modulation

manufacturers began supporting MAP. Yarbrough and Krout provide insight into modifications required on existing VAL-II robots to fully support MAP.[5]

There is some concern that MAP may be too complex for small firms or that other protocols (such as DDCMP) and other networks (such as Ethernet) may be sufficient. Another problem is that some groups (including the European CNMA implementations) are targeting their activities toward MAP 3.0, which was just released, although many products are still using MAP 2.2. Unfortunately, 2.2 is not software nor hardware compatible with 3.0, so many manufacturers have decided to support the latest version only. The major difference between the two is in the protocol. Originally, the Manufacturing Messaging Format Standard was adopted. This was changed to Manufacturing Messaging Specification in MAP 3.0.

SECS

Another communications standard, often specified for clean-room applications, is SECS. This protocol defines the formats and semantics for communicating devices used in semiconductor manufacturing. SECS-I is somewhat similar to ISO layers 2 through 5 and defines a block transfer protocol with a 10-byte header that follows the block length byte. A maximum of 244 data bytes then follows, ending with a 2-byte CRC. In these applications, RS-232 is used as the physical interface (layer 1).

12.7 SAFETY

All devices and tools are potentially dangerous: the stronger the device, the larger its reach, the more attachments it has, then the more dangerous it is. In 1981, the first worker was killed by a robot at Japan's Kawasaki Heavy Industries plant. In 1984, a robot killed a man at a Ford Motor Company factory, resulting in a $7 million judgment. Numerous injuries have also occurred.

Safety must be a prime goal of robot designers, robot users, and industrial workers. Three areas of safety must be included in all standards.

1. *Built-in safety features:* The engineer must consider safety as an important design feature. Systems should be built to fail-safe, and safety interlocks should not be easily defeated just to make a test technician's job simpler.

2. *Safety barriers:* Barriers are meant to keep people out. The company installing the robot and the company receiving the robot should both insist that adequate safety barriers be installed. At least one major robotics vendor claims in its literature that safety is the responsibility of the purchaser. Not so! It is the responsibility of the purchaser *and* the vendor.

3. *Safety training:* Anyone who will have close contact with robots should be trained in their use, particularly in related safety procedures. Someone in the factory must be given the responsibility to develop a continuing training program. This program should be given periodically to both new and experienced personnel, and should be updated as new equipment, procedures, or processes are added.

Section 13.4 discusses safety techniques that can be used.

REFERENCES

1. *An Interpretation of the Technical Guidance on Safety Standards in the Use, etc. of Industrial Robots.* Tokyo: Japanese Industrial Safety and Health Association, 1985 (available in English).
2. Prange, James M. and James A. Peyton. Standards development. *Robotics Today,* Vol. 8, No. 6, Dec. 1986, pp. 23–24.
3. Becker, Stanley E. Development of mechanical performance standards for robots. *Proc. National Bureau of Standards/U.S. Navy Workshop on Robot Standards.* June 6–7, 1985.
4. *Map Specification Version 2.1,* GM MAP taskforce. GM/APMES, Warren, Mich., 1985.
5. Yarbrough, Grahan G., and Francis H. Krout, Jr. Robot controllers: A case study of MAP, SECS and VAL-II. *Proc. Robots 11 Conference.* Dearborn, Mich.: Robotics International of SME, April 1987, pp. 20-15–20-43.

APPLICATIONS ENGINEERING

Unlike previous chapters, which have concentrated on the system and design issues involved in developing robots, this chapter discusses the application of robot technology to industry. We begin with the steps necessary for planning, selecting and installing robots. That presentation is followed by an examination of those issues which might be faced by a company considering the installation of a seam-welding robot. That example was chosen because it is illustrative of many other types of robot installations. Some special issues involved in work cell systems are covered, and we end with an examination of safety in the plant, including needs, approaches, and available techniques.

13.1 SYSTEMS ANALYSIS

Deciding about robot installation involves two stages that are separated in time and separately addressed. The first stage includes those activities to be done before the robot is ordered, and includes planning, justification analysis (is an installation warranted?), and robot selection. The second stage covers installation-related activities.

After a company decides to investigate adding robot technology to its manufacturing operations, many steps are required before any purchases should be made. Of these steps, perhaps seven are the most important: system planning; robot justification; robot selection; specification and test; installation; documentation and training; and planning for success.

System Planning

Critchlow[1] discusses 10 steps in robot system planning:

1. As part of the justification process, establish the system goals.
2. Determine system selection criteria. Are they to be based on economic, quality improvement, improved working conditions, productivity, life cycle costs, or other considerations?
3. Determine system requirements, including throughput, accuracy, and reliability.
4. Determine system specifications (weight handling, peak/average velocities, number of axes, end-effector requirements, controls required, auxiliary equipment needed).
5. Consider integration into existing production facilities. Is there a need for CAD, CIM, and other computer interfaces?
6. Encourage wide review of the specifications and the planned implementation. All levels and all departments should be included.
7. Determine documentation requirements [and ensure they are met].
8. Develop a training plan [particularly for safety].
9. Plan the organizational structure necessary to implement the operation.
10. Define acceptance test procedures and requirements.

With a well-planned system, the robot can be a powerful manufacturing aid that can both increase productivity and free humans from laborious environments. Unless a firm has strong in-house robotics experience, it will probably find that a turnkey system is preferable to a system purchased from several suppliers and assembled in house.

Robotic consultants offer valuable assistance to firms without engineering staffs, or whose staffs have limited experience in robotics. The successful purchase and installation of an industrial robot requires that the entire process be planned and carried out in a logical sequence. Robots have unique properties and companies have unique requirements. Both of these must be considered to achieve a proper match. Consultants can assist in evaluating factors such as whether it is practical, advantageous, and cost-effective to automate the manufacturing operation. They can also assist in the selection of appropriate robots and their installation, and in the training of personnel.

If a complete system, rather than an individual robot is being considered, additional factors must be considered. A complete system can range from a single robot with associated vision equipment and necessary data communications to a complete work cell, containing several robots, material handling equipment, central computer, and management software. Venkataraman[2] suggests five steps to provide a methodology for designing a flexible manufacturing system.

1. *Data collection and analysis:* Define the present product design and manufacturing processes.
2. *Manufacturing technology survey:* Determine the available alternatives to material handling, process equipment, hardware and software, and management control systems.
3. *Cost justification:* Consider both tangible and intangible cost savings resulting from benefits such as space savings, better quality, worker safety, and morale improvements.
4. *Hardware and software specification development:* Prepare specifications covering the system's intended use to ensure that selection will be based on meeting the company's minimum requirements at the least. This step is also important when deciding whether to purchase standard or custom-designed equipment.
5. *Vendor selection:* Evaluate a vendor's past experience, capacity, financial status, and after sales-support record.

Robot Justification

Questions that need resolution in the area of justification include: Is the proposed robot application one that should be robotized? For example, in welding applications a robot installation is generally justified only if the following six criteria are all met:

1. An orderly environment
2. A repetitive operation
3. Cycle time over 3 seconds
4. No judgment required by the robot
5. Uniform parts to be welded
6. Installation meets the company's investment criteria

Other questions are: Can it be cost-justified? What additional information is required? What are the steps in the implementation process? What changes will be necessary to plant facilities or operations?

Robot justification must be based on realistic information. Robots should be installed only if the circumstances—whether economic, safety, quality, or other—justify it. If it can be shown that a robot will save money, that is usually sufficient reason to justify the purchase. However, it is almost impossible to determine in advance all of the costs associated with installing the robot and all of the benefits to be gained. It becomes even more difficult to trade off noneconomic issues.

The first step in any justification analysis is to study all of the impacts, pro and con, that the robot installation will have. Once the study is complete, a trade-off list showing both sides of this issue can be presented to management for review. In many cases, the decision is not clear cut. Management must weigh the various factors, consider the uncertainties, and make the final decision.

One factor worth considering is the experience of other firms. Have robots proven worthwhile to other manufacturers? How many have regretted their decision to go ahead? One study has reported a ratio of 27 users satisfied with the installation of robots in their factories for every user dissatisfied.

Cost Considerations

Both anticipated cost savings and expected cost increases need consideration. The most obvious additional costs are those associated with the robot: the price of the robot, shipping expenses, and the cost of necessary accessories (grippers, tools, vision system). Price comparisons between vendors are difficult because robot capabilities, features, and accessories usually vary. In addition, price comparisons must consider support, particularly the level of warranty, maintenance, spare parts, and training offered by the robot supplier.

Other expenses include installation, plant modification, necessary fixtures, and related safety equipment. Costs of required engineering planning and support activities, lost production time during installation and start up, and training time and expense must also be considered. In addition to those one-time expenses, there are on-going costs, such as maintenance, as well as reprogramming and retooling costs when production requirements change. Indirect costs should also be considered. If robots are going to free workers for other tasks, the costs of retraining and/or early retirement must be evaluated.

Robotics workstations have different types of overhead than human workstations. Just as certain costs (such as installation) must be added to a robot's price to determine its true cost, others (such as fringe benefits) need not be included. It is easy to recognize that a robot will save fringe benefit costs, since robots do not get benefit packages. Other overhead expenses are more difficult to determine, such as how to factor in the cost of heating a plant. Obviously, the amount of heat necessary for robot workers will be less than that for humans, so heating costs will be lower, unless both must share the same space.

One expense often overlooked results from the increased productivity of robots. Adding a robot could increase inventory, both in final parts and semi-finished goods. Additional space is needed and material handling increases.

Countering all of the expenses of adding robots are the many cost advantages. For example, the direct economic benefits potentially include:

• Reduced labor cost

• Higher productivity

- Less scrap and rework
- Less handling costs
- Lower inspection costs

Indirect cost savings result from:

- Decreased insurance costs
- Reduced management involvement
- Improvement of product quality leading to more sales and fewer returns

It is also important to consider three noneconomic but major benefits:

1. Removal of personnel from boring or hazardous jobs
2. Compliance with OSHA requirements
3. Flexibility and growth potential

Application Considerations

A number of applications make excellent candidates for robot installations. (Chapter 14 discusses application areas and their associated requirements.) The most obvious are loading and unloading machines (i.e., die casting), handling parts in manufacturing, welding, spray painting, assembly, machining (deburring and drilling), and inspection. Since robots can support several machines at once, material handling is also an obvious candidate.

The use of robots is ideal for unhealthy and unpleasant environments, such as the painting booth. Solvents are toxic, and the air discharge from the painting nozzle can be noisy and cause ear damage. In addition to removing humans from an unsafe environment, the use of robots reduces the amount of ventilation required, the amount of paint used, and the reject rate, all of which contributes to lower costs. Another example of dangerous and dirty work that can be performed by robots is fettling, the process of removing excess metal in foundries. Robot spot welding can be quite cost-effective, especially if numerous spot welds must be done. Robots are better than humans in accuracy, uniformity, and positioning of weld.

However, in spot welding as well as in other potential robot applications, some savings may never materialize. For example, suppose a company employing only a single welder is considering robot welding. If this company must retain its welder for occasional difficult welds, or if a robot operator/ technician must be added to the staff, it is obvious that adding a robot would increase and not reduce labor costs. But, even in this example, there may be other important reasons to justify a robot. For example, productivity/through-put may be raised significantly. Product quality could go up. The ability to meet certain close tolerance welding specifications might increase market opportunities.

When considering the application, the environment that the robot will be placed in must also be considered. Some environments are rougher than others on robots, and some robots are more sensitive to environmental prob-lems. Clean-room applications (discussed in chap. 14) strongly influence ro-bot selection.

Payoff Considerations

If we look at cost factors alone, justification can be based on payback period (how soon will the robot pay for itself?), return on investment (net savings for the investment), or the discounted cash flow method (which considers the cost of money).

Most robot systems for small manufacturers should offer a payback within two years. Factors affecting payback include the availability of two-shift opera-tion, reduced scrap costs, reduced rework costs, a simpler method of inspec-tion, and higher product quality. If the robot will take two years to pay for itself, the cost of the money advanced during this period should also be considered.

Troxler and Blank[3] have prepared a justification methodology that identi-fies relevant decision factors and determines the value of the planned robot system. Their approach provides for final choices of either human workers, individual robots, or complete flexible manufacturing systems (FMS). Table 13-1 summarizes four areas that should be considered and lists three typical factors in each area that serve as trade-off indicators. For example, part orien-tation requirements are listed under tooling complexity. Parts may need no orientation if they are symmetrical or if they have been oriented at an earlier

Table 13-1. Factors in Robot Justification

Task Complexity	Tooling Complexity	Related Operations	Work Environment
Number of parts	Part orientation requirements	Synchronization	Monotonous
Number of operations	Ease of handling	Bottlenecks	Dangerous
Manipulation difficulty	Supporting fixtures	Coordination	Fatiguing

station; they may need minimum orientation, which can be done through gripper manipulation; or they may require vision system assistance to determine correct orientation. Depending on which of these conditions applies to the selected application, different requirements, and thus different costs, are imposed on the robot system.

Selecting the Robot

Selecting a robot requires more than picking a company. What specific requirements must be met? In what area, if any, is customization of the robot required? Which competing systems should be examined? What training and services do these companies offer?

The choice of robot should be made with care, as the wrong choice could impact total system costs and not provide the anticipated cost benefits. Worse yet, the wrong robot might not even be able to perform the intended task with sufficient speed or accuracy, or it might not handle that new product planned for the future.

The following topics are examples of some of the information needed for robot selection:

- Type of task

- Weight to be handled

- Special grippers or tools required

- Complexity of task

- Ease of programming or training the robot

- Work envelope needed

- Accuracies required

- Special environmental hazards present

- Expansion considerations

A few robots may offer extra features, but which of them are really required and which can be deleted, thus saving costs?

In selecting a vendor, it is important to talk to several of a vendor's customers that have successful robot installations, particularly if their applications are similar to yours. Try to see existing systems in operation, and ask appropriate questions. It is also important not to overlook the different levels of help that different vendors offer in areas such as maintenance, locally available spares, diagnostics, and operator training.

Specification and Test

Unfortunately, there are currently no standards for robot specifications, test, or calibration, although as mentioned in chapter 12, a draft performance standard was circulated by ANSI in 1988. Without complete standards, it is even more difficult to compare robots or to test the robot after it is purchased to ensure it is working satisfactorily.

The importance of the user testing robots as soon as they are delivered, and before they are put into production, was stressed in a speech to the RIA in 1986 by R. J. Piccirilli, Jr., Chrysler Motors' Director of Manufacturing. Mr. Piccirilli recounted the experience his firm had with 200 robots, purchased from various manufacturers.[4] Of the 200, 50 broke down even before being placed in test, and none of the others (including American, Japanese, and European) were able to pass a 50-hr test. Further, in tests done at vendor plants before shipping the robots to Chrysler, there were 393 breakdowns in only 48 systems, yet they were considered reliable enough to ship. Fortunately, robot reliability has increased since then.

Figure 13-1 shows the effects of various errors on robot performance. Shown are the results of joint dead zone (caused by encoder resolution or servo limitations), friction, angular/offset errors and a composite of servo hunting, offset, and angular errors. Benhabib[5] concentrates on joint error and the effect of link length changes, distortions and deflections. Critchlow[6] discusses six major items affecting positioning accuracy and repeatability:

1. The load on the robot and the force due to gravity causing downward arm deflection
2. Acceleration of heavy loads contributing to deflections
3. Drive gears and transmission systems having slack and backlash
4. Thermal effects expanding or contracting the links, particularly important in high-accuracy applications or in large robots
5. Bearing wear and "play"
6. Twisting under load of long rotary members

Many other problems or error sources can affect performance. Day[7] discusses over 20 sources of errors and divides them into environmental, parametric, measurement, computation, and application classes. Sturm[8] emphasizes the contribution of tooling, fixtures, and the work part.

Becker[9] provides a good discussion on standards for testing. Repeatability, accuracy, and resolution have had stringent definitions developed,[10] but they are not yet formally adopted by robot manufacturers.

Wodzinski[11] is a good source for evaluating and testing robots, and Roth[12] covers robot calibration research. Several companies have begun to specialize in testing robots, and they may be able to certify robot performance.

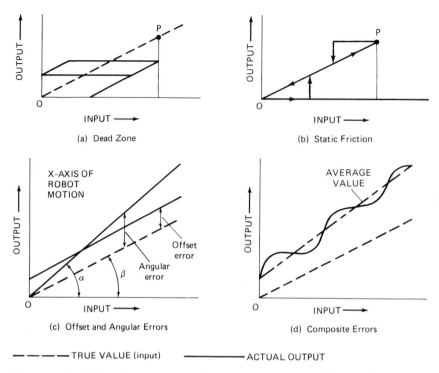

(a) Dead Zone

(b) Static Friction

(c) Offset and Angular Errors

(d) Composite Errors

— — — — TRUE VALUE (input)   —————— ACTUAL OUTPUT

Figure 13-1. Effect of various types of errors. (*From W. H. Holzbock,* Robotic Technology: Principles and Practice. *New York: Van Nostrand Reinhold, 1986, p. 5; copyright © 1986 by Van Nostrand Reinhold)*

Installation

A partial list of questions in this area includes: What preparatory work is necessary in the workplace? How will installation of the robot be handled? What provisions should be made for test and checkout? How much help will the vendor give? How should training of personnel be handled?

As part of the applications engineering tasks, a plant layout of the immediate area should be prepared, using scale drawings. The layout must include obstacles that might interfere with the robot arm in its work area, safety precautions for personnel, and provision for introducing raw materials and removing finished products. Changes in facilities are usually necessary, and they are the responsibility of the firm purchasing the robots. These changes might include moving part of the production line to accommodate the robot, installating safety equipment, and providing a supply system to the robot. Other modifications might involve provision for utilities, (electrical power and water/air if necessary), requirements for spare parts, and maintenance access.

If a standard model is ordered, robot delivery can be rapid; some firms claim to ship the day after they receive the order. However, actual installation time (including necessary testing) for a single robot usually takes a minimum of one week, and can take as long as a month in some of the more difficult installations. Plans must be made to have someone available at all times to coordinate and supervise installation activity.

Issues that might arise include scheduling of delivery and installation around production requirements, scheduling of personnel training, and installing needed safety equipment. Plans need to be made to ensure that the robot performs in an effective manner. Last-minute details such as supplying material to the robot and scheduling its use must be handled. It is also wise to schedule production start-up on a gradual basis and to make sure necessary maintenance activities (both routine and emergency) and spare part procurement have been considered.

Documentation and Training

Documentation and training are two fundamental issues that often do not get enough attention.

Too many manuals provide insufficient information, especially if the robot must interface with equipment from other manufacturers. Unfortunately, once the sale is made, the amount of help available from the seller is too often limited, and in any case it is up to the user to maintain his own equipment (or contract out for maintenance). In order to maintain equipment, update it with new features, or integrate it into systems to be installed later, clear and complete documentation is essential.

Four separate sets of documentation are generally required: installation, operation, programming, and maintenance. If insufficient documentation is available from the selected vendor, perhaps a different vendor should be chosen.

In order to use robots properly and safely, many levels of personnel must be trained. Insufficient training for key people who must operate and maintain the units, and insufficient education of all individuals who will have even limited contact with the robot, have led to many problems. A lesson can be learned from the early history of computers. In those days, the training of maintenance personnel or janitors was often overlooked, with the occasional result of having the computer system turned off ("there was no one using it") and the subsequent loss of system data.

With robots, insufficient training has led to injury of personnel. Robots may look like they are shut down, and yet move suddenly, endangering an unsuspecting or poorly trained janitor. Some of this training should be available from the manufacturer, but the user is the one who is ultimately responsible.

Be sure never to install robots without providing sufficient safety training for all personnel who will be associated with the robot. (The topic of safety is covered in detail in section 13.4.)

Planning for Success

Robot installation failures can often be traced to a lack of preinstallation planning. Lack of planning may be due to inexperience, overreliance on sales claims, or a desire to hurry the installation.

Planning Checklist

The following checklist should help to flag problem areas before they develop. This advice is general and primarily aimed at the first- or second-time user, but it should still prove helpful for experienced individuals.

1. Which application will be selected for the first installation in your plant? Be careful in this selection. In general, pick the simplest application that warrants automation for your first robot.

2. Has the responsibility for in-plant decisions been clearly designated? A robot installation has ramifications in every department. Be sure that the individual in charge has enough authority and experience to consider all ramifications. Representatives from all departments should also be included in various planning meetings.

3. Is this application unique? For example, does it need specialized tools or use robots in a way not previously done? Never combine a brand new application area for robots and a brand new user of robots. Leave experimentation to the more experienced firms. If the robot supplier cannot supply the specialized tooling, fixtures, and other peripheral devices you need, you had better reconsider your choice of application (or vendor).

4. Will the robot vendor assist in all phases of the work? With limited in-house experience, you will need more help than you anticipate. This first installation should be done by an experienced robotics systems company that has made similar installations and that will work with you from before the order until after training and installation. If you have any doubts about the firm or the application, retain a good robotics consultant.

5. Are all parts of the installation completely planned and scheduled, and are the plans approved? Be sure to plan the entire operation before making committments and have the plan reviewed by as many people as possible,

including the systems house that will provide the robot. They may suggest steps you are overlooking.

6. What types of training and documentation will be provided? Examine available documentation and training courses offered by the manufacturer or system house and make sure they are thorough and included in the price. Make sure several of the employees attend these training courses. You must have more than one in-house person who is familiar with your installation.

7. How will maintenance be handled? Leave maintenance to experts. Most in-plant technical people are not equipped to handle the repair of their first robot. Either contract with the system house that installed the robot for maintenance (including preventative maintenance), or select one of the growing number of firms specializing in robot service.

Experience

Robot installations require many types of engineering expertise. Some factories have a complete selection of engineering skills. Smaller firms must choose between bringing in a robotics consultant or relying on the vendor for technical decisions and after-sales support. In some cases, components needed by the robot were built in-house, but the firm was inexperienced both with robots and the effects of various materials and tolerances, resulting in problems. To capitalize on experience, use third parties that specialize in supplying the added system components you need.

Four unexpected problems have been flagged by firms after their first installation: (1) the high cost of engineering development, (2) installation-related requirements, (3) problems found in integrating the robot with conventional manufacturing, and (4) inadequate support available from the manufacturers after the sale. Unfortunately, some manufacturers provided insufficient applications engineering to the user, and these users found they had insufficient experience to handle all the problems that developed.

Use a consultant to determine whether robots should be used, which applications to do first, and which robot to select. Always start with the simplest robot application.

Maintenance

If in-house maintenance is attempted, the approach may result in temporarily replacing the robot with a human, while the maintenance department tries to determine the problem. Once the line is running again, there is no longer any great pressure to fix the robot and robot downtime can often end up much longer than necessary. It is usually much better to contract with an experienced outside service firm.

13.2 SYSTEM EXAMPLE

System considerations are more than the sum of individual factors. This section uses welding to illustrate some actual questions involved in justifying, planning, and installing a robot. Welding has traditionally been one of the most critical and costly manufacturing processes, and one of the most difficult to automate. Yet the benefits of automating welding are obvious: higher product quality, lower rework and rejection rates, lower materials and energy costs, and freeing personnel from a dangerous, and sometimes boring, occupation.

Welding robots were first used in spot welding, where the location of the weld was well defined. Seam tracking techniques have now been introduced, and intelligent robot welders using advanced seam trackers are now available. Robots for welding small oil storage tanks (weighing about 250 lb) will be taken as an example.

In one current factory manufacturing home fuel oil tanks, manual operations are used for the necessary seam welding. The company wished to determine the advisability of automation, considering the difficulty of finding and retaining skilled welders.

To begin the determination, a study of current operations was undertaken. For the seam welding operation, metal is cut to the appropriate size and rolled into the proper shape. The long seam along the tank is spot welded to hold it in place, and then the end pieces are spot welded on. The tanks are then divided among four manual welding stations that together produce 80 welded tanks a day, or an average of $2\frac{1}{2}$ tanks per hour. This operation involves complicated seam tracking and large objects. Five approaches were considered:

1. Use a robot welder in close cooperation with a human (a robot-assisted approach)
2. Use a limited seam tracking system with fixtures providing the necessary precise control over the position of the parts
3. Use an advanced seam tracking system with some method (such as a turntable) for roughly positioning the parts
4. Use an advanced seam tracking system with sufficient robot motion (such as through a gantry) to reach all areas of weld
5. Use several robots to speed up the welding and allow the entire area to be covered without moving the tanks.

Picking the wrong system may mean spending more money than the company needs to, since some advanced welding robots can perform additional (unneeded in this application) tasks, such as recognizing the size of the tank being welded. This information can easily be entered into the robot by an operator, so the feature is not required. On the other hand, the system might be too limited to be satisfactory or require excessive manual help and intervention. For example, some seam following systems can only deal with a single or limited range of seam types, some can only work with low-current welding arcs, and some cannot deal with workplace distortions due to heat.

Intelligent seam tracking systems currently available can provide quality welds if the initial fit is reasonably correct, and if it is able to compensate for thermal-induced part warpage during welding. Many techniques are applicable to seam tracking; some use contact sensors, some noncontact sensors, and others employ a vision system. Vision systems can be used in two-pass systems, where the robot first determines the path of the proposed weld and then performs the weld. This approach eliminates the difficulty of seeing under the bright light produced by the welding and allows better planning and control of welding speed.

An alternative that allows faster throughput is the single-pass system. In this method, three approaches can be used to track the weld: lead systems (the camera or another sensor looks ahead), laser systems (welding with lasers), and through-the-arc systems (the vision system uses the arc as illumination).

In the example described, a GE robot welding system was selected. This GE system (and many other systems) tracks and welds at speeds from 20 to 100 in/min. Compared with perhaps 12 in/min from a human welder, this speed significantly increases productivity.

13.3 WORK CELL SYSTEMS

A slogan in many factories a few years ago was "automate or liquidate." As a result, many FMS were developed. Some of these systems proved to be overkill for the needs.[13] Others, such as at Deere's Waterloo tractor factory, were unsuccessful, perhaps even computerizing confusion and inefficiency.

The trend is still toward automation but with integration and simplification at the same time. Thus, flexible manufacturing cells, or work cells, are being used more often, and complete factory automation systems, or even "islands of automation," are being postponed. The work cell approach is currently growing at the fastest rate, helped by the addition of MAP (see chap. 12) to standardize the interface between robots and the interface between cells. The robot can be treated as part of a work cell or tied to other machines via a network. Work cells are a much smaller investment (perhaps $500,000) compared with a factory-wide FMS.

All activity in a work cell must be controlled by a single source. One of the robots in the work cell may be used as a master controller, or more often a separate intelligent cell controller is in command. Sometimes, a simple relay logic controller is used.

When the robot is the work cell controller, it bears an extra burden since it now must interface with and control external sensors, machine tools, conveyor belts, and other robots. Most robot languages and I/O were developed for stand-alone applications, so it is difficult to efficiently add other cell-related tasks. Hence, robot controllers are primarily used in applications involving only a few external devices. In almost every case, complex sensor processing (such as from a vision sensor) must be done before the robot receives the data, since robots have neither the language set or processing speed necessary to do complex processing.

It is more common to have a separate, central computer controlling the work cell, as shown in Figure 13-2. This computer (possibly connected to a master factory computer system) controls several numerical control (NC) machines, a couple of robots, a conveyor belt, and, in some cases, connects to an automated guided vehicle (AGV) to move raw materials and finished goods between work cells.

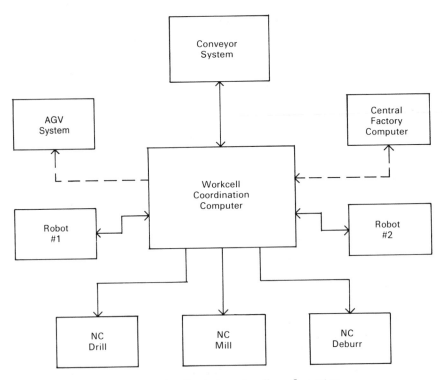

Figure 13-2. Typical work cell configuration.

In one work cell system developed for their dishwasher manufacturing facility, General Electric included four SCARA-type robots in conjunction with a vision system, conveyor line, and numerous small specialized sensors. The cell was controlled by a programmable controller and a control information center. The work cell completes the assembly of the dishwasher and fan subassembly.

Five workstations were provided in this work cell. At the first workstation, a robot accepts incoming parts (shaft/rotor assembly) from the conveyor and places them in an empty pallet. The pallet travels to the second station where a robot picks up a hub, adds it to the shaft, and places a washer on the hub. The third robot takes the piece and adds a fan blade and a circular sleeve to the shaft. At the fourth workstation, a mechanical press is actuated to fasten all of the components on the robot shaft. At the last workstation, a fourth robot unloads the finished subassembly and places it in front of a vision system that inspects the completed work and accepts or rejects the part.

General Electric has also opened an automation center in Frankfurt, West Germany, that demonstrates automated manufacturing cells on an international scale. In this center, a Swedish AGV transports parts to GE robots. These robots supply parts to Italian- and Swiss-built machining centers. Then the robots unload the finished components, which are placed back on AGVs for transport to a warehouse. The center is used for training, demonstrations, and application development.

A final example of a manufacturing cell is an application at International Harvester's Farmall Division. One robot, two turning centers, and a gauging station are used. Parts enter the cell as rough castings, and leave it completely machined and inspected. The sequence follows:

1. The robot takes a part from the pallet on the conveyor and loads it on the first turning center. It also removes the part just machined.
2. The robot moves the part it just removed from the first machine to the automatic gauging station. If found within tolerance, the robot takes it to the second turning center.
3. The robot removes a finished part from the second turning operation and loads the part just gauged.
4. The robot takes the finished part to the gauging station. If the part is good, it is returned to the pallet, and another part is picked up.

13.4 SAFETY IN THE PLANT

What causes accidents? In one study of accidents, 20 out of 32 accidents were directly caused by poor workplace design, primarily a lack of proper access control. Another seven were due to poor robot design, which allowed unexpected movements under failure conditions. Poor training and miscellaneous factors caused the other five. Thus there are three basic steps to improve worker safety:

1. Design a safe workplace
2. Install a safely designed robot
3. Properly train all workers

Safety must be a primary goal of robot designers, of robot users, and of the individual workers themselves. Unfortunately, the increasing use of robots has not been accompanied by completely safe workplaces. Even though we know more about robots, accidents still occur because familiarity tends to result in carelessness. All individuals having any involvement whatsoever with the robot, including individuals who sweep the floor in the area, should receive training in safe procedures.

According to Graham and Millard[14] three factors contribute to unsafe workplaces. First, as robots are used for more complex applications, their actions become more unpredictable. Second, a cage or barrier is easy to provide around a single robot, but much more difficult to use around a group of robots, such as in a work cell. Third, mobile robots bring problems to more locations in the plant.

Safety is also a function of the robot's operating modes. There are at least three modes for running robots, and the safety considerations are different in each one. In the normal run mode, an operator is not usually present. Provisions can be made to shut down the system automatically if someone enters the robot's work area. Supplies can be brought by conveyor or by another robot. If the operator must provide supplies or perform tool changes, the system should be shut down before the operator approaches the robot.

In the teach mode, an operator will be near the robot, but should stay clear of the work envelope. In this case the robot will be under local control, which should both restrict the speed of the robot as well as ensure that the robot cannot be put into an automatic, operating mode while the pendant is plugged in. In addition, the provision for deadman switches and the availability of an emergency stop button should further improve safety. Unfortunately, the design of some robots allows the operator to bypass the slower speed teach mode and to run the robot at full speed.

Maintenance operations provide the highest exposure to accidents. Although the robot is normally shut down during maintenance, some tests may be necessary during any repair work. If the robot is malfunctioning, some of the built-in safety features may also be disabled. Therefore, untrained personnel should never try to repair a robot, and all testing should be done with personnel away from the robot.

Safety Techniques in Workplace Design

Safety in the workplace has been extensively studied. One of the first steps was to recognize that there are three or four zones, also defined as levels of protection, around a robot and that safety practices should be different for each. These levels are[15]

Level 1   Perimeter protection
Level 2a Within the perimeter but outside the reach envelope
Level 2b Within the reach envelope but outside the robot's normal
          work envelope
Level 3   Within the robot's normal work envelope

As can be observed, the danger to the operator increases at the higher numbered zones. Different techniques are more effective at different levels. For example, the obvious approach to level 1 protection is to fence off the area. Wire partition panels (Fig. 13-3) made of heavy gauge wire mesh are structurally strong, yet open enough for unrestricted airflow and visual monitoring of the operation. Panels may be rearranged when manufacturing requirements change. Doors built into the panels can be locked to restrict access, and can have an electrical interlock that cuts off power to the robot if the door is

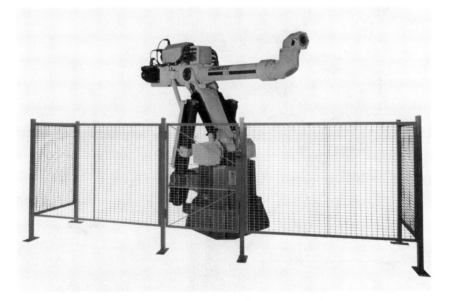

Figure 13-3. Wire screen protective panels. (*Courtesy of Wire Crafters, Inc., Louisville, Ky.*)

opened. Wire panels are usually supplied in OSHA safety yellow. For maximum protection, safety barriers should be higher than eye level.

Other approaches at level 1 are photoelectric beams and watchmen. Photoelectric beams, or light curtains (often using IR wavelengths), if interrupted, stop the robot. Figure 13-4 illustrates one commercially available light curtain in place, and Figure 13-5 shows the sensing equipment.

Occasionally a robot must be operated without all of its safety features in place, such as during initial testing before the installation is complete and some operations during maintenance. A watchman should be used, as required by the Japanese safety standard (chap. 12), to prevent unauthorized personnel from entering the area and to watch for the safety of personnel in the area. A watchman must have direct access to the emergency stop button.

Level 2 protection is provided by pressure mats or IR movement sensors, level 3 (the most difficult) may use ultrasonic sensors mounted on the robot, although limit switches can also help.

AGV Safety Issues

The AGV offers some additional concerns about safety. Although industrial robots can (and do) kill, they at least stay in one place. The AGV moves

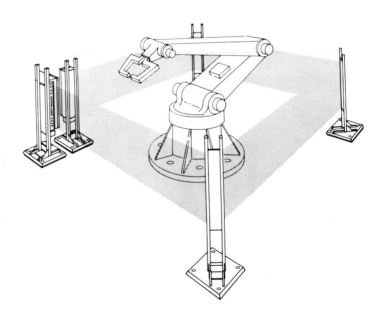

Figure 13-4. Light curtain surrounding a robot. (*Courtesy of Dolan-Jenner Industries*)

Figure 13-5.   Light curtain units. *(Courtesy of Dolan-Jenner Industries)*

around, often where people are. Although the safety record on AGVs is quite good, their use is increasing and they are being given additional autonomy as to where they can travel. Mobile robots are beginning to be designed with a zero turning radius. Although this may be desirable in certain applications, it offers a new type of safety problem. In the past, a worker could watch an AGV go by and know that any turn required a large turning radius, so the robot could be seen while turning. If the AGV can turn on a dime, it is possible for it to make a turn just as it reaches an individual, and move sideways into him or her.

Most AGVs are equipped with bumpers that upon making contact, will shut off the unit. However, these bumpers have little capability for give when they hit something and the AGVs cannot stop on a dime, so there is still a possibility of injury. Tracey suggested using a foam-type bumper to provide a soft impact and prevent any damage.[16]

Special Safety Issues

Soft materials can now be cut at high speed through water jet cutting techniques. Even light metals have been cut with water jets containing abrasive materials. This technique requires a different type of safety barrier, due to the high-pressure water jet (as much as 50,000 psi). One type of barrier used in an application to cut automotive floor rugs is steel and Mil-Spec plexiglass.[17] This type of application requires baffles for noise, a method of removing the resultant moisture and particle cuttings, a backup power supply, and some way to shut off the water and release pressure automatically in case of a power failure.

Finally, even the best of plans can go astray. In one example of an unexpected safety problem, a safety system using a light curtain was defeated when a light on the roof of a building reflected from a car in the parking lot into the equipment.

REFERENCES

1. Critchlow, Arthur J. *Introduction to Robotics.* New York: Macmillan, 1985, p. 421.
2. Venkataraman, V. S. Design and Implementation of a Flexible Manufacturing System. Paper presented at Electro 86, IEEE, 1986.
3. Troxler, Joel W., and Leland Blank. Justifying flexible automation: A system value model. *Proc. Robots 11 Conference.* Dearborn, Mich.: Robotics International of SME, April 1987, pp. 3-9–3-37.
4. Piccirilli, Jr., R. J. Robotics: A users perspective. *Robotics Engineering,* Vol. 8, No. 4, April 1986, pp. 6–10.
5. Benhabib, B., R. F. Fenton, and A. A. Goldenberg. Computer-aided joint error

analysis of robots. *IEEE Journal of Robotics and Automation,* Vol. RA-3, No. 4, Aug. 1987, pp. 317–322.

6. Critchlow, *Robotics,* pp. 78–81.

7. Day, Chia P. Robot accuracy issues and methods of improvement. *Proc. Robots 11 Conference.* Dearborn, Mich.: Robotics International of SME, April 1987, pp. 5-1–5-23.

8. Sturm, Albert J. Robot testing and evaluation. *Robotics Engineering,* Vol. 8, No. 12, Dec. 1986, pp. 4–10.

9. Becker, Stanley E. Development of mechanical performance standards for robots. *Proc. National Bureau of Standards/U.S. Navy Workshop on Robot Standards.* June 6–7, 1985.

10. Acherson, Scott D., and Daniel R. Harry. Theory, experimental results and recommended standards regarding the static positioning and orienting precision of industrial robots. *Proc. National Bureau of Standards/U.S. Navy Workshop on Robot Standards.* June 6–7, 1985.

11. Wodzinski, Michael. Putting robots to the test. *Robotics Today,* Vol. 9, No. 3, June 1987, pp. 17–20.

12. Roth, Zvi S., Benjamin W. Mooring, and Bahram Ravani. An overview of robot calibration. *IEEE Journal of Robotics and Automation,* Vol. RA-3, No. 5, Oct. 1987, pp. 377–385.

13. Palframan, Diane. FMS: Too much, too soon. *Manufacturing Engineering,* Vol. 98, No. 3, March 1987, pp. 33–38.

14. Graham, James H., and Donald L. Millard. Toward development of inherently safe robots. *Proc. Robots 11 Conference.* Dearborn, Mich.: Robotics International of SME, April 1987, pp. 9-11–9-21.

15. Kilmer, R. D. Safety sensor systems for industrial robots. *Proc. Robots 6 Conference.* Dearborn, Mich.: Robotics International of SME, April 1986.

16. Tracey, P. M. Automatic guided vehicle safety. *Proc. Robots 10 Conference.* Dearborn, Mich.: Robotics International of SME, April 1986, pp. 9-1–9-11.

17. Foster, John A. Robotic water jet cutting of automotive carpeting. *Proc. Robots 11 Conference.* Dearborn, Mich.: Robotics International of SME, April 1987, pp. 15-15–15-19.

APPLICATION-ORIENTED REQUIREMENTS

This chapter examines some design issues and requirements that are functions of the application being automated.

14.1 APPLICATION-ORIENTED REQUIREMENTS

As we saw in chapter 2, there are many ways to classify robots. Each classification method has further subdivisions. The type of application determines the classification method to be chosen. For example, the application often determines whether you need a robot with eight DOF or can accept one with four and one half DOF. Application requirements cannot be absolute because numerous variations occur within each application class. For example, problems in assembling engines are not the same as those found assembling components on printed circuit boards. Nevertheless, there are many common requirements within the same application class. This section reviews requirements by class for:

- Sealing systems

- Welding systems

- Assembly systems

- Drilling

- Inspection systems

- Material handling

- Spray painting

- Grinding

- Deburring

- Wire wrapping

Sealing Systems

Sealing systems are used in the manufacture of many items. In the automotive industry, sealing is used for waterproofing or soundproofing joints. A robot that performs this task must be able to reach the area to be sealed and must lay down a precisely controlled bead of sealant, often over a complex path. Figure 14-1 shows a robot sealing the inside of a car body by reaching down and through the window. This example illustrates two requirements often imposed on sealing robots. First is great flexibility; that is, the robot must be able to

Figure 14-1. Robot placing sealer inside car body. *(Courtesy of Cincinnati Milacron. Note: Safety equipment may have been removed or opened to clearly illustrate the product and must be in place prior to operation.)*

approach its work area via numerous paths. This flexibility sometimes requires as many as eight DOF. The second requirement is a short forearm, to allow the robot to work with a small tool in confined places. Both of these are needed for maneuverability.

Sealing applications are also characterized by the complexity of their work envelopes. Typically, the robot must follow a complicated set of curves. Hence, the robot must be able to operate under a controlled path mode and to have an accurate method of speed control. Speed control is required to match the sealant output with tool tip velocity. Since the velocity affects the size of the bead, velocity control becomes quite important. It must be set and held to, perhaps, a 10% tolerance. Speed is reduced when a thick bead is to be put on. Minimum speed can be as low as 5 in/s, although many sealing applications have minimum speeds of 20 in/s. Maximum speed is perhaps 50 in/s.

Position accuracy is not as important in these applications. Path deviations held to within ± 1 mm are usually satisfactory. The load on the robot is also not heavy (from 5 to 50 lb), depending on wrist and gun design.

Seam Welding Systems

Most welding systems now available are second-generation robots, which can handle only selected types of seam welding because of their limited seam tracking capability. Hence, the piece to be welded has to be accurately positioned with expensive fixtures. In addition, the only method available to handle minor irregularities in the seam is the "weave" welding process, which is primarily applicable to butt-type welds.

Third-generation welding robots are now being marketed. These robots can handle more complex applications. They may use vision systems or contact sensors to follow the seam. Laser welding is also being used in some third-generation robots, with this weld being of higher quality and more accurate.

Tracking is an important design consideration in a seam welding robot system. If the seam has no sharp turns (radius greater than 4 in), many current welding systems can track by contact and need no vision equipment. When tracking is required, through-the-arc seam trackers perform well, particularly when welding speed is not a problem. Available systems can position the torch to within 0.01 to 0.02 in of the seam.

The type of weld contributes to the difficulty of automating the weld. End welding is more difficult than overlap welding; if relatively long seams must be welded, some seam-following technique is necessary. Butt welding is usually less difficult, since a "weave" process may often be employed. In a weave-type weld, the robot moves in a slight vertical oscillatory motion as it follows a horizontal seam. T joints, inside welds, box welds, and thin metal or aluminum welds are all difficult for seam tracking systems. However, general-purpose welding systems have been designed to automatically handle different types of metals.

Smaller welding robots have different limitations. Some can handle only a single or limited range of seam types. Others have problems with heat produced by high current arcs, which induces part warpage and inaccuracies in the weld. To prevent workpiece distortion, many robots use a cooling system of circulating water inside the electrodes. Systems without cooling can usually work only with low-current welding arcs, which limits the size of the pieces that can be welded and the welding speed. The heavy current can also generate a lot of electrical noise during current surges, which in many applications causes interference. Thus, the robot often must be placed on a separate power transformer, or some other method of electrical filtering provided.

Welding robots cannot do every welding task yet, but in many applications they compare favorably to humans. A human welder can produce 10 to 12 in of weld per minute. Robots easily surpass this, welding at rates from 40 in/min to 100 in/min. The maximum rate is 200 in/min. Robots are also more accurate, with a typical repeatability of under ±1 mm.

Robots used in welding need more peripheral items than in many other applications. Already mentioned were the welding gun and electrical isolation. A power source is needed to hold the welding voltage to ±2% during the weld. In addition, there is often a seam tracking system, various fixtures and adapters to hold the work piece, and a water cooling system. Completing the accessories is a wire feed system that can hold wire speed to 0.2% of the value set, as well as checking for a stuck wire.

Welding robots do not need much flexibility because they cannot weld in tight places. Usually four and one half DOF are satisfactory, with some applications needing five or six DOF. Some applications do need extensive reaching capability. Large transformer cases, for example, have needed up to 500 in of weld. If the area to be welded cannot be brought to the robot, the robot must be a gantry type, or must have extended reach capabilities.

Assembly Systems

Assembly applications require the robot to pick up a part, often from an automatic feeding bin, and place it correctly into a partially completed product. In many cases, some method of holding the part in place must also be provided by the robot. In other cases, the parts either snap in place or are attached at a later stage in the assembly process.

Only limited types of assembly can be done with present robots because many assembly tasks require skills that robots do not yet possess. For example, the robot may have to find the part in a bin. Although the general bin-picking problem has been solved, a vision system will be needed, thereby adding to the cost of the robot. Fortunately, most assembly systems have parts that have been separated and prepositioned.

A related problem is locating the position to place the part. In hard automation, part positioning depends only on the accuracy of a positioning fixture. In more complex applications, a vision system with an interface to the robot may be needed. In the most complex systems, special collision-avoidance software capability may also be required, especially when two robots team up on an assembly task.

The simplest assembly applications can be done under hard automation. In this case, the robot only handles part of the assembly, and manual intervention is often necessary to finish the task. Many electronic assembly applications can be done satisfactorily with a four DOF robot, such as the SCARA, and a human adds a few specialized parts to the board to complete the assembly.

Other assembly tasks can be very difficult to automate. Not only are the tasks difficult to describe in a program, but each step may require pauses (and perhaps tests) to ensure that the previous step was completed satisfactorily. Errors and problems that develop during assembly need special consideration. Disassembly, which may be required as one step in an assembly process, can cause problems of its own.

Speed in most assembly applications is very important, if the robot is to provide a reasonable return on investment. However, objects with greater size or weight must usually be handled at slower speeds.

Drilling

Drilling systems do not have to handle heavy loads nor do they need many DOF, but they must have flexible end-effectors and be capable of precise part placement. Multiple and changeable end-effectors and/or tools are always required in this type of application. Drilling systems must often consider the task scheduling problem: Production rates and resulting accuracies can vary as the order of drilling is changed.

Inspection Systems

Robots and their associated vision systems are quite useful in the inspection and test areas. Some small robots can position very accurately, 0.0001 in over a small area, and the better vision systems can measure to 0.001 in. Thus, the two can be used to inspect the position accuracy of parts placed by other robots. This capability is important, since the accuracy of the testing device must be greater than the error to be measured.

Cooperation between robot and camera can take many forms. Bumpers have been inspected by having the robot pick up the completed bumper and hold it for the view of an inspector. A robot can move a camera to inspect closeups of assembly parts. Currently one of the most important uses of vision inspection systems is to inspect printed circuit boards (Fig. 6-13). These systems are faster and more accurate than human inspectors, so they are cost-effective and more reliable. Robots have also been used to check the continuity of PC board traces.

Material Handling

Material handling can be anything from machine loading and unloading (which historically was the first use of robots) to material transport between workstations. Moving parts to storage areas and stacking completed assemblies (palletizing) are examples of material handling jobs. Three important requirements are a suitable gripper, load capacity, and reach. Although any type of robot can be used in small machine loading jobs, the larger palletizing tasks often require gantry robots to handle the reach and load capacity required.

Machine handling operations depend on the use of a suitable gripper to pick up the object, the ability to handle the necessary weight, and (on occasion) a need to provide force feedback to ensure that the part will not be damaged. Figure 14-2 shows a robot moving a PC board with a force feedback gripper.

Material transport requirements are similar to those for machine loading, but robots must often handle several objects at one time. Three types of robots are used for transport. Within most work cells, material transfer is usually done by one or more fixed robots. Figure 14-3 shows three robots working together in a work cell. For heavy loads or for movement between nearby work cells, overhead gantries are often the best choice. For transfers around a factory, the AGV is the least expensive approach.

Spray Painting

Spray painting is possibly the most demanding robot process, and product reliability (paint quality) is very important. Off-line programming has not been successful, and the robot must be led through the motions. Electric robots are not as good for lead through tasks, and they also generally require exposure-proof housing. The first spray painting systems used hydraulic robots, but they generally cannot provide as good a quality paint finish as electric or pneumatic robots, although they are still used.

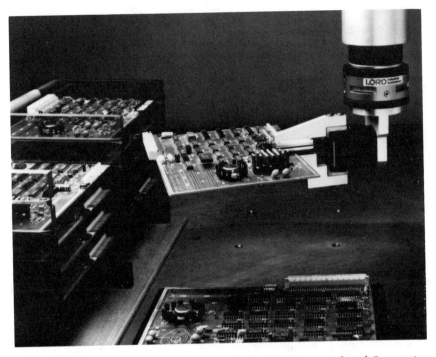

Figure 14-2. Material handling using force feedback. (*Courtesy of Lord Corporation, Industrial Automation Division*)

Both air spraying and electrostatic sprays may be used. With the latter, high voltage (75–150 kV) attracts the paint droplets to the surface. Paint droplets in the air are easily ignited, hence systems may use hydraulic or pneumatic drives to avoid ignition. If electrostatic spray is used, there must be a restriction on the amount of current provided by the high voltage system to reduce the possibility of a spark. There are also obvious safety problems and quality control problems due to the electric field that tends to concentrate the paint along the edges.

Spray painting requires continuous-path robots, with the flexibility of a six- or seven-axis robot. Since the paint will spread out anyway, repeatability and accuracy are not as important. A $\frac{1}{8}$-in (3 mm) repeatability value is often satisfactory. Speed is typically 1–2 m/s. Velocity should be held to 10%.

The robot needs very flexible wrists, such as multiturn or 420° of travel. The gun must be easy to clean. There may also be a problem with old color flakes dropping off the gun and contaminating the new paint job.

The work envelope must be large enough to reach the entire object to be painted. If necessary, the robot could be on rails. Since paint booths are

Figure 14-3. Three robots cooperating in a work cell (one has been installed upside down). *(Courtesy of GMF)*

usually small, a small robot is used, although it must be able to carry about 20–25 lb.

Grinding

Two approaches are used for grinding: either the robot holds the tool against the part, or it holds the part against the tool. Part (or tool) orientation and grinding depth accuracy (often down to 0.002 in) are important. The tool is used to remove excess material from the part, and in accomplishing this task the robot must often follow a complex contour.

Deburring

Deburring is similar to grinding, with the primary difference being that grinding is done to known specs whereas deburring is used to remove burrs of unknown size. Thus these robots must be highly rigid due to the high torque and forces required, yet the wrist must be soft to follow height variations in the burr. One approach uses a force sensor to compensate for these conflicting requirements.

Burrs are generally of irregular height, which cannot be foreseen. The robot must not remove too much or too little of the burr. Drilling and deburring require high repeatability, about 0.2 mm. Adaptive robots use sensors to detect the size and location of burrs in real time, although many deburring robots only use programmable servo-controlled robots.

Wire Wrapping

In wire wrapping and harness manufacture, the robot deals with light loads (less than 5 lb), but requires the full dexterity of a six-axis robot.

14.2 CLEAN-ROOM ENVIRONMENTS

Clean rooms are particularly important in semiconductor manufacturing. Figure 14-4 shows a specially designed robot in use in a semiconductor wafer etching facility. As VLSI chips become more complex, the need for higher yields demands ultraclean environments to prevent contamination of the silicon wafer base. For example, it was acceptable just a few years ago to allow a hundred 0.5-micron particles per cubic foot of air in most clean rooms. Now the tightest standards have been lowered to a single 0.5-micron particle per cubic foot. Even in a class 10 clean room, the contaminant levels may be high enough to limit yields to only 60%. As higher-complexity chips are designed and manufactured, the requirement for ultraclean air will become even more stringent. Federal Standard 209C[1] provides guidelines for air quality testing and gives limits on the size and number of contaminant particles.

Clean-room classes are defined as the number of 0.5-micron particles present per cubic foot of air. Thus, a class 100 clean room can have no more than 100 0.5-micron particles; a class 1 clean room can have no more than 1 0.5-micron particle of that size. Since actual contaminants come in many different sizes, there are also limits set on smaller and larger particles. For example, a class 10 clean room can have no more than ten 0.5-micron particles per cubic feet of air, no more than one tenth of a 3.5-micron particle (on average), and no more than 350 particles of size 0.1-micron.

Laminar airflow in the clean room keeps the air clean by quickly removing dust, dirt, skin, or metal particles. Quick as the air system may be, it is the initial appearance of these particles that can cause a problem. If a single

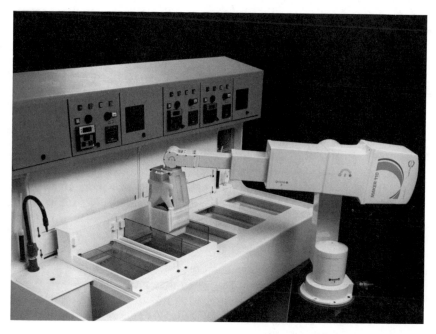

Figure 14-4. Clean-room robot. *(Courtesy of U.S. Robots)*

particle falls onto a wafer, the wafer can be easily contaminated, even if most particles are removed while airborne. Even when wearing clean-room suits, people give off many particles per minute. Thus, air changes as often as twice a second may be needed to keep the air environment within limits. Humans are the major source of contaminants due to the air they exhale, minor skin cells that slough off, and dirt and dust particles they bring in. Hence, the semiconductor industry is finding that robots are better for ultraclean applications. However, unless they are specially designed, robots can also contaminate a clean room.

Class 10 environments are difficult enough. New chips, such as a 1M DRAM (dynamic RAM) will require class 1 clean rooms. As long as neither a human nor a robot enter it, a class 1 clean-room environment is not too difficult to produce. The challenge is to design the robot to limit its contaminants. Although humans can work in a class 10 environment, it appears that only robots can work in class 1 environments. Not only does the human contaminate the silicon material, but some of the chemicals used in the clean room contaminate the human. For example, the number of miscarriages apparently due to contaminants has increased, and other long-term health effects may appear as well.

Robots designed for clean rooms must meet the following requirements:

1. Their joints must be sealed so that metal particles normally found due to wear of the joint cannot escape into the atmosphere.
2. They need an internal cavity exhaust system to remove through negative pressure any particles that may come off the robot.
3. Because they will generally be used for handling delicate parts under tight tolerances, these robots need a high degree of accuracy and repeatability. For example, a part chamfer may be less than 0.005 in, requiring very accurate alignment.
4. Because only a single robot may be put in a clean room, to limit contamination, the robot must be highly adaptable, able to handle a variety of tasks, such as material transport, assembly, inspection, and test functions. An important, often-overlooked element is that the robot may need a tool changing capability.
5. Robots must be small. Large robots require large clean rooms, which are much more expensive, and create more potential contamination. Smaller robots also disturb the airflow less when they move.
6. The degrees of freedom must be limited so that there are fewer joints that lose metal particles. Therefore smaller SCARA-type robots are often selected over more general-purpose robots.
7. Robot speed must be minimized to protect the delicate silicon wafer and, more importantly, to not stir up the air during arm motion. One test showed that doubling the robot's speed increased the number of contaminating particles sixfold.
8. The robot must be reliable. The silicon wafers cost as much as $10,000 each, so the robot must limit damage to the wafers it handles. It is also important for the robot to fail safely under a loss of power or during malfunctions.
9. Most links and joints must be below the level of the wafer, thus allowing most particles from the robot to be caught in the laminar airflow and never reach the wafer. Obviously joint design is critical.
10. The exterior finish must be very smooth, and special paints must be used.

14.3 MOBILE ROBOT REQUIREMENTS

As mentioned in chapter 8, almost no autonomous mobile robots are in general use in factories, and only a few experimental models are commercially available. The requirements discussed here, therefore, are based on what we believe autonomous robots must meet before they can be readily accepted. A few applications may be able to relax one or two of these requirements.

Operating Area

How far must the robot go, and under what terrain conditions, to satisfy most mobile applications? Operating area limitations obviously limit the types of tasks that the robot can perform and the locations it can get to. Yet some limitations are necessary, unless we can design a robot the size of an ant to do the work of a horse. In the near term, we can limit mobility to indoor environments, although there is a need for outdoor autonomous vehicles, and some models are being developed for this environment.

Indoor environments include manufacturing and office buildings, and in these locations it is desirable that the robot go anywhere a human can. A wheeled vehicle will be unable to reach some areas (such as climbing a ladder to the roof), but it is at least desirable that the robot travel anywhere a human in a wheelchair could. Thus, the mobile robot should be able to go through standard doors, over minor raised portions of the floor (such as door sills), and down normal corridors and aisles between factory workplaces. Eventually, it is desirable that they operate elevators, but that is farther in the future. For now, if the robot needs to go to a different floor, a person will have to escort it.

Size

If a mobile robot is to travel through these areas, it cannot be much wider than about 28 in. (Some standard doors are often only 30 in wide, with the actual clearance being 1 in less, and some aisles only provide a 36-in clearance.) The robot's height should be no greater than about 72 in, if it is to go through openings with low overhead clearance (typical doors are 78 in high). It has been suggested that a robot's "vision" sensors be at least 36 in high, to allow the robot to see over desks, tables, and workbenches.

Payload Capability

The robot must be able to carry some equipment or parts with it, although few applications would need an intelligent mobile robot that can carry heavy loads. For light office applications, the robot should be able to carry a minimum of 40 lb. More typical factory applications might require a 200-lb capacity. If much heavier payloads are needed, a conventional AGV or an auxiliary cart, which the robot could push or pull, should probably be used.

Gripper/End-Effector

Although in some instances the robot only needs to observe (such as a night watchman application), and in others people will load and unload it (such as some material transport applications), the general requirement is that the robot have an arm with some type of gripper or end-effector in place that can handle the entire payload capability, or at least load and unload in blocks, so perhaps one third of the maximum payload capacity could be considered the minimum.

Speed

The robot need not travel faster than a human would in a factory setting (that would be dangerous); perhaps it should normally travel at half a human's speed, about 0.3–3 mph (0.4–4.4 ft/s), with the slowest speeds used only during maneuvering in tight places.

Operating Time

Mobile robots will have to carry their energy source, usually a battery, with them. The longer the robot's operating time, the more energy that the robot will need. There are trade-offs in operational time and payload. Larger batteries mean less payload capacity of the robot.

After a study of mobile robot applications, we have concluded that there are two types of time constraints. For many applications, the robot should operate for one full shift (8 hr) and then have 16 hr to recharge its batteries. In some applications, the robot may have to operate for 16 hr (a night watchman, for example); but rarely will it be required to operate continuously. Therefore it can be recharged periodically during its shift. Provision for periodic and automatic recharging is highly desirable.

Accuracy

The robot needs to meet a tight accuracy (perhaps $\frac{1}{4}$ in) when it is docking at one of its stops or providing material to another robot. When it is traveling between locations, it can have a much larger position error, say ± 2 in, unless it had to travel in tight places (a narrow door, for example). The robot may then have to hold its position to within $\frac{1}{2}$ in. Currently available AGVs offer $1\frac{1}{2}$-in positioning accuracy during travel, but they are helped by their guide path. Experimental mobile robots have demonstrated a $\frac{1}{2}$-in accuracy after extensive travel, using dead reckoning alone.

Collision Avoidance

AGVs know where other vehicles in the same system are because a central computer system schedules them and tracks their progress. They handle unexpected obstacles (a wastebasket, a person standing in the aisle) by gently bumping into it and stopping. But this approach is not acceptable for the general purpose mobile robot. It must have a collision-avoidance system to detect obstacles, with provision for going around them (local path planning). In one office analysis it was determined that the robot should start paying attention to obstacles at a 5-ft distance, reduce speed at a 3-ft distance, and turn or stop at $1\frac{1}{2}$-ft distance. Perhaps the most difficult obstacles for the robots are items hanging down from the ceiling or protruding from benches, as well as the barely visible electrical cords and air hoses.

Navigation

The principle stumbling block to widespread use of mobile robots has been technological, specifically their ability to navigate autonomously. In other words, the robot must maneuver through an unknown and changing physical environment. Investigators in the field have subdivided this problem into three subproblems, all currently difficult to do in completely unstructured environments: collision avoidance, global mapping, and path planning.

We have already discussed collision avoidance. Global path planning is the ability of the robot to determine where it wants to go and what path it should take. It is always possible, for some applications, to use lead through to teach the robot the correct path.

Mapping is the ability of a robot to look around its environment, determine where obstacles are placed, and map its area (usually two-dimensional). The robot must update the map as conditions change or as it obtains new information; it must also be able to position itself in a previously prepared map of the local environment. Some of these functions might be done at a central computer site, with instructions sent to the robot through a communication link.

For any mobile robot to be usable, it must, as a minimum, be able to accept a map generated elsewhere and to locate itself on that map as it travels around. A more desirable situation is for the robot to be able to generate and update its map by itself.

Floor Conditions

Floor conditions include problems associated with sloping floors, door jams, and even not leaving wheel marks. The robot must also have clearance to turn, especially for wide-angle turns, and remain upright. It must not run off edges (such as loading docks).

REFERENCE

1. *Airborne Particulate Cleanliness Classes for Clean Rooms and Clean Zones,* Federal Standard 209C. Washington, D.C.: U.S. Government Printing Office, Sept. 1986.

PART
V

FUTURE CONSIDERATIONS

As technological improvements progress, robots will be used in new and diverse applications. What are some of these new techniques and where will they lead us over the next 10 years? These final chapters attempt to provide a realistic view of the future, based upon what is known today. Chapter 15 explores the trends in robot technology development. Chapter 16 looks at a few newer technologies in more detail. Chapter 17 discusses some future applications for robots, many of which have already reached the demonstration stage. It concludes with a forecast of robots to come. Part of this forecast was provided by JIRA, the Japanese Robotic Association, and covers the types of robots they expect to see introduced over the next 5 to 10 years.

TRENDS IN ROBOTIC SYSTEMS

There are perhaps nine major trends in robotic systems:

1. System integration
2. Emphasis in networking robots
3. Switch to off-line programming
4. Move into mobility
5. Use of intelligent sensors
6. Dominance of the electric drive
7. Robot complexity
8. Increased modularity
9. Move into new application areas

After reviewing current research, this chapter discusses the first eight trends and then gives a prognosis of robot capabilities within the next 10 years. Chapter 17 discusses the last trend.

15.1 CURRENT RESEARCH PROJECTS

United States

Robotics is a multidisciplined activity. Many universities and corporations are performing research supporting advancements in robotics. We have already mentioned some of this research activity in other chapters. This section presents a number of specific and potentially significant examples.

Fundamental work in speech recognition and visual systems has led to a robot at Bell Laboratories that responds to voice commands to move objects. This experimental robot can also track a Ping-Pong ball and catch it in midair. The robot thus demonstrates that speech recognition is practical for robots, that something as small as a 2-in ball can be recognized and tracked (while in motion) by a vision system in real time, and that arm motion can be made fast and accurate.

Research on a dexterous hand is being conducted at many universities. At the University of Southern California, studies of human hand movements are being used to develop hands with fewer degrees of freedom (and therefore they are less expensive) that are tailored to specific tasks.

There is much research on robots that walk (see chap. 16). Researchers at Carnegie-Mellon University and later at MIT have shown that it is possible to develop a one-legged robot that can solve in real time the balance and kinematic problems necessary to keep a hopping, single-legged robot stable. Other researchers are studying two-, four-, and six-legged robots.

Legged locomotion has been commercially applied to an outdoor environment. International Robotic Technologies has developed a robot window washer that crawls along the surfaces of buildings by means of vacuum grippers on its six legs. Weighing only 44 lb, the robot carries its own wipers and washing fluid and can wash large glass windows in skyscrapers.

Martin Marietta is also applying current robot technology to an outdoor environment. In cooperation with other companies and universities, Martin Marietta is developing an autonomous vehicle that can travel along a roadway by visually tracking the road edges. The company is also supporting research at the University of Florida to develop a robot orange picker that can recognize ripening fruit on trees, reach the fruit, and remove it without damage.

One limitation of mobile robots is a good collision-avoidance system—one that not only recognizes an object in its way but can also figure out how to get around it. Advancements have been made, however. Bruce Krogh has developed a novel method, called the *potential field approach,* to path planning and collision avoidance in an unknown environment. Avoidance vectors are assigned to obstacles, and attraction vectors are assigned to goal points. The resultant vector then provides the desired path.

Purdue University has developed, and is using, an automated robot for organic chemical synthesis. This robot improves the low yield they were previously getting in some of their processing steps. The robot has a gripper to hold glassware and a syringe to dispense chemicals. In a related application, Phillips Petroleum has performed eight chemical laboratory procedures with robots. In one procedure a work cell microswitch was occasionally sticking in the closed position. A robot was used to periodically check the switch. If the switch is stuck, the robot reaches over, takes hold of it, and pulls it free.

Robots that can accept speech input, drive along a highway, pick oranges from trees, move around on one leg (or more), perform chemical processes, detect and correct mechanical malfunctions, and wash windows is an impressive list of accomplishments indeed!

Japan

The Japanese have developed many types of advanced robots and are applying their research in many areas. (Table 15-1 lists 28 such areas. Here we discuss only one.) One of the most impressive demonstrations of current robot technology the author has seen is WABOT-2, the keyboard-playing robot from Wasada University (Tokyo).[1] It looks quite human (see Fig. 15-1) and has an amazing number of advanced capabilities. By using both arms and legs, WABOT can sit down at an organ and play very well. It can select from its own repertoire of prelearned songs, or it can play new songs by reading sheet music supplied to it. WABOT can even recognize some songs by hearing them sung. It then accompanies the singer on the organ without missing a note. To accomplish these feats, the robot has five different subsystems: limb control, vision, conversation, singing tracking, and supervisory.

We talked with two of WABOT's designers and learned a lot about its inner circuitry. The limb control subsystem operates 2 arms, 10 fingers, and 2 feet. The vision subsystem allows the robot to read sheet music in printed or handwritten scores. With its speech subsystem the robot can recognize speech from other individuals and speak to them. The speech system recognizes and interprets sentences and asks questions about ambiguous information. (Naturally, it converses in Japanese.) The singing tracking subsystem adjusts to the singer's voice, abstracts the melody from the words, and recognizes the music—a most impressive robot.

Europe

The French have implemented a Plan Productique, which is a set of nationally supported R & D programs specifically dedicated to third-generation robotics.[2] One of these programs, ARA, covers the development of advanced teleoperated robots. Another program, RAM (multiservice autonomous robots), is aimed at developing a series of special applications, including robots for nuclear plants, mining operations, underseas, forestry, and plant cleaning. The French have announced their intention to be world leaders in most of these technologies.

West Germany is a leader in using robots. Among the research topics they are pursuing are speech recognition (particularly at the Technical University of Munich and the University of Braunschweig), mobile robotics/autonomous

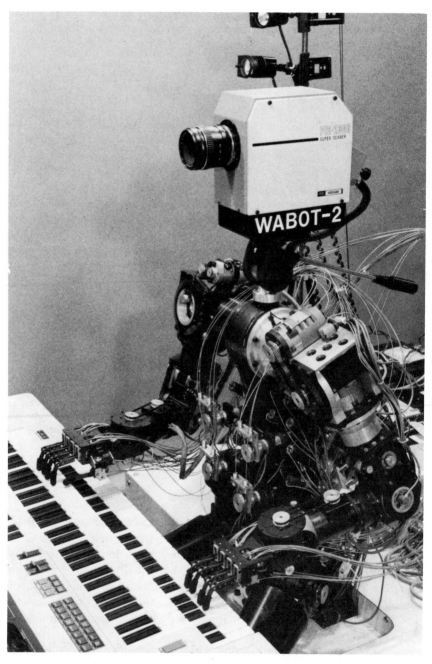

Figure 15-1. WABOT-2 playing an organ. (*Courtesy of Kato Laboratory, Wasada University*)

navigation at Karlsruhe University, and vision inspection systems at the Fraunhofer Institute for Production.

The Italians are working on the integration of visual data with ultrasonic data. Robots receive vision data from two charge-coupled device (CCD) cameras and use ultrasonic data to provide accurate range information and to assist in image processing. They are also developing a type of robotic "skin" at the University of Pisa.[3] The "skin" is made from the polymer polyvinylidene fluoride (PVDF), which can detect pressure, slip, and temperature changes.

One of the many areas being examined in England is robotic vision. Since scene analysis for vision systems demands high computational power and lengthy processing, many laboratories are developing parallel architecture approaches to handle a number of visual frames simultaneously. A practical parallel vision system is Visive, developed by Sowerby Research Center of Bristol. The University of Hull is looking into specialized vision sensors and parts inspection.

15.2 SURVEYS AND PREDICTIONS

There have been a number of studies done on robot trends and related predictions made. This section summarizes some of that information.

United States

In a 1985 survey of opinions from leading robot manufacturers,[4] general agreement was reached on a number of trends over the next 10 years, including user growth and the availability of two-dimensional and three-dimensional vision systems, which will allow robots to be used in new applications. In fact, some in the industry are predicting that future systems will not be called robots if they do not have vision capability. Six-axis robots equipped with vision and other sensors and employing multiple end-effectors will be at the center of flexible assembly lines. These robots will have expanded off-line programming capabilities and compatibility to computer networks, especially MAP.

In a 1985 report, Ayres[5] presented four of his predictions:

1. Two types of robots will be increasingly used: simple and highly sophisticated. Simpler robots will be limited to hard-automation-type applications.
2. Robot prices will decrease due to lower technology costs and the resultant economies of scale as more robots are built.
3. The development of lighter-weight materials and the use of parallel linkages will improve the robot performance-to-weight ratio.

4. Multiaxis force sensing, the increased use of tendon drives, and the development of general-purpose hands with high-resolution force sensing skin will increase the robot's flexibility and ability to handle complex tasks.

Japan

The Japanese have already demonstrated advanced robotic capabilities. Observing where they are and where they are going is, therefore, particularly significant.

The Japanese Industrial Robot Association (JIRA) conducted a research study[6] in 1985 to determine the current level of Japanese robot technology, the types of research being conducted, and the expected maturity date of this research. Approximately 200 topics were investigated, with survey questionnaires sent to most robot specialists in Japan. Table 15-1 summarizes the results of this survey. Two dates are shown for 28 of their robot-related programs. The first date indicates the anticipated start time of the research; the second indicates the expected time when practical results should be available.

Several points from Table 15-1 should be emphasized. First, only sensor improvements can be expected in the near future; computer-based enhancements are further off. Second, in the mobile robot field, legged robots are anticipated as being available in five years. What is not indicated is that wheeled mobile robots are already available from many Japanese companies. One of the most optimistic predictions is that better robot operating systems and more sophisticated robot language are predicted for the next two to four years. However, we see little movement in this direction unless the user community forces it on the manufacturers (similar to the way MAP developed).

Worldwide

A world conference on robotics research, looking five years into the future,[7] was held at Lehigh University in 1984, and a second conference was held during the summer of 1986. Speakers indicated that a major limitation in robot kinematics has been in finding accurate, robust, and, perhaps most important, computationally efficient models. General issues raised included the need for better modeling and control of dynamic motion, better integration of sensor data into robot systems, and improved hardware and software standards. One area of research that was stressed was that of piezoelectric and silicon-based sensors, which are replacing more conventional conductive elastomers. Looking especially good are Kylar film and PVDF film. Two of the research leaders are D. de Rossi and P. Dario from the University of Pisa. Areas in which the participants saw major research continuing include mobile robotics, robot modularity, and high-precision robots, particularly for micro-surgery.

Wikol[8] predicts that future robots will be

- Modular

- Computer-controlled

- Faster

- User-friendly

- Intelligent

- Cheaper

- Compact

- Strong

Note that the first five predictions are a direct result of better application of computers to robotics.

15.3 TECHNOLOGICAL TRENDS

It is difficult to predict what improvements will be practical during the next five years. This section presents eight areas in which major changes are currently underway and in which continued improvements are expected.

System Integration

Improvements in sensors and sensor integration, and an increase in robot intelligence, satellite robots, and work cell concepts are leading to a completely integrated factory. Currently, several robots can work together within automated manufacturing cells and on automated production lines. The growth of system integration will result in an increased role for robot system houses, and manufacturers will have to provide more system-level assistance, especially as applications become more sophisticated, the need to integrate various sensors with the robot increases, and robots continue to be integrated into a complete factory network.

Companies specializing in robot systems (as opposed to robots as components only) will become increasingly important to interface new applications with available robots. Users will support this systems trend in two ways. First, during the design for production cycle, they will increasingly design products

Table 15-1. Period of Materialization of Robot Technology

Field	87	88	89	90	91	92	93	94	95

SENSING

Visual
- 3D Recognition/Sophisticated Vision
- 3D Measurement
- Super Small Cameras
- High Speed Tracking
- Color Recognition

Tactile
- Contact, Pressure, Strength Sensing
- Slip Sensing

Aural
- Abnormal Sound Discrimination
- Continuous Voice Recognition, Independent Spkr

MECHANICAL

Actuator
- Small High Output Servomotor
- High Output to Weight Ratio Actuators

Arms/Hands
- Arms with Funicular Freedom
- Fingers with Human Responsiveness

Locomotion
- Mobile robot able to adjust for obstacles
- 4/6 Legged Robots
- 2 Legged Robots

COMPUTER

Cooperative
control
— Harmonic Control of Fingers
— Coordinated Control of Multiple Robots

Remote control
— Visual/Strength Information
Sophisticated Remote Control

Distributed
control
— Sophisticated Hierarchical Control

Intelligent
robots
— Component Fitting Algorithm
Problem Solving Systems
— Learning Functions
— Accept Commands through Speech

Information
processing
— Robot Operating System
Sophisticated Robot Languages
— Integration with CAD/CAM

Source: After Kanji Yonemoto, Ichiro Kato, and Kensuke Shima. Technology forecast on industrial robots in Japan. *Proc. 15th International Symposium on Industrial Robots.* Tokyo: Japanese Industrial Robot Association, 1985, p. 53, with permission.

that are most able to utilize the advantages of robot production. Second, as new plants are built or older ones are expanded, the advantages of robot production cells and automated lines will be realized, and the factory layout arranged accordingly.

Thus, we see a welcome trend occurring in which complete robotics system sales and servicing will be made available. When robots were first introduced, many manufacturers limited their responsibilities to providing the robot, and left system responsibility to the user. However, most first-time (and even second-time) robot users need a lot of support in planning, plant modifications, training, safety systems, installation, production changes, and preventive maintenance to use robots efficiently. Several robot system integrators have appeared to help fill this gap, to custom design special robots for unique applications, and to integrate advanced vision systems and other intelligent sensors into the system. Other specialized firms now offer robot testing and robot servicing.

Robot Mobility

As hardware and software techniques improve, mobility will be increasingly applied. Used first for transport purposes, its role will then expand to security, inspection, and minor repair. Mobile robots will further contribute to the move to all-electric robots and direct-drive motors. Although legged robot research will continue to be supported, we see commercial mobile robots being based on wheels for at least the next 10 years.

Improvements in robot mobility will occur in three areas.

1. Extension of current AGV technology. The University of Georgia, among others, is adding collision avoidance and local path planning to AGVs, thus extending their capabilities.
2. Capability for autonomous navigation. The distinction between mobile robots and autonomous mobile robots is based directly on this capability. A robot without navigation capability is just a movable vehicle. An autonomous robot can act completely independently and find its own way to its assignment.
3. Development of better and faster three-dimensional vision systems for collision-avoidance systems and other uses.

In summary, improvements in robot mapping, collision avoidance, navigation programs, ultrasonics, self-contained power, and better vision systems will all support the move toward completely autonomous mobile robots.

Off-Line Programming

Many robots are still being taught what to do by a technician walking the robot through the necessary steps via a teach pendant. Although a viable approach for facilities with only a few robots or for facilities that use robots only for certain types of applications, it will become less and less viable for larger and more complex installations.

Computer systems have supported off-line programming in other areas for years, and this technology, along with advanced graphics workstations and specialized robot simulation languages, is now being applied to robot program development. This trend is so important to the development of future robotics applications that both the number of systems employing off-line programming and their available capabilities must increase. Advanced programming capability is also particularly important for custom products or batch manufacturing.

Languages (e.g., Karel and ROPS) have been developed to facilitate off-line programming and simulation languages have been written that allow the programs to be tested at off-line workstations by simulating the robot. Therefore, current programs may be fully tested before they are loaded into a robot on the production floor.

Robot Communications and Networking

A few years ago, robots were stand-alone devices with little communication with other robots or to a factory-level computer. With the current trend to work cells, where several robots must work together, there is an increasing need for robot networking capabilities. The development of the MAP network standard and its support by many companies in the United States and Europe should help robot intercommunications.

A second area, which is still in its early stages so that it cannot be called a trend, is the addition of voice communications to robots. Not all robots need voice communications, but in many areas it would help, and there are potential applications that will never fully develop until two-way voice communications with robots is easily performed. Currently, limited speech understanding may be found in some inspection systems.

Intelligent Sensors

All types of advanced vision, tactile, and ultrasonic sensors are being developed. With processing capability added directly to the sensor, they are becoming more intelligent. They are also being directly integrated into the robot system.

At an IEEE seminar in 1983, Dr. Larry Leifer explored some of these trends. Five years later, we see this area expanding. He stressed the many new sensors (force, position, chemical, etc.) being developed based on silicon substrates. Since these sensors need the support of signal processing technology, he foresaw the next step: bringing the signal processes into the sensor. Another advantage of the silicon-based approach is the increased pressure sensitivity and the decreased thermal sensitivity. Work in this area is being pursued at the University of Michigan.

Dr. Leifer also predicted that tactile display resolution would soon rival that of current vision systems. For example, experimental capacitive sensors have been produced with a 16 × 16 array of 1 mm spacing, a resolution approximating that of a human finger.

In addition, completely new sensor technologies are in the research stage. One example of a new technology not yet applied is the development of a robot nose. Dr. Siegel at Carnegie-Mellon University hopes to combine various sensing elements into a single electronic sniffer and to produce a unique output signal for each type of smell. One problem in this area is that even such a simple smell as a lemon peel aroma has 400 separate characteristics.

Electric Drive

The current trend in robotics is to use electric motive power in place of hydraulic, even though hydraulic actuators are more powerful for a given size. Electric systems have five major advantages:

- They are less noisy.

- They do not leak hydraulic fluid or contaminate the workplace.

- They provide higher accuracy, greater repeatability, and finer movement.

- They do not need a warmup period.

- They interface more effectively with advanced sensors and advanced servo drive systems.

There is also a growing trend to direct-drive motors. In these robots, transmission elements between the motor and the arms are eliminated, thus increasing efficiency and accuracy. Carnegie-Mellon is a leader in this area, with their first model built in 1981.[9] Their latest version of direct-drive offers an impressive slew speed (the robot arm can travel from any point to any other point within 1 s).

Other types of robots are still needed, however. Hydraulic and pneumatic robots are still sold and can be the best choice in some applications, particularly spray painting.

Robot Complexity

Many limited robots, such as pick-and-place, are still being built, and there will continue to be applications in which they are satisfactory and cost-effective. Nevertheless, the trend in all areas of robotics is for more complex systems, due to requirements for robots to work with other robots, to be used in more complex applications, and to use advanced peripherals and multiple grippers. Advanced peripherals include more sophisticated vision systems, integration of advanced tactile sensors, multiple-tool-carrying end-effectors, and ultrasonic collision-avoidance systems.

Modularity

Modularity includes the ability of the robot to interface a wide variety of peripherals and to add such specialized programming modules as voice recognition, network communications, and work cell robot coordination. Modularity also refers to the ability to provide separate software application modules developed by third-party vendors. For example, the personal computer's modularity—where many vendors can offer specialized hardware boards that can be plugged in and specialized software that can be integrated with other packages—has made it highly desirable in the business community. Robot manufacturers are just starting to see the need for modularity. In fact, if personal robot manufacturers had offered an open architecture, it is quite likely there would have been fewer failures.

Current research in modular programming include such developments as the JIRA suggested software standard and the effort put into Sharp, a robot language with three major planning modules. Unfortunately, more progress in software applications modules cannot be made without adoption of some type of software standards, or more vendors adopting a common robot language. Modularity will hopefully permit interchanging parts and programs between robots. In the long term, both hardware and software modularity will be important steps in maturing the technology.

Although progress is being made in interface standards, some vendors are still clinging to their own "standards" or simply resisting any standards at all. Communication standards in particular have seen too much disagreement by some of the larger computer firms, who seem to follow the "not invented here" syndrome.

15.4 PREDICTIONS

What will be the state of the art by the turn of the century? We close this chapter with our own predictions. Among others we see the following:

1. One overriding trend is growth in vision systems. In five years almost all advanced robots will contain vision systems. Both two-dimensional and three-dimensional systems will play important roles in many applications, not just in inspection. Vision systems may grow much greater than the 25% predicted. After-sales support from vision vendors will also increase. A common robot language must evolve, and vision system interfaces and requirements may lead the way.
2. The growth in vision systems will be supported by faster processing speeds and increased preprocessing capabilities, both of which will reduce the burden on the image recognition system.
3. Hierarchical approaches to pattern recognition will be used, especially in stereo images.
4. Parallel architecture will start moving out of the research labs and into practical products.
5. Neural networks will provide the next big improvement in image processing architecture.
6. In 10 years most intelligent robots will be mobile.
7. Industry will move away from dedicated robots and toward flexible automation. Stand-alone robots will become less important, and robot controllers will interface with more than one processor, especially for multiple robots working in automated manufacturing cells or on completely automated production lines.
8. Integrated complete solutions will replace simply selling robots with advanced technology. Robot firms will become systems-solution oriented rather than robot-sales oriented. Robots that can better interface with other types of robots will become increasingly important, rewarding companies willing to standardize.
9. Precision will become more important. High precision will refer to models with accuracies and repeatabilities under 1 micron. These robots should be available within five years.
10. Off-line programming will soon be done on engineering workstations with advanced graphics simulation of robots and direct input from CAD and CAM data bases.

REFERENCES

1. Fujisawa, E., T. Seki, and S. Narita. Supervisory system and singing voice tracking subsystem of WABOT-2. *Proc. 85th International Conference on Advanced Robotics.* Tokyo: Japanese Industrial Robot Association, 1985, pp. 489–495.
2. Feldmann, M. Third generation robotics in France. *Proc. 85th International Conference on Advanced Robotics.* Tokyo: Japanese Industrial Robot Association, 1985, pp. 3–6.
3. Dario, P., et al. A sensorized scenario for basic investigation of active touch. *Proc. 85th International Conference on Advanced Robotics.* Tokyo: Japanese Industrial Robot Association, 1985, pp. 145–152.
4. Niebruegge, Douglas, et al. Trends and transitions in the robotics industry. *Robotics Age,* Vol. 7, No. 11, Nov. 1985, pp. 22–27.
5. Ayres, Robert U., et al. *Robotics and Flexible Manufacturing Technologies: Assessment, Impacts and Forecast.* Park Ridge, N.J.: Noyes Publications, 1985.
6. Yonemoto, Kanji, Ichiro Kato, and Kensuke Shima. Technology forecast on industrial robots in Japan. *Proc. 15th International Symposium on Industrial Robots.* Tokyo: Japanese Industrial Robot Association, Sept. 1985, pp. 51–58.
7. Coleman, Arthur, Robotics research: The next five years and beyond. *Robotics Age,* Vol. 7, No. 3, Feb. 1985, pp. 14–19.
8. Wikol, Murray D. Flexible automation implementation in Europe, Australia, Japan and North America. *Proc. Robots 11 Conference.* Dearborn, Mich.: Robotics International of SME, April 1987, pp. 3-39–3-48.
9. Stauffer, Robert N. Researching tomorrow's robots. *Robotics Today,* Vol. 9, No. 5, Oct. 1987, pp. 27–35.

NEW TECHNOLOGY

We looked into many areas of robotics-related research in chapter 15. Five of these areas are important enough to future robot design to warrant more discussion: natural language processing, speech recognition, legged locomotion, collision avoidance, and neural network computing.

16.1 NATURAL LANGUAGE PROCESSING

Understanding the meaning of words, at least within the context of industrial and commercial use falls within the province of natural language processing.[1] Although this subject can be readily identified with speech recognition, it also deals with text entered into a computer through a keyboard or by optical scanning.

The distinction between natural language understanding and speech recognition is straightforward. A speech recognizer converts sounds patterns received into a word within its available vocabulary. A natural language processor then takes the word and interprets it in the context of its rules and other words received. The language processor must distinguish between words that sound alike but have different meanings. Natural language processing rules can also be used to distinguish among several candidate words from a speech recognizer. Therefore, most complex speech recognizers include some type of language processing.

Goals

The first attempted use of natural language processing was translation from one language to another. Understanding the meaning of words, at least for translation, seemed to be straightforward, and work on natural language understanding occurred during the 1950s. The initial result was frustration and failure. Problems arise due to the numerous meanings that a word has, compounded by idioms and implied meanings. For example, an invitation to attend a presentation on land opportunities in Florida implies to most people that sales attempts will be made. How would a robot understand this when the word *sales* is never mentioned?

If we artificially restrict the meanings of a word and eliminate idioms and implied meanings, understanding sentences becomes much easier. It is quite practical today for robots to understand the limited vocabulary needed in some factory inspection tasks. The goal of language understanding is to expand this capability to include larger vocabularies and multiple word meanings, to allow much more ambitious applications of voice input to robots.

Problems

Understanding words with several meanings, depending upon the context, is the most difficult problem. The problem is compounded when the words come from a speech recognizer rather than as typed input, since the system cannot distinguish homonyms (i.e., did the robot hear *pale* or *pail*?). Although some advanced speech recognizers can address this problem, it is really a language processing function. To illustrate the level of complexity, Table 16-1 lists 12

Table 16-1. Alternate Word Meanings of "Rise"

Alternate Meanings	Example
Increase	The temperature is on the rise.
Extend upward	We live in a high-rise building.
Advance	He will rise in the ranks.
Come into view	The town rises in the distance.
To begin	From this spring a river rises.
To erect	On this land a building will rise.
Cheerful	Her spirits began to rise.
Anger	That got a rise out of him.
Expand	With yeast, the dough will rise.
Revolt	Will they rise against him?
High ground	On the rise is the town.
Height	The step rise is 8 in.

meanings for the word *rise*. Although some of the meanings are related, no single word (or meaning) could be substituted for *rise* and cover all these uses.

Approaches

Language understanding systems must determine the meaning of the word from its context. A built-in dictionary can give a set of possibilities, but only from the context can the robot choose among them. In fact, we understand words in the same way, and we can often guess at the meaning of a new (to us) word solely from its context. We use different clues to do this, and computers have been programmed to use the same clues.

One clue is *syntax,* which can be used to determine a word's meaning through grammatical usage. For example, the robot might be told to "use orange paint" or "reject a blemished orange." The language processor could separate these meanings because the first use of *orange* is as an adjective, and the second is as a noun. The language processor uses parsing techniques to divide sentences into their basic parts of speech. Parsing is particularly difficult with spoken words, since we often do not talk in complete or grammatically correct sentences.

But syntax alone cannot distinguish all word meanings. A second clue is *semantics*—that is, does the meaning of the word make sense in a particular context? For example, the homonyms *pair* and *pear* are both nouns, so they cannot be distinguished syntactically. However, through semantics the robot could distinguish the meanings of the words in the statements "separate the pears by size" and "place pairs of shoes together." Thus the sense of the sentence often allows the correct choice to be made.

A third method is *pragmatics,* which refers to discerning the meaning of a word in different contexts from experience. When the robot learns new meanings, it can use pragmatics (i.e., its experience) to understand the difference between grasping an egg and grasping a heavy tool. A related factor is association. We naturally associate certain words with others. The presence of one can help to determine the meaning of the other. For example, the robot might associate *close* or *open* with *gripper.*

Prosodic features are another strong clue to a word's meaning. They include additional information contained in a spoken word, such as stress, timing, transition, and pitch. Rising pitch often indicates a question. A pause in speaking usually sets off a clause or the end of a sentence. The manner in which stress is applied to syllables can help distinguish *con'·duct* from *con·duct'*. Note that prosodic clues require more information from the speech recognizer system than just its best estimate of a word's spelling. For example, in the statements "close the gripper," and "move close to the object," the robot must use the pronunciation differences of *close* to determine its meaning.

Humans gradually obtain a vast amount of experience, which allows us to quickly understand words through their context. Yet computers have no similar experience, and it is impractical to enter all of a human's vocabulary into a robot. A robot must use most, if not all, of the clues to detect the meaning of words. The best approach is probably through learning, just as we did. Artificial intelligence (AI) techniques are being developed to allow the computer to learn from its past mistakes, and, by asking intelligent questions, to understand the meaning of many words.

Status

Various language processors are available, some that operate from typed input and others with spoken input. As long as the set of language is quite restricted (multiple meanings of the same word are not allowed, total vocabulary size is not too large), even some of the more limited speech recognizer systems are quite adequate.

Human factors studies have been used to improve the human-machine interface, and one important recommendation concerns the computer's handling of rejects. Studies have shown that soft rejects are tolerated in the workplace much better than hard rejects. Suppose we want to enter the phrase *move left five feet.* If the computer did not understand one of the words and was designed for hard reject, it would beep or say "please repeat" constantly. On the other hand, if the system was designed for a soft reject, the computer would make an educated guess, such as "did you say *move left nine feet?*" Experience with current speech systems shows that most of the time, the computer guesses correctly. Even if the guess is wrong, the operator immediately knows which word is causing the difficulty and can more readily clarify the instruction.

16.2 SPEECH RECOGNITION

Voice has always been the most important medium in human communication, and the electronic transmission of voice has been essential to business since the invention of the telephone. With the advent of the robotics age, it is only natural that there be research into voice communications with robots. For maximum value, the communications must be both ways.

Work in speech synthesis was begun in the 1930s by Homer Dudley of Bell Laboratories, but supplying the robot with quality speech was not possible until speech synthesis chips were developed by Texas Instruments in the 1970s. Today, the capability of natural sounding speech is almost taken for granted; even elevators can announce the floor by using speech synthesis.

In speech recognition, however, there has been less progress. Two major reasons are that we do not completely understand how a human hears sounds and interprets them as words, and that everyone speaks with slightly different pronunciations, tempo, and accents, thus producing distinctly different sound waves.

Definition

A speech recognition system accepts the sounds produced by a speaker and converts these sounds to words within its vocabulary range. Within this broad definition are several subclassifications. Continuous speech recognizers can accept speech at a normal rate and need no artificial pauses to separate words. Multiple-speaker systems allow the system to understand speech from many different individuals, and are not limited to a single speaker.

Goals

The goal of a speech recognition system is to understand normal human speech. Normal human speech is what a person not trained on the system would speak at his or her normal rate, with a standard vocabulary, an average accent, and under conditions of normal background noise.

Problems

If the voice recognition system operates with no background noise and needs to recognize only 50 words from a single trained speaker speaking in an isolated manner (pauses between words), then many current systems provide this capability. When more advanced capabilities are needed, recognition problems include

- Large vocabularies (over 200 words)
- Connected speech (normal conversation)
- Multiple speakers
- Near real-time response (less than 2 s)
- Extreme noise conditions (factory environment)

Each condition creates its own set of problems. Many speech recognition systems can only accept speech from people they have been trained with (speaker-dependent systems). Often the speaker must enunciate clearly and speak with pauses between words (isolated word systems). Any noise will degrade the response, and if other sounds are picked up by the microphone it is sometimes impossible to recognize speech. In noisy environments the operator might have to strain his voice to ensure a high signal-to-noise ratio.

Current vocabularies are generally limited and must be carefully selected to eliminate words that sound like others being used. This process is increasingly difficult as the size of the vocabulary increases. Large vocabularies have three difficulties: (1) potential confusion between similar sounding words; (2) large memories for storing large vocabulary templates; (3) longer processing times to search for the correct word.

Another problem with speech systems is false acceptance of a word. This problem is far more serious than not understanding a word because the computer can always ask for a repeat. But if the word is misunderstood and no clarification is requested (such as mistaking a nine for a five), the system may never notice this error. Therefore some speech recognition systems provide a verbal feedback (repeat each word) to eliminate misunderstandings.

Approaches

Most voice recognition systems follow one of three signal processing methods: (1) signal processing techniques (such as via Fourier frequency analysis), (2) speech modeling (such as with an LPC chip), and (3) speech reception techniques (analog of the human ear). The last approach (trying to build a circuit that emulates the ear), though offering the most promise, has not been fully implemented. Signal processing techniques determine the amount of energy in different parts of the speech spectrum and compare this data with stored spectral patterns.

A technique that has received a lot of attention is linear predictive coding (LPC). Originally developed for speech synthesis, the LPC chip uses parameters found in speech waveforms as a basis for word recognition. These parameters include formant frequencies, sound energy, zero-crossing rates, and short-term spectrums. The parameters are measured during time "windows," typically 10–50 ms of speech. The resultant set of time-varying parameters may be further modified to normalize the pattern length.

Most systems are trying to increase vocabulary size, because the more words a system can recognize, the more potential applications there are. Although some applications can accept limited vocabularies (<100 words), most applications require much greater vocabularies.

For example, the Defense Advanced Research Project Agency (DARPA), which sponsored one set of speech research in the early 1970s, is now setting an eventual goal of 10,000 words for an office environment, with a goal of

5,000 words by 1989. Most uses for speech recognition with industrial robots do not require as large a vocabulary, and 500 to 1,000 words should be adequate.

A second area in which progress is being made is the robot's ability to recognize continuous or connected speech rather than isolated words or phrases. There are two problems here. First, without an artificial pause, the robot may not be able to tell where one word stops and the next word starts, especially if the word has a stop consonant in the middle. For example, *rubber* has a natural pause between the two syllables. This problem may be solved, at least partly, by a language processor that feeds back information to the speech processor.

Second, most systems cannot recognize speech in real time, and a continuous speech input would quickly overload the computer. Incidently, systems that accept phrases rather than individual words are not continuous speech systems and are actually easier to implement than isolated speech, since there is a much larger pattern to match. The real-time problem is being solved by using faster hardware and improved algorithms.

Another area under extensive development is that of recognizing speech from many speakers, not just the few users the system has been trained on. This capability is necessary for many reasons. For example, the robot should obey a stop command regardless of who gives it. The simplest technique for pattern recognition is based on template matching—comparing certain characteristics of the speech to stored templates. The characteristics are compared one by one against each possible vocabulary entry. The resultant matches are scored (a likelihood value is assigned to each potential match), and the highest score (over a threshold) is selected. Large vocabularies require long processing times and large memories to hold all the templates.

If a separate set of templates must be stored for each speaker, the system could quickly become unwieldy. Several techniques are used to get around this situation. If there are only a few speakers, a separate set of templates could be on line for each. If there are many speakers, they can identify themselves before each session and only templates previously stored for the identified speaker would be needed. However, if the speakers were new and no templates had ever been made for them the situation is much more difficult. They could be asked to say half a dozen predefined words, and the system could then estimate which stored template matched this speaker most closely and use it.

Another difference between speakers is the speed of their speech. Speech rate varies for any word according to the context of the speech, its importance, and the activity of the speaker. Time warping techniques have been used to adjust the templates to compensate for the speech pace, but they are only an averaging method, they require additional processing time, and the resulting distortion can cause other problems in recognition.

The techniques we have discussed are primarily used for complete word recognition. Another approach is to use systems that recognize phonemes. Phonemes are the smallest elements of speech, only part of a syllable. (For example, the one-syllable word *thrill* has the four phonemes θ, r, i, and l.) These systems attempt to recognize words by breaking them into individual phonemes, recognizing the phonemes, and finding the meaning of the resulting phoneme combinations in a dictionary.

This concept is promising because most languages have only 40 to 50 phonemes, compared with about 50,000 words. (The English language has 47 phonemes, 32 consonant sounds, and 15 vowel sounds. Note that there are more phonemes than letters because many letters have several pronunciations, and there is one phoneme for each pronunciation, not for each letter.)

One problem with phonemes is that their pronunciation often changes as a function of the preceding and succeeding phonemes. When allophones (slight variations in phoneme sounds) are used, the number of combinations may be 1,000. Allophones are generally only necessary with very large vocabularies.

A phoneme approach can also break down when accents affect the sound of the phoneme and when parts of words are left out, a problem known as elision (the English drop "*h*'s", and Bostonians drop final "*r*'s"). A more difficult example for a recognition system would be a statement like "glah to me cha" rather than "glad to meet you."

Some types of natural language understanding are often used as feedback to the speech recognition system. In such cases and to reduce the number of templates, systems may use rules of grammar that govern the sequence of words or application-specific information to determine that only a limited number of words is acceptable after certain key phrases. For example, after the phrase *part number* only numerals may be allowed. Although these approaches help at the word-matching level, they are of little value at the phoneme-matching level.

Deroualt[2] reports that a good strategy for phoneme-based English language processing is to take the longest match available in the vocabulary. Thus, if *cat, your,* and *catcher* were all in the vocabulary, *catcher* would be preferred over the combination *cat your,* since it is the longer match.

Status

The accuracy of speech recognizers depends on the difficulty of the task (it decreases with larger vocabularies) and the techniques used, but is usually from 90 to 98% for most commercial systems. (Note that it is always lower than company advertising literature claims, due to noise in the environment and user inexperience.) A typical spoken clause contains 6 to 12 words. With a 94% recognition rate, the system would, on average, miss one word in every two clauses.

Speech recognition systems have a threshold, which can often be user set, to determine the level in which they are sure of the interpretation of a word. Below that level, they will still be able to guess the correct word most of the time, but the system should request a verification ("Did you mean *nine?*").

Like other fields, speech recognition has had its casualties, with some of the pioneers out of business or pursuing other applications. In a 1983 study,[3] 10 companies were producing speech recognition products. A 1987 study[4] identified 12 companies. But there were only two names in common between these studies.

Ten predictions were made in 1982 for progress over the next five years, and eight have come true. Two examples are reduced price and larger vocabularies. System prices were $10,000 to $100,000, but are now $5,000 to $40,000. Board-level prices were $200 to $2,000, but are now $150 to $1,300. Vocabulary capabilities were approximately 100 words under the best conditions (single speaker, isolated speech, low noise), but are now approximately 1,000 words under the same conditions.

Applications

Limited vocabulary, isolated word, speaker-dependent speech recognition is used in factories in inspection and inventory applications. One example of a voice recognition system built to operate in the noise and dirty environment of a factory is Votan's complementary metal-oxide semiconductor (CMOS) RAM-based system with no moving parts. The total memory system is in RAM rather than on disk.

Direct speech input into robots is farther in the future, although a system that cuts initials into glassware from a spoken input has been demonstrated.

16.3 WALKING VEHICLES (LEGGED LOCOMOTION)

Mobility is a highly active research topic. Most of this work is being done with wheeled vehicles because they are less expensive and more efficient than legged vehicles, but they are not generally usable over rough terrain or for climbing stairs.

How important is movement to a robot? In an IEEE short course in 1983, Dr. Larry Leifer said that only because we are mobile have we been able to learn as much as we have. Travel is one of the main distinctions between plants and animals. He added that robots must be mobile in order to explore new areas and thus significantly increase their skills and abilities. In addition to being an aid to learning, in many applications the robot needs mobility so that it can move to where the work is.

The most versatile locomotion system uses legs rather than wheels. Various problems in control and balance have shown that the most stable configurations are six-legged, although progress is being made on two- and four-legged models.

Goals

The goal in legged locomotion research is a design that is stable, reasonably energy efficient, and easy to control. Progress has been made at several universities in the United States and Japan, but we do not seem to be near to obtaining a commercial version of any of these research vehicles.

Problems

Problems exist in four areas: static balance, dynamic balance, power efficiency, and limb control algorithms. Static balance is balance with the robot standing still. A robot with three or more legs has static balance. A two- (or one-) legged robot must rely on dynamic balance, or balance under motion, which is much harder to accomplish due to the generally high center of gravity of a legged robot. Even a three- or four-legged robot becomes potentially unstable as it moves its legs and requires dynamic balance to move. If three legs can always be kept on the ground at one time, and the center of gravity of the robot is within the leg contact points, static balance is sufficient. This can be accomplished with many six-legged robots.

Efficiency is a problem in legged units, since the power used to raise and lower the leg is mostly wasted, as far as forward motion is concerned. Limb control has been a problem because three types of control are being attempted at once: dynamic balance, forward motion, and turning. These three tasks are each difficult, which explains why few researchers try to provide a natural looking gait as well.

In addition, design choices must be made in two other areas: the number of legs and the gait. In general, more legs means better balance, but control and coordination problems are more difficult. Gait refers to the order in which the legs are moved and to the number of legs on the ground at one time. Some gaits are more stable than others.

Approaches

Most investigators have chosen either a two-legged robot for size, maneuverability, and ease of control, or a six-legged robot for stability and availability of different gaits. One of the most popular six-legged gaits is the alternating tripod gait, in which three legs remain on the ground at all times.

Although most models use electric drive, the models developed by Professor Ichiro Kato of Wasada University have generally used hydraulic systems. Researchers are also looking into the possibility of combining manipulator and locomotion functions in the same leg, to give a walking robot added capability without requiring a separate arm.

Status

One-Legged Robots

Marc Raibert, while at Carnegie-Mellon, developed a one-legged robot that could dynamically balance while hopping around (similar to a pogo stick).[5] His main purpose was to develop the necessary algorithms for dynamic balance. The unit used a remote computer to solve the motion equations, so a tether cable was required between robot and computer. His approach would automatically compensate for unbalancing forces, so even a hard shove against the side of the robot would not knock it over; the robot would just hop off in the direction in which it was pushed. This work is being continued at MIT.

Two-Legged Robots

Several two-legged robots are under development. The Born model, built in Japan, is illustrated in Figure 16-1. It can walk over rough terrain and go up and down stairs.

Professor Ichiro Kato of Wasada University has developed several experimental walking machines. One model, called the WHL-11 (Wasada-Hitachi Leg) and developed in cooperation with Hitachi, is a true walking robot, with links resembling human limbs. The robot has six DOF per leg, compared with seven in humans. It walks slowly, taking 10 s per step. With a step of 20 in, the speed is thus limited to about 0.1 mph. Joint drive is hydraulic. Unlike many U.S. models, this robot is completely autonomous, containing its own hydraulic pump, electric power, and computer system. A different model, the WL-10RD also developed by Kato, takes only about 1.5 s for each 17-in step. It weighs about 84 kg and operates under an onboard Z8002 microcomputer.

The University of Portland is studying the theoretical problems of distributed process control as applied to walking machines. Initially using the 68000, their latest biped now is based on Inmos transputers (a specialized chip with onboard RAM and other features). Balance is obtained by comparing the actual position with the computer intended position and doing 500 of these calcula-

Figure 16-1. Experimental two-legged robot.

tions per second. They have been able to separate the automatic reflexes from motion control, thus simplifying motion control and making the gait appear much more natural.

Four-Legged Robots

There has been less interest in four-legged robots because they are not as stable as six-legged ones, but they are more complex than two-legged robots. Carnegie-Mellon did develop a four-legged horse that could trot. The University of Tokyo designed a four-legged robot called Titan II.

Six-Legged Robots

Ivan Sutherland at Carnegie-Mellon is working on a six-legged hydraulically driven crawling machine. It travels 2 mph while carrying a person.

Ohio State is building a more ambitious machine, called the adaptive suspension vehicle (ASV); see Figure 16-2. This robot weighs $2\frac{1}{2}$ tons, is 15 ft long, 5 ft wide, and 10 ft tall while walking, and can travel 5 to 8 mph. It carries an operator and a 500-lb payload. With an infrared optical radar system, the ASV can detect $\frac{1}{2}$-in-wide objects 30 ft away.

Oregon State University is studying the leg motion of insects to provide a theoretical foundation for next-generation walking machines. Insect legs are not identical to each other, and some can also be used as manipulators.

Kaneko[6] has described a six-legged walking robot using an alternating tripod gait. Named Melwalk-III, this unit attempted to reduce energy consumption by reducing leg weight and by using a single vertical-drive motor, with motive power supplied to all legs through a pantograph type of linkage system.

One famous six-legged walking robot is ODEX-I, developed by Odetics. This robot, which looks like an octopus, can walk to a pickup truck and pick it up! Besides being strong, it is dexterous enough to climb into the truck and small enough to fit into it. ODEX-I has been referred to as the first "functionoid." It weighs 370 lb and can stroll at 4 mph.

Figure 16-2. Adaptive suspension vehicle. *(Courtesy of Ohio State University)*

16.4 COLLISION AVOIDANCE

The robotics industry can be characterized as an emerging high-tech industry. However, the mobile robotics industry is still in its early stages. Improvements in areas such as onboard processing and cost reductions have spurred research into various mobile robot applications. Progress is being made in global path planning and autonomous navigation. However, there is at least one area that still needs attention—collision avoidance.

Research is progressing in all facets of autonomous navigation, and some of the currently available hobby robots offer a limited type of collision detection system. Nevertheless, there is no current robot commercially available that can autonomously detect the presence of an obstacle through noncontact sensing under all conditions and, more importantly, take the appropriate action by going around the object.

Definition

Collision detection and avoidance are two of the four requirements imposed on any robot that is going to move autonomously. It must know where it is and where it wants to go (global-level mapping). And it must know how to reach the goal, considering known obstacles in the way (global-level path planning). Next, as it travels, it must recognize additional, unexpected objects in its path to prevent collisions (collision detection). Finally, it must navigate around these obstacles to continue toward its goal (local path planning). This section addresses these last two questions, which together can be considered collision avoidance.

Goals

The first goal of collision avoidance is to ensure that the robot does not contact an object because of errors in its navigation or because a new object has suddenly appeared in its path. When the robot finds its direct path to a goal blocked, it must find a method to circumvent the blocking object, if possible, by using some type of local path planning. A collision detection system by itself can alert the robot to a potential collision. Separate software is needed to provide a local, alternative path. Collision detection systems must operate in real time while the robot is in motion. Local path planning can be done with the robot stopped.

Problems

Some of the many reasons for the lack of a complete and effective collision-avoidance system are sensor limitations, system time and computing constraints, and path geometry considerations. Collision detection sensors suffer from three types of problems. The first is insufficient coverage. Because of the cone pattern found in most obstacle detection systems, there will always be areas near the robot that the sensor cannot see. Multiple sensors can alleviate some of this problem.

A second problem is to determine the exact location of an obstacle and the exact width of an available opening. Sensor inaccuracies often add inches to the required clearance distance that the robot must have. Another problem is found when attempting to determine an alternative route around the obstacle. The limited coverage of the sensors does not readily support side views, and the placement of the obstacle often blocks part of the optional path from view. With the robot traveling at a reasonable speed, it may not have enough time to detect a potential collision and react to it unless a separate computer is assigned the task of collision detection (and perhaps subsequent path planning).

A final difficulty is due to the three-dimensional problems inherent in collision avoidance, while a two-dimensional approach is often taken for path planning. The robot travels in two dimensions, and one approach often used for global path planning is to shrink the robot to a point and extend the boundaries of all objects by the radius of the robot. This approach allows fewer complex calculations for determining which path will permit the robot to clear any obstacles. Unfortunately, this procedure is usually insufficient in a collision-avoidance situation, unless the robot and objects that may collide with it are smooth and symmetrical. Although acceptable for a research environment, this situation is not what one would expect in an industrial setting. Objects may hang down from a ceiling, protrude from a wall, or rise from the floor, and yet be so placed as to not be seen from any collision-avoidance sensor.

A collision detection system would ideally detect any penetration into a three-dimensional rectangular space directly in front of it and would ignore objects outside of this space. The rectangle's height and width would be identical to the robot's, and the depth would be sufficient to allow the robot time for object detection and reaction under worst-case conditions.

Approaches

There are some excellent papers on the collision-avoidance problem. Canny[7] describes the three basic types of collisions between a moving robot and fixed objects, and considers mathematical methods to predict a collision-free path.

He divides potential collisions between a moving object and an obstacle into three classes:

1. When the moving object face collides with a vertex of the fixed obstacle
2. When a vertex of the moving object contacts a face of the fixed obstacle
3. When an edge of the moving object contacts an edge of the fixed object

Other papers include Ramirez[8] and Thorpe.[9] Ramirez describes an algorithm to handle collision avoidance through the use of stratified zones that surround objects. Thorpe's paper reviews several approaches for generating collision-free paths and recommends one that first approximates the path and then refines the approximation.

Collision detection systems can be designed to err on the safe side; that is, if an object is too near the path of the robot it might stop and examine the situation further. However, unless the environment has plenty of open spaces, the path planning algorithms around the object will need to be more accurate. Some of the issues found in this area were covered in chapter 7.

The type of collision-avoidance system required depends on the rules under which the robot is operating. A mobile robot generally travels under one of four conditions:

1. *Fixed path:* The robot is following a fixed path, such as a buried wire, in which the only action the robot should take on detecting something is to stop.

2. *Known environment:* Following a fixed, predetermined path where collision avoidance will be used to move the robot temporarily from this path and then return to it. This condition is a natural extension of AGV work and is being looked at by various researchers, including the University of Georgia.

3. *Changed environment:* Following a predetermined path between two locations that will vary between runs because the robot can determine a better path based upon its own map of the area or because changes occur in the environment. Examples of changes include doors opened and closed, or movement in goal point location.

4. *Unknown environment:* Exploring a new environment in which no current map exists, and the robot must explore and prepare its own map. For this condition, the goal might be considered to be complete a map of the area. In this case, the collision detection system will only alert the robot to a potential collision. The global mapping program would have to determine alternative paths.

Many current systems are built around multiple sonar sensors for obstacle detection. Crowley[10] reports on a robot with 24 sensors in a ring around its body. Everett[11] reports using seven sensors in the shape of a cross. The robot Hilare uses 14 sensors. Information from these sensors is then plotted on an internal map, and, if the robot's path will take it into this area, alternative path planning is then undertaken.

Once an object has been detected, and a path planned around it, some type of navigation is necessary while the robot is following its alternative path. Dead-reckoning systems, especially those using proportional, integral, differential (PID) algorithms can be very accurate. Cybermation offers dead reckoning to less than $\frac{1}{2}$ in over good floors after very extensive and complex movements of the robot. Most dead-reckoning systems are accurate enough to bring the robot back to its original path.

Status

Both AGVs and mobile robots are supplied with some type of contact bumper that, if activated, will stop the robot. It might be considered a fail-safe backup.

Various research projects have used sensors to map objects in a room. Once the objects appear on the map, the robot can plan a path around them. Efforts at Stanford and Carnegie-Mellon have used vision and ultrasonics to map the area. In most cases this approach was based on the robot stopping every few feet to make another map update. It is not clear that this approach would be satisfactory if the robot is rapidly moving along a known path.

Less effort has been devoted to the related area of local path planning. One promising technique is the composite local mode approach, discussed by Nitzan.[12]

In one MIT project Anita Flynn[13] studied the use of redundant sensors to improve the accuracy of a robot's navigational capability. That study assumed that the robot had a priori a map of the room for planning and monitoring its movement. Although this approach did not address the problem of the presence of new and unexpected obstacles, this subject was planned for a second-phase effort.

The Naval Surface Weapon Center addresses the problem of obtaining sufficient information for mobile robot collision avoidance by using a reconfigurable single lens. The lens has a wide-angle, extremely narrow depth of field and can focus at speeds up to 30 frames per second. By focusing during successive fields at specific distances, it can eliminate backgrounds. This approach thus rejects all items unnecessary for digital processing and only passes on data for processing from objects in the range of interest.

A different approach to collision detection is the use of proximity detectors. These devices can detect when an object is within 6 in. By the use of multiple infrared proximity sensors with different directions of view and restricted fields of reception, both the presence of a potential collision and its direction can be determined. Proximity detectors in general can usually only tell if an object is within a minimum range, not how close it actually is, so the robot must clear objects by at least the range of the proximity detector.

Applications

Saito[14] describes using a gyrocompass to keep a concrete-finishing robot aligned with a preprogrammed route. With this technique, a 0.3% position error can be obtained. The system has no local path planning to handle potential collisions, but relies on an obstacle-free preplanned route.

Previous autonomous robots include the Stanford Cart, the Carnegie-Mellon Rover, and the Hilare from France. This latter robot uses IR beacons in corner reflectors to determine where it is, and three-dimensional vision, laser range finding, and ultrasonics to control movement. Hilare integrates the data from 14 ultrasonic emitter-receivers to detect potential collisions in the range of 2 ms.[15] Vision is also provided from a TV camera, laser range finder, and infrared emitter-receiver.

Investigators at the University of Karlsruhe, West Germany, use vision systems, ultrasonic sensors, proximity sensors, and tactile sensors on their mobile robot.[16]

Crowley[17] refers to three types of obstacles typically found in a real-world environment: stationary/static, stationary/dynamic, and moving obstacles. Stationary/static refers to things like walls or factory workbenches, which are always present at the same position. Stationary/dynamic refers to objects that are stationary when seen by the robot, but whose position could have changed from the last time it was seen, such as an open or closed door or an office chair that had been moved. Dynamic refers to objects that are usually in motion, such as people.

16.5 NEURAL NETWORK COMPUTING

Definition

An artificial neural network is a type of computer system that uses a large number of elements in parallel and that adjusts the interconnections of those elements based on information presented to it. In other words, it has the capability to learn—an ideal arrangement for pattern recognition. Artificial neural networks are being examined to determine if they can be used to improve a computer's visual and speech recognition capabilities.

Work in this area began during World War II, and the concept of the Perceptron showed some promise during the 1950s. However, not until the

1980s did research, especially by Hopfield,[18] lead to major improvements. This activity was spurred both by new analog VLSI techniques and progress in understanding the human brain. Neural nets thus offer perhaps the only method of obtaining the necessary level of complex processing that can duplicate some human pattern recognition feats.

Goals

Current computer architectures are limited when trying to solve pattern recognition problems. Two major limitations are the serial method used by computers to solve problems, thus requiring long processing times to recognize complex images, and the fact that the programming techniques are better suited to mathematical computations (conventional languages) or symbolic processing (AI languages) rather than pattern recognition processes. Neural network techniques are poorly suited to mathematics but directly applicable to learning patterns.

One goal of neural networks is to develop a computer that uses processes similar to that of the brain (which is very good at pattern recognition) and to use this system to solve such currently difficult tasks as understanding speech and recognizing objects.

A second goal being pursued by such companies as Neurogen, Neural Systems and Hecht-Nielsen Neurocomputers is to provide the adaptive learning necessary for real-time robot manipulator control.

Problems

Neural networks are still in such an early stage that it is difficult to list all of the problems. One problem applicable to vision is the apparent necessity of having a network with more total nodes in it than picture patterns being recognized. Originally, Hopfield felt that the number of patterns that his approach could handle, when used as a content addressable memory, would be less than 0.15 times the number of nodes in the net. Other research work indicates that it can handle much greater numbers. Hitachi Corp. has developed a combined neural net front end, an S820 supercomputer and fuzzy logic to simulate a system with 1 million neurons.

Since these networks allow learning from experience, in which exemplars (an advanced template) are stored and modified on a real-time basis, another problem is the possibility of an unstable exemplar pattern if there are too few differences between its pattern and another pattern. For example, it would be possible for a valid exemplar pattern for an "O" to break down under noisy conditions and join with a "C" pattern.

Another problem is a requirement for multiple passes through all the data before a decision is valid. For this reason, the greatest potential for these nets is the high-speed processing that can occur with implementation in VLSI architecture.

Approaches

Lippmann[19] has an excellent overview of neural network computers. Some of the material here is adapted from his paper. Other information came from other references and conversations with experts. Lippmann discusses six types of network approaches: the Hopfield net, Hamming net, Carpenter/Grossberg classifier, the Perceptron, a multilayered Perceptron, and the Kohonen self-organizing feature maps.

Although they all show promise, we will describe only the Hopfield net. Hopfield's[20] approach to modeling the brain's neural circuits seems to give AI researchers a better method of solving certain types of pattern sensitive problems and developing an associative memory. Hopfield has extended the original work in one type of neural network model, the nonlinear graded response neurons. Much early work on neurons assumed they were on or off (they fired or they did not fire). Rosenblatt used this approach in studies of the Perceptron, using feedforward principles. Later studies, such as those by Hartline, used linear models to explain edge detection in the eye. Circuit networks were then designed based on the graded response model. These networks were quite applicable to pattern recognition problems.

Hopfield made two simplifying assumptions in circuit design. He neglected electrical effects that are a function of neuron shape, and he neglected some small internal time delays. He then modeled a network to include cross-connections and nonlinear elements. Such networks operate in parallel, often with nonlinear elements. Parallel processing permits an increase in overall speed and several alternatives to be examined simultaneously. Nonlinear elements allow weights to be assigned based upon the computer learning from its past experiences.

In Hopfield's approach, the optimization problem is mapped into a neural network such that the various configurations correspond to possible solutions. When the circuit is given a new set of inputs, it finds a new stable set of values that minimizes the amount of energy required. Thus, the solution with minimum network energy is chosen, resulting in optimal pattern selection.

Neural networks allow learning on a continuous basis. In older approaches the system was taught once, and everything else was compared to that stored template. This learning capability is extremely important to speech and visual processing, since small variations in otherwise identical signals will always occur. This ability is based on three factors: large connectivity, analog response, and reentrant connections. The resulting computations bear no resemblance to Boolean logic, or to standard computer architecture. To keep the number of connections at a minimum, we can use a hierarchical approach.

Figure 16-3 shows the steps involved in the recognition of a noisy "3" through the use of a Hopfield net. A network containing 120 processing elements was trained using eight exemplar patterns. The input was corrupted by randomly reversing each bit with a probability of 0.25. After seven iterations, the recognition was perfect.

In summary, neural networks do not follow the conventional sequential

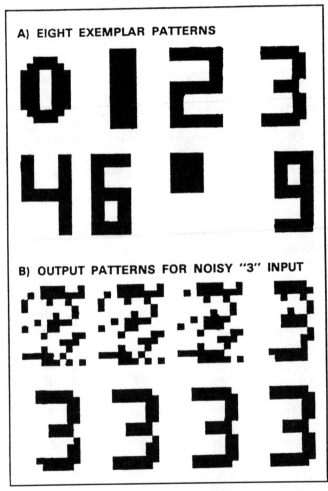

Figure 16-3. Neural net example. (*Courtesy of Richard P. Lippman*)

processing of normal (von Neumann architecture) computers. Their non-linearity allows problems to be solved that could not otherwise be solved.

Status

Perhaps the only circuit which has been fully developed to date is an analog-to-digital converter that can operate within a microsecond. Neural nets have also been applied to the problem of text-to-speech conversion and to solving the multicity traveling salesman problem.

As an example of what can be done in the associative memory area, memories based on 32 neurons have been built. They are particularly applicable to vision problems because of the large amount of data involved and the potential for data overload. In the speech area, Sejnowski[21] developed a speech synthesizer using back-propagation that showed 78% accuracy with a 439-word vocabulary.

One result of this research is the renewed understanding that analog preprocessors may be needed for any real-time pattern generation hardware. Analog processing techniques, discarded in the 1950s in favor of digital processing, appear to again offer unique capabilities.

Some researchers hope that existing speech and pattern recognition algorithms can be used in these networks. However, others believe that it is the ability of these networks to try new approaches that offers the most potential. DARPA, already funding speech recognition, will fund neural network research starting in 1989. Initial funding will be at the $33 million level for evaluating neural networks against conventional technology.

Applications

Although these techniques can be used in many applications, including manipulator control, their greatest potential seems to be in pattern recognition, especially speech and image processing. The potential benefits include a greater degree of robustness, because the failure of a single node will not cause the whole network to fail, as well as their ability to continually improve their performance through learning.

What Hopfield has showed is that there is an approach using these neural simulated models that can be used to optimize problem solution. He gives an example of determining what three-dimensional object comes closest to providing an observed two-dimensional shadow. This problem-solving approach is further applicable to edge detection, stereo ranging, and motion detection. Hopfield[22] believes the recognition of phonemes in speech is theoretically similar to some vision processing. The biggest difference is the large variability between phonemes. In this respect, Hitachi's approach using sound fractals to distinguish between phonetic elements is a major step forward.

REFERENCES

1. Waldrop, M. Mitchell. Natural language understanding. *Science*, Vol. 224, No. 4647, April 27, 1984.
2. Deroualt, Anne-Marie, and Bernard Merialdo. Natural language modeling for phoneme to text transcription. *IEEE Transactions on Pattern Matching and Machine Intelligence*, Vol. PAMI-9, No. 6, Nov. 1986, pp. 742–749.
3. Poole, Harry H. Computer recognition of speech. *The Handbook of Computers and Computing*, A. H. Seidman and I. Flores, eds. New York: Van Nostrand Reinhold, 1984, pp. 186–197.
4. Wallich, Paul. Putting speech recognizers to work. *IEEE Spectrum*, Vol. 24, No. 4, April 1987, pp. 55–57.
5. Raibert, M. H., and I. E. Sutherland. Machines that walk. *Scientific American*, Vol. 248, No. 1, Jan. 1983, pp. 44–53.
6. Kaneko, M., et al. Basic experiments on a hexapod walking machine (MELWALK-III) with an approximate straight-line link mechanism. *Proc. '85 ICAR International Conference on Advanced Robotics*. Tokyo: Japanese Industrial Robot Association, 1985, pp. 397–404.
7. Canny, John. Collision detection for moving polyhedra. *IEEE Transactions on Pattern Analysis and Machine Intelligence*, Vol. PAMI-8, No. 2, March 1986, pp. 200–209.
8. Ramirez, C. A. Stratified levels of risk for collision free robot guidance. *Proc. 15th International Symposium on Industrial Robots*. Tokyo: Japanese Industrial Robot Association, Sept. 1985, pp. 959–966.
9. Thorpe, Charles E. *Path Relaxation: Path Planning for a Mobile Robot*, Technical Report CMU-RI-TR-84-5. Pittsburgh, Pa.: Carnegie-Mellon University, April, 1984.
10. Crowley, J. L. Navigation for an intelligent mobile robot. *IEEE Journal of Robotics and Automation*, Vol. RA-1, No. 1, March 1985, pp. 31–41.
11. Everett, H. R. A multielement ultrasonic ranging array. *Robotics Age*, July 1985.
12. Nitzan, David. Development of intelligent robots: Achievements and issues. *IEEE Journal of Robotics and Automation*, Vol. RA-1, No. 1, March 1985, pp. 3–13.
13. Flynn, Anita M. *Redundant Sensors for Mobile Robot Navigation*, Technical Report #859. Cambridge, Mass.: MIT Artificial Intelligence Laboratory, 1985.
14. Saito, M., et al. The development of a mobile robot for concrete slab finishing. *Proc. 15th International Symposium on Industrial Robots*. Tokyo: Japanese Industrial Robot Association, Sept. 1985, pp. 71–78.
15. Julliere, M., L. Marce, and H. Place. A guidance system for a mobile robot. *Proc. 13th International Symposium on Industrial Robots*. Chicago: Robotics International of SME, April 1983, pp. 13-58–13-67.
16. Dillman, R., and U. Rembold. Autonomous robot of the University of Karlsruhe. *Proc. 15th International Symposium on Industrial Robots*. Tokyo: Japanese Industrial Robot Association, Sept. 1985, pp. 91–101.

17. Crowley, J. L. *Dynamic World Modeling for an Intelligent Mobile Robot Using a Rotating Ultra-Sonic Ranging Device,* Technical Report CMU-RI-84-27. Pittsburgh, Pa.: Carnegie-Mellon University, Dec. 1984.

18. Hopfield, John J., and David W. Tank. Computing with neural circuits: A model. *Science,* Vol. 233, Aug. 8, 1986, pp. 625–633.

19. Lippmann, Richard P. An introduction to computing with neural nets. *IEEE ASSP Magazine,* Vol. 4, No. 2, April 1987, pp. 4–22.

20. Hopfield and Tank, Computing with neural circuits.

21. Sejnowski, T., and C. Rosenberg. *NETtalk: A Parallel Network That Learns to Read Aloud,* Technical Report JHU/EECS-86/01. Baltimore, Md.: Johns Hopkins University, 1986.

22. Tank, D. W., and J. J. Hopfield. Simple neural optimization networks: An A/D converter, signal decision circuit, and a linear programming circuit. *IEEE Transactions on Circuits and Systems,* Vol. CAS-33, No. 5, 1986, pp. 533–541.

NEW APPLICATION AREAS

Previous chapters address robot technology as it exists today and research activities that are moving forward. This chapter examines some of the current and future uses for robots.

17.1 TASKS FOR ROBOTS

Robots can do many types of tasks, and their future potential is almost unlimited. However, robots are expensive and have limitations, so there is a natural tendency to only use them where they are more suitable than human labor. In addressing their applicability, Nitzan[1] says that robots should do tasks that are (in order of priority)

1. Lethal to a human (e.g., a high radiation environment)
2. Harmful to a human (e.g., handling toxic chemicals)
3. Hazardous (e.g., firefighting)
4. Strenuous (e.g., loading heavy items)
5. Noisy (e.g., riveting)
6. Boring or dull (i.e., assembly)

To this list, we would add tasks that require

7. Environments incompatible with human presence (e.g., class 1 cleanroom part handling)
8. High levels of attention (e.g., PC board inspection)

9. High repeatable accuracy (e.g., precision drilling)
10. Trained people (who may be difficult to find)

17.2 CURRENT APPLICATIONS

Robots can be used for a much broader range of applications than the ten standard applications discussed in chapter 14, and this section looks at some of the nonstandard uses.[2]

AGV Systems

AGVs are one of the few types of mobile robots now available, and current models have many capabilities. Murata offers a series of AGVs as part of their Robo-Family line. These robots are used in automated warehouses and for carrying parts to assembly areas within manufacturing plants. One of their larger models is the RoboCarrier, a type of robot train, which can carry 5,000 kg (11,000 lb) and travels on a rail.

Eaton Kenway also offers a model on rails, the ML 200 small parts storage/retrieval system. It retrieves and returns bins to their designated storage places, allowing operators to find and retrieve parts. Storage bin capacity is 500 lb each, with a maximum of 30,000 bins available.

Several autonomous AGV systems are under development. One is an autonomous parts delivery cart, a combined project between Tektronix and Stanford University. Although the system is based on optical tracking to follow various standard paths, its autonomy shows up at branch points and when objects block the main path. It can also recognize various proximity stations along the route.

Public Safety Telerobots

Telerobots are becoming increasingly important in several areas. One application is public safety. Teleoperated systems are used for police, fire, and bomb disposal work. The Pedsco Mark-3 (Fig. 17-1) is a low-cost ($20,000) system built by a Canadian firm and used by the New York City Police Department. This robot can be used to get a closeup picture of something that may be dangerous, or it can retrieve a small object, such as a bomb.

One model is equipped to fight back. An English firm has developed RO-Veh (remotely operated vehicle), which fits in the trunk of a small hatchback car and is controlled through a 300-ft cable. It has numerous attachments, including a standard TV camera, a thermal image camera, floodlights,

Figure 17-1. NYPD teleoperated robot. *(Courtesy of New York City Police Department)*

x-ray equipment, water hose, loudspeaker, gripper, and armaments—a bank of slug-throwing bomb disrupters, a five-shot shotgun, and a device that can drop explosives equipped with remote detonators.

Telerobots are also being developed to fight fires. The RO-Veh model can find smoke victims through its thermal image camera and drag them to safety. Another fire-fighting robot is being developed by the U.S. Navy[3] at Aberdeen Proving Ground, to be used on flight decks. Dubbed the A/S32P-16, the vehicle is similar to a heavy-duty industrial truck. It carries a 375-gal tank containing aqueous film-forming foam (AFFF). The vehicle is driven to the scene of a fire. The driver then leaves the vehicle, carrying a control box, connected to the robot by a 250-ft umbilical cable, for remotely operating the fire-fighting apparatus.

Marine Applications

Robots have been used in marine applications, both as underwater teleoperated robots and as stand-alone mobile robots. A typical example of a teleoperated robot is the Scorpio, shown in Figure 17-2. This robot is used in underwater exploration and recovery, and has both a video camera (seen in the lower right of the picture) and a multi-jointed arm (seen in the lower center).

The Oceanographic Institute at Woods Hole, Massachusetts, has also developed a small research robot, Jason Jr., to explore underwater sites. It was used

Figure 17-2. Underwater robot. (*Courtesy of Straza Division, AMETEK*)

in 1986 to investigate the causes of the *Titanic* shipwreck. A model used by the French Navy for undersea remote viewing systems is the Eric-II, which can be used to a depth of 18,000 ft. Robotic Systems International of Sidney, British Colombia, created a robotic manipulator set of arms that was used during the *Challenger* disaster to recover wreckage from the sea.

Some marine robots are not teleoperated. One such robot, developed by the French firm Nomed Shipyards, can paint boats, remove barnacles, perform high-pressure washing, and inspect structural welds with gamma rays. The robot has three legs with electromechanical adhesion pads and vacuum grippers attached to allow it to walk up steel ship hulls at the rate of 8.2 ft/min, which allows it to clean over 50,000 ft^2 of hull a day. The robot carries an arm, end-effector, and an ultrasonic obstacle detection system. It receives its power via a low-voltage cable that also contains fiber-optic data links to its remote computer.

Mobile Robots

The Japanese have invented a concrete-slab-finishing robot for the construction industry.[4] The robot can travel over the newly poured cement and trowel it smooth. Another mobile robot under development by the Japanese and several U.S. firms is a mobile security guard, or night watchman.

It is now possible to use mobile robots with an accurate dead-reckoning navigation system (Cybermation) for transporting parts around a factory. Another mobile robot being developed is a general-purpose model from Cyberotics. This robot includes a novel wall-following and door-finding system. Unfortunately, we are still a few years away from commercially available mobile robots with fully operable collision-avoidance and navigation systems.

Agriculture and Food Industry

One potentially very important area is the food industry. The robot farm worker will be able to relieve humans from backbreaking seasonal jobs, and the robot factory food worker can inspect produce without being bored. Several firms have developed vision-based inspection systems for the food industry. They are being used to detect broken yolks, sort and grade cucumbers, and inspect candy. One example is the citrus grading machine developed by Sunkist Corporation which handles about eight pieces of fruit per second and can grade oranges or lemons according to size, color, blemishes, and frost injury.

Another sorting application combines a conveyor belt system with machine vision for handling and sorting fruits and vegetables. Optasort, of Reedley, California, has developed a system that holds individual produce in a cup.

Once the product has been identified and sorted by size through data from the visual system, the cup is tilted at the appropriate location into the proper size sorting bin. Their system handles everything from small shrimp to bulky pineapples.

The British company PA Technology has teamed with the West German manufacturer Otto Haensel Machine Company to develop a chocolate-decorating robot. This machine combines a vision system and a manipulator-controlled nozzle to decorate 60 pieces of chocolate each minute. The robot is able to recognize and reject odd sizes of chocolates. The system costs about $150,000.

Suzumo Machinery Industry of Japan introduced robots in California in 1985 that make shari (rice cakes), which are the basis of sushi dishes. The robot costs $12,000. At Arthur D. Little, Donald L. Sullivan is developing a robot meat inspector that looks at a product, such as a slice of bacon, and determines if it has too much fat.

A robot is being developed to recognize types of fish and measure their lengths. Based on this information, it then determines the fish's approximate weight. This robot is planned for use aboard boats to keep a running account (in pounds) of each type of fish caught.

Robots will even be used down on the farm. Researchers in Japan have produced an automatic grain harvester that can be placed in a field and shown, by an operator, where the boundaries are. The combine will then harvest grain until finished, keeping track of where it had previously cut by the location of cut patches. Another system offering promise, but still in development, is a robot cattle feeder.

A unique robot was developed in Australia.[5] A few years ago farmers contacted several large robotics firms and asked if it were possible to develop a robot that could economically shear wool without hurting the sheep. They were told it could not currently be done, so the University of Western Australia developed and tested a sheep-shearing robot on its own. After the sheep is placed in a holding pen, the robot is able to shear the entire sheep. The robot has a unique collision-avoidance system that controls the angle of the cutter and its closeness to the skin. Figure 17-3 shows this practical robot, which has successfully sheared thousands of sheep.

Still in the laboratory stage is a robot that can transplant seedlings. Currently too slow for most applications, the robot performs at one fifth the speed of humans and misses about 1 seedling out of every 36.

Pipe-Cleaning Robots

Robots have been developed to maintain large pipes. A French firm has developed a three-wheel pipe maintenance robot that crawls along the outside of

Figure 17-3. Sheep-shearing robot. *(Courtesy of University of Western Australia)*

nonburied pipes searching for leaks. Both eddy current and ultrasonic sensors are used to detect breaks in the pipe.

Japan also has a three-wheeled pipe cleaning robot (MOGRER) that actually maneuvers inside the pipe. The robot uses a scissorslike arrangement to enable its legs to expand and contract, thus allowing it to work with pipes of various sizes. It also carries a CCD camera for remote viewing.

Speech Recognition

Practical robots that can understand speech are just beginning to be introduced. The organ-playing robot WABOT-2 (see chap. 15) has speech recognition and speech synthesis capabilities. An English robot known as Robuk employs speech recognition to obtain the information necessary to engrave initials in glassware. This system of voice recognition is good enough to work in a noisy factory environment. After the operator tells the robot the appropriate initials and the quantity, the system engraves the glassware with a Cincinnati Milacron T3-726 robot.

Many cars are available with speech synthesis systems. How far in the future is a car that can listen to you? Perhaps not very far. A voice-operated passenger car has been developed by Nissan Motor Co. Although there are no current plans for export, Nissan states it adds $4,000 to the price of a typical car. Not to be outdone, Renault is also developing a speech recognition system for its passenger cars.

Aid to the Handicapped

Speech recognition is also being used as an aid to the handicapped. A voice-controlled wheelchair manipulator was developed in the late 1970s by NASA's JPL and evaluated at the VA Prosthetics Center in New York City.

Larry Leifer at the VA Hospital in Palo Alto is involved in a project that uses a robot arm to respond to commands from a paralytic. In one demonstration, the arm opened a refrigerator, removed food, placed the food in a microwave, closed the door, set the timer, and turned it on. The robot then waited until the food was cooked, took the food tray out, and served it, all under direct verbal command of the operator. The commands were high-level, task-oriented rather than low-level, movement-oriented.

Dr. Leifer is also running tests on a robot nurse with a 3-ft-long robot arm on a mobile base. A handicapped person can talk the machine through such complicated tasks as cooking a meal, serving the food, playing board games, turning the pages of a book, and picking up the phone.

Meldog Mark 1 is a robot guide dog for the blind. It uses sonar sensing to match its master's speed as it travels by his or her side. This dog, shown in Figure 17-4, warns the person if he or she leaves a predetermined safety zone or if obstacles are in the way. Intelligent disobedience has even been built into the dog. Unsafe commands are ignored.

Aid to Surgeons

Robots are beginning to assist the surgeon in the operating room. At Robots 11, Dr. Kwoh, of Long Beach Memorial Hospital, demonstrated a robot that assists during brain biopsies. The patient's head is firmly held in position by a steel ring. A CAT scan of the brain defines the three-dimensional location of the suspected tumor. An operator designates the tumor location on a terminal connected to the CAT system, and the robot arm then moves into position, directly over the required entrance point in the skull. Even the depth of the biopsy needle is specified. For safety reasons, current systems require a surgeon to actually drill the small hole and perform the biopsy. However, the surgeon does it through a tube positioned and held in place by the robot.

Figure 17-4. Experimental robot guide for the blind.

Another area of robot use by surgeons was pioneered at the University of Tennessee. This system uses a micromanipulator made by Laser Industries of Tel Aviv to remove a tumor through carbon dioxide laser surgery.

Perhaps the best current candidate for telesurgery (using a remote-controlled robot) is the eye. The earliest work in this area was done in 1979, when a device constructed at the UCLA School of Medicine was tested at the Jules Stein Eye Institute under Dr. Manfred Spitznas.

Miscellaneous Applications

A specially developed French robot is being used to locate and identify color-coded wires within a telephone cable. This robot has a color vision system. By using a single camera, a reflecting prism, and two mirrors, it obtains the necessary three-dimensional information to precisely locate wires.

A robot developed in the Peoples Republic of China uses advanced vision systems and advanced tactile sensors to recognize and pick up low-contrast and fragile items, such as glassware and plastic parts, from a moving conveyor belt. The vision system locates the part, and the tactile system provides slippage information to allow the fragile parts to be picked up with a minimum of force.

One interesting project in Sweden is a specialized robot developed by ASEA that glues and assembles windshields. First, the glass base is positioned within 1.5 mm. Then the glass is cleaned, the surface is primed, and glue is applied. Optical sensors, a glue flow sensor, and a gripper sensor, to ensure the part is in place, are used.

A university in Korea has developed a robot that can pick up and recognize the shape of a piece of a jigsaw puzzle.[6] It then orients and places the piece in the proper place and selects another piece. The robot uses a technique based on shape recognition (through edge detection) and does not recognize the actual photographic scene image, so several pieces of the same shape would be used interchangeably.

The Zero toy robot, manufactured by an English firm, is a line-following and line-drawing robot. It can recognize a line drawn on a piece of paper and follow the line as it extends around the paper. Basically very simple, the image detector has a single-diode light source and two or three light sensors. Mobility is provided with independent motors on two of its three wheels. This robot illustrates what can be accomplished with only one integrated circuit (serving as its brain) and a few other parts. It can potentially be expanded into a new type input device that could automatically enter drawings into a computer.

17.3 FUTURE ROBOTICS APPLICATIONS

Japanese Predictions

JIRA conducted a study to determine expected results of its technology research and future application areas of this technology.[7] Chapter 15 discussed their technology findings. This chapter examines some of the many new applications they estimate to be available from 1992 to 1999.

Table 17-1 summarizes 28 new applications for robots in 14 different fields. Lines represent the expected development time as well as when the robot may be available. The applications were divided into three groups. Those robots performing relatively simple, repetitive work are capable of being developed with current technology. They would have low intelligence, limited mobility, and require limited gripper motion. Examples of some in development include fertilizer disseminating robots, egg inspection robots, and ship tank cleaning robots. The second group of robot applications would begin development in 1989, after some technology improvements in mobility, hand manipulation, and limited intelligence become available. Examples are garbage collection, nuclear plant inspection, and tree felling. In the medical area, the Japanese predict that robots that can provide some nursing functions and others that can transport patients in wheel chairs will be available by 1993. The third group requires near-human-level skills in mobility and hand manipulation and more intelligence than can be found now in the laboratory. It is expected that technology improvements will allow robot development in this group to begin in the early 1990s. Examples include models for sea rescue, electrical and communications line stringing, and space station construction.

In summary, the Japanese are predicting numerous new applications for robots in a wide range of industries, and these new robots should be available

Table 17-1. Period of Materialization for New Applications

Field	86	87	88	89	90	91	92	93	94	95	96	97	98	99
Agriculture					Fertilizer Dissemination Robots									
								Crop Conveyance Robots						
Livestock							Stable Cleaning Robots							
				Egg Inspection/Bagging Robots									
Forestry									Tree Felling Robots					
							Lumber Collection Robots							
Marine development and fishery									Robots to Extract Sea Bed Minerals					
												Sea Rescue Robots		
Construction, mining							Scaffold-building Robots							
							Pit Face Work Robots							
Transportation, cargo handling					Stock Handling Robots									
							Ship Tank Cleaning Robots							
Gas and water utilities								Robots for Inspection of Buried Gas Lines						
								Robots for Internal Pipe Inspection					

Electric power
and telephone — Line Stringing
— Transmission Line Inspection Robots

Nuclear power
plants — Welding Inspection Robots
— Nuclear Plant Inspection

Aerospace
applications — Space Station Construction Robots
Astronaut Rescue Robots

Medical and
welfare — Nursing Robots
Wheelchair Robots

Waste disposal,
cleaning — Sorting Non-combustibles
Garbage Collection Robots

Fire fighting,
disasters Robots to Aid at Fire Scenes — Fire-fighting Robots

Service,
other — Filing Robots for Libraries
Robots for Inspection of Oil Tanks

Source: After Kanji Yonemoto, Ichiro Kato, and Kensuke Shima. Technology forecast on industrial robots in Japan. Proc. 15th International Symposium on Industrial Robots. Tokyo: Japanese Industrial Robot Association, 1985, p. 56.

in Japan within 5 to 10 years. The author has seen firsthand some of this development work and concurs in their prediction.

Application Assessment

Factory applications will continue to be the main use of robots, and they will be expanded as more accurate, lower-cost, and better vision systems appear. Standard industrial robots can also be used in other environments. For example, a basic robot could make sandwiches in a fast-food shop or do some construction activities, and an AGV could empty wastepaper baskets or deliver mail.

A much larger range of applications will open up only after the development of practical, low-cost, autonomous mobile robots that extend the physical range of robots. These first mobile robots need not have a global mapmaking capability; they can easily be led over the desired route, or a local map could be downloaded to them. But they must have collision-avoidance and local path planning capabilities and (for some tasks) better speech recognition, vision systems, and higher intelligence.

Based upon our own observations and projections from the Japanese, French, and U.S. experts, we expect the first true autonomous mobile robot with collision-avoidance capability to be available in 1991. Thus, sufficient improvements should be available within a couple of years to allow mobile robots to start moving out of the university laboratory and into business. When that happens, many potential applications will open up.

In one brainstorming session among individuals with different backgrounds, many potential applications were examined. The ideas are summarized in Table 17-2. Although no constraints were placed on the initial brain-

Table 17-2. Future Applications for Robots

Factory applications	Security, trash pickup, plant maintenance
Domestic applications	Dishwashing, floor cleaning, window washing
Fast-food restaurants	Order taking, cooking, cleaning tables
Library robots	Filing books, retrieving books, answering questions
Entertainment robots	Serving hors d'oeuvres, telling stories, playing music
Construction robots	Iron work, rough carpentry, concrete smoothing
Hospital robots	Patient transport, food servers, nurses' aide
Automated farm	Animal feeding, crop planting, harvesting
Lawn care	Watering, grass cutting, maintaining golf greens
Department stores	Stock handling, automated customer carts, automated checkout

storming, several constraints were placed on items before they were listed in the table. These constraints included general practicality of the approach, current status of the required technology, perceived need, and sufficient interest to make it commercially viable.

With a few exceptions, this list of 30 potential applications does not match the projected applications given by JIRA, nor some of the work being done under the French-supported research. It is, however, based upon known interest by various companies and known needs by various users. In addition, there is at least one application in each of the 10 areas that is currently in the development or product demonstration stage. For example, Borenstein describes his prototype robot nurse in a communication published during 1988.[8]

REFERENCES

1. Nitzan, David. Development of intelligent robots: Achievements and issues. *IEEE Journal of Robotics and Automation,* Vol. RA-1, No. 1, March 1985, pp. 3–13.
2. Poole, Harry H. A World View of Advanced Robotic Applications. Paper presented at IEEE Advanced Robotics Seminar, North Kingstown, R.I., Oct. 1986.
3. Tarr, Tomas F., William E. Cutlip, and Sharon M. Hogge. Navy fire fighting truck performance enhancement through remote control. *Proc. Robots 10 Conference.* Dearborn, Mich.: Robotics International of SME, April 1986, pp. 5-67–5-98.
4. Saito, M., et al. The development of a mobile robot for concrete slab finishing. *Proc. 15th International Symposium on Industrial Robots.* Tokyo: Japanese Industrial Robot Association, Sept. 1985, pp. 71–78.
5. Kovesi, D. Collision avoidance. *Proc. 85 ICAR: International Conference on Advanced Robotics.* Tokyo: Japanese Industrial Robot Association, Sept. 1985, pp. 51–58.
6. Oh, S. R., J. H. Lee, K. J. Kim, and Z. Bien. An intelligent robot system with jigsaw-puzzle matching capability. *Proc. 15th International Symposium on Industrial Robots.* Tokyo: Japanese Industrial Robot Association, Sept. 1985, pp. 103–112.
7. Yonemoto, Kanji, Ichiro Kato, and Kensuke Shima. Technology forecast on industrial robots in Japan. *Proc. 15th International Symposium on Industrial Robots.* Tokyo: Japanese Industrial Robot Association, Sept. 1985, pp. 51–58.
8. Borenstein, Johann, and Yoram Karem. Obstacle avoidance with ultrasonic sensors. *IEEE Journal of Robotics and Automation,* Vol. 4, No. 2, April 1988, pp. 213–218.

APPENDIXES

These appendixes provide the reader with sources of additional information in a number of areas. The first appendix is a list of major robotic system manufacturers in the United States, Europe, and Japan. The second appendix provides a partial listing of major universities currently active in robotics research, with most of these located in the United States. Appendix three offers a list of the major international robotics organizations with their addresses.

ROBOTIC SYSTEMS MANUFACTURERS

We have tried to list most major robot manufacturers. Part 1 covers firms with manufacturing plants or major sales offices in the United States; Part 2 lists foreign addresses.

Part 1: United States Companies

Accuratio Systems Inc.
1250 Crooks Road
Clawson, MI 48017
(313) 288-5070

Action Machinery Co.
1725 NW 24th
Portland, OR 97210
(503) 278-2987

Adept Technology
150 Rose Orchard Way
San Jose, CA 95134
(408) 432-0888

AKR Robotics
35367 Schoolcroft
Livonia, MI 48150
(313) 261-8700

American Cimflex Corp.
121 Industry Drive
Pittsburgh, PA 15275
(412) 787-3000

Anorad Corp
110 Oser Ave.
Hauppauge, NY 11788
(516) 231-1995

Applied Robotics
10 Northern Blvd.
Amherst, NH 03031
(603) 883-9706

ASEA Robotics, Inc.
16250 West Glendale Drive
New Berlin, WI 53151
(414) 785-3400

Automatix, Inc.
1000 Tech Park Drive
Billerica, MA 01821
(508) 667-7900

Cincinnati Milicron Inc.
Industrial Robot Division
4701 Marburg Ave
Cincinnati, OH 45209
(513) 841-6200

Conco-Tellus Inc.
100 Tower Drive
Burr Ridge, IL 60521
(312) 789-0333

Control Engineering
34375 W. 12 Mile Road
Farmington, MI 48018
(315) 533-1293

Cybermation Inc.
5457 JAE Valley Rd.
Roanoke, VA 24014
(703) 982-2641

Cybotech, Inc.
P.O. Box 88514
Indianapolis, IN 46208
(317) 298-5890

Devilbiss Company
300 Phillips Avenue
Toledo, OH 43692
(419) 470-2169

Eaton-Kenway
515 East 100 South
Salt Lake City, UT 84102
(801) 530-4000

FMC Corporation
3400 Walnut St.
Colmar, PA 18915
(215) 822-4300

GCA Corporation
One Energy Center
Naperville, IL 60566
(312) 369-2110

GMF Robotics Corp.
Northfield Hills Corp. Center
5600 New King Street
Troy, MI 48098
(313) 645-1517

Graco Robotics Inc.
12898 Westmore Avenue
Livonia, MI 48150
(313) 523-6300

Henry Mann Inc.
3983 Mann Road
Huntingdon Valley, PA 19006
(215) 355-7200

Hirata Corp of America
3901 Industrial Blvd.
Indianapolis, IN 46254
(317) 299-8800

IBM Corporation
Advanced Manufacturing Sys
1000 N.W. 51st Street
Boca Raton, Florida 33431
(305) 998-1701

Intelledex Inc.
4575 SW Research Way
Corvallis, OR 97333
(503) 758-4700

International Robotmation/
Intelligence
2281 Las Palmas Drive
Carlsbad, CA 92008
(619) 438-4424

C. Itoh & Co. (America)
24085 Research Drive
Farmington Hills, MI 48024
(313) 855-1290

Kawasaki Heavy Industries
375 Park Ave.
New York, NY 10152
(212) 759-4950

Kuka Welding Systems
40675 Mound Road
Sterling Heights, MI 48078
(313) 977-0100

Liebherr Machine Tool
1465 Woodland Drive
Saline, MI 48176
(313) 429-7225

Manca-Ameco Automation
165 Carver Avenue
Westwood, NJ 07675
(201) 666-4100

Microbot Inc.
453-H Ravendale Drive
Mountain View, CA 94043
(415) 968-8911

Mobot Corporation
110 Technology Parkway
Norcross, GA 30092
(404) 448-6700

Modular Robotics Industries
814 Chestnut Street
Rockford, IL 61105
(815) 964-8666

NEC America, Inc.
8 Old Sod Farm Road
Melville, NY 11747
(516) 753-7000

Netzsch Incorporated
119 Pickering Way
Exton, PA 19341
(215) 363-8010

Oriel Corporation
250 Long Beach Blvd.
Stratford, CT 06497
(203) 377-8282

Panasonic Industrial Co.
One Panasonic Way
Secaucus, NJ 07094
(201) 348-7000

Pickomatic Systems
37900 Mound Rd.
Sterling Heights, MI 48077
(313) 939-9320

Prab Robots, Inc
P.O. Box 2121
Kalamazoo, MI 49003
(616) 329-0835

Precision Robots, Inc.
6 Cummings Park
Woburn, MA 01801
(617) 938-1338

Reis Machines, Inc.
1150 Davis Road
Elgin, IL 60123
(312) 741-9500

Robotic Systems & Controles
3206 Lanvale Avenue
Richmond, VA 23230
(804) 355-2803

Seiko Epson Corp.
683 W. Maude Ave.
Sunnyvale, CA 94086
(408) 773-9797

Spine Robotics
1243 Chicago Rd.
Troy, MI 48063
(313) 589-9085

Thermwood Corp.
P.O. Box 436
Dale, IN 47523
(812) 937-4476

Tokico America Inc.
15001 Commerce Drive N.
Dearborn, MI 48120
(313) 336-5280

Toshiba/Houston International
13131 West Little York Road
Houston, TX 77041
(713) 466-0277

Westinghouse Automation Div.
200 Beta Drive
Pittsburgh, PA 15238
(412) 963-4000

United States Robots, Inc
2463 Impala Drive
Carlsbad, CA 92008
(619) 931-7000

Yaskawa Electric America
3160 MacArthur Blvd.
Northbrook, IL 60062
(312) 291-2340

Part 2: Foreign Companies

ACMA Robotique
3–5 Rue Denis Papin-Beauchamp
P.O. Box 7724
95250 Cergy-Pontoise Cedex
France

AKR Robotique
6 Rue Maryse Bastié
91031 Evry Cedex
France

ASEA Electronics Division
72183 Vasteras
Sweden

Atlas Copco Tools AB
Sickl Alee 5
10460 Stockholm
Sweden

Automelec S.A.
Case Postale 8
Rue des Poudrieres 137
2006 Neuchâtel
Switzerland

Axis
Tavarnelle Valdispesa, 50028
Italy

Berger Lahr GmbH
Breslauerstrasse 7
7630 Lahr
Federal Republic of Germany

Bisiach E. Carru SpA
Corso Lombardia 21
10078 Venaria Reale (Torino)
Italy

Robert Bosch GmbH
KruppstraBe 1, Postfach 300268
7000 Stuttgart 30
Federal Republic of Germany

Broetje Automation
August-Broetje-Str. 17
2902 Rastede
Federal Republic of Germany

Camel Robot S.R.L.
Via S. Guiseppe 10
121040 Gerenzano/Varese
Italy

Citroen Industrie
35 Rue Grange Dame Rose Zi Vélizy
92360 Meudon-La-Foret
France

Cloos Schweisstechnik GmbH
Postfach 121
6342, Haiger
Federal Republic of Germany

Commercy Soudure
P.O. Box 79
55202 Commercy
France

Compagnie de Signaux et
d'Entreprises Electriques
17 Place Etienne Pernet
Paris 75738
France

Compania Anomina de Electrodos
Infanta Carlota 56
Barcelona
Spain

Daewoo Heavy Industries
6, Manseog-Dong
Dangitgu
P.O. Box 7955
Inchon 160-00
Korea

Dainichi Kiko Co, Ltd.
Kosai-Cho
Nakokoma-Gur Yamanashi
Japan 400-04

DEA SpA
Corso Torino 70
10024 Moncaliere Turino
Italy

Elco Robotics Ltd.
PO Box 230
Ramat Gan 52101
Israel

Eshed Robotec Ltd.
Comercial Center
Ramat Ilan 51905
Israel

Euromatic
8 Rue du Commerce
68400 Riedisheim
France

Fanuc Ltd.
3580, Oshino-Mura
Yamanashi-Pref
Japan

Feedback Industries Ltd
Park Road
Crowborough, Sussex TN6 2QR
England

Feinmechanische Werke Mainz
GmbH
Postfach 20 20
6500 Mainz 1
Federal Republic of Germany

Fuji Electric Co. Ltd.
Shin-Yurakucho Bldg.
12-1, Yurakucho 1-chome
Chiyoda-Ku, Tokyo 100
Japan

Gaiotto Impianti SpA
Statale Milano-Crema Km 27
26010 Vaiano Cremasco
Cremona
Italy

GEC Robot Systems, Ltd.
Boughton Rd.
Rugby, Warwickshire CV21 1BD
England

Hirata Industrial Machines Co.
5-4 Myotaiji-machi
Kumamoto 860
Japan

Hitachi, Ltd.
4-6, Kanda-Surugadai,
Chiyoda-Ku, Tokyo
Japan

IGM GmbH
Strabe 2A, Halle M8
2351 WR. Neudorf
Austria

Industria
28 Av. Clara
94420 Le Plessis Trévise
France

IROBOS
Tiroler Str. 85
8962 Pfronten
Federal Republic of Germany

ITMI s.a.
Chemin des Clos-Zirst
38240 Meylan
France

Jungheinrich
Friedrich-Ebert-Damm 184
2000 Hamburg 70
Federal Republic of Germany

Kawasaki Heavy Industries Ltd.
4-1, Hamamatsu-cho 2-chome
Minato-ku, Tokyo 105
Japan

Kotobuki Industry Co., Ltd.
2, Minamimatsuda-Cho
Higashikujo
Minami-Ku, Kyoto
Japan

Kuka (IWKA Keller & Knappich)
P.O. Box 43 13 49
Blücherstrasse 144
8900 Augsburg 43
Federal Republic of Germany

Lamberton Robotics Ltd.
26 Gartsherrie Road
Strathclyde ML5 2DL
Scotland

Machine Industry Ivo Lola Ribur
Zeleznik
Belgrad 11250
Yugoslavia

Marrel Hydro
BP 56
42160 Andrézieux Bouthéon
France

Matra Automation
3, Avenue du Centre
78182 Saint-Quentin-en-Yvelines
Cedex
France

Matsushita Electric Ltd
1006, Oaza Kadoma
Kadoma-shi, Osaka 571
Japan

Messma-Kelch-Robot GmbH
Postfach 1208
Heinkelstrasse 37
7060 Schorndorf
Federal Republic of Germany

Meta Machines
9, Blacklands Way
Abingdon Industrial Park
Abingdon, Oxford
England

Microbo SA
Avenue Beauregard 3
2035 Corcelles
Switzerland

Mitsubishi Electric Corp.
2–3, Marunouchi 2-chome
Chiyoda-ku, Tokyo 100
Japan

Motoda Electronics Co., Ltd.
32-9 Kamikitazawa 4-chome
Setagaya-ku, Tokyo 156
Japan

Murata Machinery, Ltd.
136, Takeda-Mukaishiro-cho
Fushimi-ku, Kyoto Pref.
Japan

Nachi-Fujikoshi Corp.
20, Ishigane
Toyama-pref
Japan

NEC Corporation
33-1, Shiba 5-chome
Minato-ku, Tokyo 108
Japan

Nissho Iwai Corporation
4-5, Akasaka 2-chome
Minato-ku, Tokyo 107
Japan

Olivetti Spa
Via Torino 608
10090 S. Bernado D'Ivrea
Italy

Parker Hannifin Rak Schrader
  Bellows
48 Rue de Salins
25303 Pontarilier Cedex
France

Pendar Robotics
EBBW Vale
Gwent NP3 55D
England

Promatic
Gesellshaft fur Automation und
  Handling MBH
Blutenweg 2
7465 Geislingen-Binsdorf
Federal Republic of Germany

Reis GmbH & Co.
Postfach 1160
Im Weidig 1-4
8753 Obernburg
Federal Republic of Germany

Robotic Systems International
9865 W. Saanich Rd.
Sidney, BC V8L 3SI
Canada

Sankyo Seiki Mfg Co. Ltd.
17-2 Shinbashi 1-chome
Minato-ku, Tokyo
Japan

Scemi
61 Rue de Fumas
38300 Bourgoin-Jallieu
France

Sciaky SA
119, Quai Jules Guesde
94100 Vitry-sur-Seine
France

Simmering-Graz-Pauker AG
MariBrehmstrasse, 16
1071 Vienna
Austria

Sony Corporation
7-35, Kitashinagawa 6-chome
Shinagawa-ku, Tokyo 141
Japan

Spine Robotics AB
Flojelbergsgataen 14
43137 Molindal
Sweden

Stemens AG
Postfach 120
Wenerkwerkdamm 17–20
1000 Berlin 13
Federal Republic of Germany

Thorn EMI Robotics
Norreys Drive
Maidenhead SL6 4B7
England

Tokico Ltd.
6-3, Fugimi 1-chome
Kawaski-ku, Kawasaki-shu
Kanagawa-pref
Japan

Toshiba Corp.
1-1, Shibaura, 1-Chome
Minato-ku, Tokyo 105
Japan

Trallfa Nils Underhaug A/S
Postfach 113
4341 Byrne
Norway

Tsugami Robotics Co. Ltd
5-11-10, Shiba Minato-ku,
Tokyo
Japan

Voest-Alpine AG
Postfach 2
4010 Linz
Austria

Volkswagenwerk AG
Industrieverkauf VD-3
3180 Wolfsburg 1
Federal Republic of Germany

W+M GmbH
Postfach 5948
Daimlerring 2
6200 Wiesbaden
Norderstedt
Federal Republic of Germany

Yamaha Motor Co, Ltd.
2500 Shingai Iwata-Shi
Shizuoka-pref
Japan

Yaskawa Electric Mfg. Co.
2346, Fujita Yahata-nishi
Kitakyushu-shi
Japan

Zahnradfabrik Friedrichshafen AG
Postfach 2520
Lowentaler Strabe 100
7990 Friedrichshafen 1
Federal Republic of Germany

MAJOR UNIVERSITY ROBOTICS LABORATORIES

Stanford University (Stanford, CA), a pioneer since the mid-1960s, performs research in force sensing and motor control, high-level languages, and 3D inspection and vision systems.

The artificial intelligence lab of MIT (Cambridge, MA), also a pioneer, does research in vision and force sensing, gripper design, and high-level languages.

Carnegie-Mellon University Robot Institute (Pittsburgh, PA) is a leader in mobile robotics technology. They are also doing research in vision systems and hand and arm design.

Other American universities active in robotics include

Arizona State University	Tempe, AZ
George Washington University	Washington, DC
Georgia Institute of Technology	Atlanta, GA
Lehigh University	Bethlehem, PA
Ohio State University	Columbus, OH
Purdue University	West Lafayette, IN
Rensselaer Polytechnic Institute	Troy, NY
University of California	Santa Barbara, CA
University of Cincinnati	Cincinnati, OH
University of Florida	Gainesville, FL
University of Illinois	Urbana, IL
University of Maryland	College Park, MD
University of Rhode Island	Kingston, RI
University of Texas	Austin, TX
University of Utah	Salt Lake City, UT

Foreign universities include

Concordia University	Montreal, Canada
Leningrad Polytechnical Institute	Leningrad, USSR
Kyoto University	Kyoto, Japan
Milan Polytechnic University	Milan, Italy
National Technical University	Athens, Greece
Technical University of Compiegne	Compiegne, France
University of Aachen	Aachen, Federal Republic of Germany
University of Brussels	Brussels, Belgium
University of Edinburgh	Edinburgh, Scotland
University of Hull	Hull, England
University of Karlsruhe	Karlsruhe, Federal Republic of Germany
University of Nottingham	Nottingham, England
University of Tokyo	Tokyo, Japan
University of Warwick	Warwick, England
University of Western Australia	Perth, Australia
Wasada University	Tokyo, Japan

INTERNATIONAL ROBOTICS ORGANIZATIONS

Association Espanola de Robotica (AER)
Rambla de Cataluna 70 3° 2ª
08007 Barcelona
Spain

Association Francaise de Robotique Industrielle (AFRI)
61 Avenue du President-Wilson
94230 Cachan
France

Australian Robot Association
GPO Box 1527
Sydney, New South Wales 2001
Australia

Belgian Institute for Regulation and Automation (BIRA)
Jan Van Rijswijcklaan 58
B-2018 Antwerp
Belgium

British Robot Association (BRA)
28-30 High Street
Kempston
Bedford MK42 7AJ
England

Contactgroep Industriele Robots (CIR)
C/O Hoogovens Ijmuiden BV,
Research Lab 3J22
1970 CA Ijmuiden
The Netherlands

Danish Industrial Robot Association (DIRA)
Marselis Boulevard 135
DK-8000 Aarhus C
Denmark

International Federation of Robotics (IFR)
c/o Swedish Industry Association
Storgatan 19
Box 5506
S-11485 Stockholm
Sweden

IPA Stuttgart
Nobelstrasse 12
7000 Stuttgart 80
Federal Republic of Germany

Japan Industrial Robot Association (JIRA)
Kikaishinko Bldg
3-5-8, Shibakoen
Minato-ku, Tokyo 105
Japan

Robotic Industries Association (RIA)
900 Victors Way
P.O. Box 3724
Ann Arbor, MI 48106
United States

Robotics International of SME
One SME Drive
P.O. Box 930
Dearborn, MI 48121
United States

Robotics Society in Finland
P.O. Box 55
00331 Helsinki
Finland

Societa Italiana Robotica Industriale (SIRI)
Etas Periodici Tecnici
Via Mecerate 91
20138 Milano
Italy

Swedish Industrial Robot Association (SWIRA)
Storgatan 19
Box 5506
S-11485 Stockholm
Sweden

GLOSSARY

This glossary lists more than 100 terms commonly used in robotics or peripheral areas. The definitions are not precise; they simply provide a general understanding of the terms. Applications or comparison with other definitions are often included.

acceleration   The change in velocity as a function of time. Acceleration of an object requires force to overcome the object's inertia. Being a second-order effect, acceleration may not be considered in some equations of robot motion.

accuracy   The difference between the desired or expected position of the robot and its actual position. If the robot can approach the desired position from anywhere, then accuracy provides worst-case data for the robot. Accuracy information is often stated as absolute accuracy. See *repeatability*.

actuator   The power transducer or "motor" used to convert energy to rotary or linear robot motion. Can be electrical, pneumatic, or hydraulic.

algorithm   A formula or procedure written as part of a program to solve a problem or compute an equation.

alphanumerics   All the symbols of the alphabet (A–Z), the numerals (0–9), and special symbols, such as comma and period. Used with operator terminals.

ambient light   Background or area light that produces general illumination, as compared to spot lighting or other specific object lighting. Can sometimes interfere with direct lighting.

analog-to-digital (A/D) converter Digitizes analog sensor signals so that they can be entered into a computer. A/D converters are a potential source of three types of errors: quantization linearity, quantization resolution, and aliasing errors.

ANSI (American National Standards Institute) An organization devoted to developing standards for U.S. industry in many areas. See *RS-232*.

application layer The top or seventh layer in the ISO/OSI interface specification, which allows application tasks to use network utilities. See *MAP*.

articulated The ability of the robot to maintain end-effector orientation during movement of a single joint.

artificial intelligence (AI) The ability of a computer to perform functions usually associated with a human level of intelligence. Typical functions are speech recognition, visual comprehension, and self-maintenance.

ASCII (American Standard Code for Information Interchange) An 8-bit code to digitally represent information. The first half of the code (bit 8 = 0) has been standardized. The second half (bit 8 = 1) often varies with vendors.

assembly language A programming language written close to actual machine (computer) code level, in which each instruction corresponds to a single computer instruction. See *high-level language*.

automated guided vehicle (AGV) A vehicle equipped with guidance equipment for following a path in the floor. It carries raw material and finished goods around a factory between workstations.

backlash The amount of free play in gears and other transmission devices. Backlash provides errors in the final link position, since the resulting position would depend on travel direction.

binary image A binary image provides only two levels of signal (on and off). The image is similar to a black-and-white television-type image that was presenting silhouettes.

Bode plot A graph of system gain and phase plotted against frequency. Often used in servo system design because it portrays the characteristics of the servo system, especially any tendency to instability.

bridge A transparent device used to connect to a local area network. It allows new users to be added to the network without affecting current users.

Cartesian coordinate system Representation of positions in three-dimensional space, based upon linear travel in three mutually perpendicular axes, *X, Y,* and *Z*.

CCITT (Comite Consultatif Internationale de Telegraphie et Telephonie) An international organization devoted to developing international communications standards.

center of gravity A physical point on an object that represents the center of all mass of the object. The point through which all force vectors operate.

centrifugal force A force associated with rotary motion. Centrifugal force acts in a radial direction away from the center of rotation.

charge-coupled device (CCD) One type of MOS chip used for vision sensing in solid-state cameras. Slightly more sensitive than the CID chip.

charge injection device (CID) One type of MOS chip used for vision sensing in solid-state cameras. Slightly less expensive than the CCD chip.

closed loop A servo system where the output is compared to the input and a correction made to eliminate any error (difference) between the signals.

compliance The ability of the wrist or gripper to provide some "give" in the horizontal direction to allow close-tolerance parts that are not perfectly aligned to mate more easily.

computer vision The ability of a computer to interpret visual information received from sensors to locate and recognize parts.

controller A device or computer chip that controls the operation of part of the robot system, such as one link. On occasion, the controller can refer to the computer that controls the complete robot.

data link layer Layer 2 in the ISO/OSI interface specification, which establishes and manages the error-free communications path between network attachment points (nodes). See *MAP.*

degree of freedom (DOF) The number of independent coordinate planes or orientations that a joint or end-effector can move in. DOF is determined by the number of independent variables in the system.

degree of mobility (DOM) If a system has added joints that do not provide any extra degrees of freedom to the end-effector, but only provide alternative methods of positioning the end-effector, these extra joints offer additional degrees of mobility.

difference of Gaussian (DOG) A method of image processing based on taking the difference of two images that have passed through a Gaussian filter. Particularly good for edge detection.

direct drive The motor is connected directly to a joint with no intermediate gears or other type of transmission linkage. Direct drive systems remove backlash errors.

edge The line in a picture that separates a desired object from its background. Various approaches to edge detection are used by computer vision systems.

efficiency The ratio of power out of a robot to the power supplied. It usually involves energy conversion; for example, one definition of robot efficiency is mechanical power obtained divided by electrical power supplied.

encoder A sensor that converts angular or linear position data into digital form. Optical-based encoders are the most common.

end-effector   A gripper or tool placed on the end of a robot arm and used to perform work on an object.

error signal   A feedback control signal that represents the difference between a robot's actual position (or speed) and its commanded position. The error signal is amplified and used to correct the robot's motion.

expert system   An intelligent computer program that uses knowledge stored within it and logical inference to solve problems. It generally contains the combined wisdom of different human experts, hence its name.

field   In television, a single scan through the entire picture area, usually skipping every other line.

flexible manufacturing system (FMS)   An arrangement of robots, numerical control machines, and other equipment interconnected by a transport system. A central computer controls the system.

frame   In television, scanning through every picture element in a picture, often composed of two fields.

frame grabbing   A snapshot or single frame of visual data taken from the continuous output of a visual sensor.

gantry robot   A robot suspended from an overhead framework that can move over the workplace. Often used for transporting heavy materials.

gateway   A connection between networks. It provides the necessary physical connection and protocol conversion.

gradient   A variation in intensity across an image, often due to the edge of an object, shadows, or noise. A steep gradient usually signals the edge of an object.

gray scale image   A black-and-white image that includes shades of gray, such as is found in a television picture. A typical picture may have from 2^5 (32) to 2^8 (256) different intensity levels available.

hard automation   A type of robot manufacturing that requires precise position of the parts, since the robot is not able to locate parts and adjust for position errors.

high-level language   A language written at the user level rather than the individual machine command level. It must be converted to machine language through a compiler program or an interpreter before it can be used. One instruction in a high-level language often converts to many machine-level instructions.

hydraulic system   A robot that uses hydraulic actuators as the power source. Hydraulics employ fluid under high pressure to move objects.

image enhancement   Improving the quality of an image by removing from the picture the effects of blur, noise, nonlinearities, or other distortions.

inertia   An object's resistance to a change in velocity. For translational motion, it is a function of object weight. For rotational inertia it is a function of object size, shape, and weight.

inference engine "Intelligent" software that can draw conclusions from facts presented to it based upon rules it has been given.

intelligent robot A robot that can sense and react to changes in its environment.

inverse kinematic solution The equations to determine joint movements for reaching a predefined location.

ISO (International Standards Organization) An organization approving and supplying all types of engineering standards.

joint Part of a robot arm. The intersecting point between two links. A joint allows the two links to be moved with respect to each other. Usually rotary, there are also slide joints.

Kell factor A degradation factor that must be used in any scanning system to determine effective vertical resolution. It results from discrete scanning lines that do not precisely line up with vertical picture elements. Usually considered to be 70%.

knowledge-based systems Computer systems employing special types of data or knowledge representation that allow this knowledge to be linked to other related data. It is this link between data that makes this system so valuable, compared to the types of isolated data found in most normal data base systems.

lag A phase delay through servo circuitry. Too much lag can cause an oscillation and system instability, unless compensated for.

local area network (LAN) A communications network tying together computers and related devices within a limited area, allowing each device to communicate with each other, thus sharing memory and resources. Local area networks are generally limited to shorter distances, such as less than a mile.

local coordinate system A coordinate system based upon a point on the robot gripper, often the tool center point. The local coordinate system will move as the robot arm moves.

manipulator The arm of a robot—the part that moves and operates on a workpiece.

manufacturing automated protocol (MAP) A set of standards to allow communication between computers and smart robots of many different vendors over a wideband local network. Based on the ISO draft proposal DP7498.

MOS (metal-oxide semiconductor) One method of manufacturing integrated circuit chips.

network layer Layer 3 in the ISO/OSI interface specification, which provides message addressing and message routing services.

numerical control Originally a method of supplying digital control signals to machine tools through a punched paper tape, the term now refers to computer control of a machine tool or robot.

open loop   A motor control system that does not use external sensors or feedback to provide position or velocity correction.

orientation   The relationship between the coordinate system of an object and that of a local framework. Generally, the object is on some angle with respect to the local coordinate system. If an object has complete freedom of orientation, three independent angles are required to specify orientation.

OSI (open system interconnect)   A set of communication standards that allows vendors of products at all levels to readily interconnect and pass information.

pattern recognition   The ability of a vision system to recognize a part from the information provided by its sensors and compared to its stored image of that part.

phase angle   The angle by which the input signal phase differs from the output signal phase.

physical layer   Layer 1, or the bottom layer in the ISO/OSI interface specification. Physical hardware necessary to interface to the network, as well as the necessary communications drivers.

pick-and-place robot   A nonservo robot used for very simple assembly or transport tasks.

pitch   One of the three angles used to specify end-effector orientation (along with roll and yaw). Pitch provides the up-and-down motion, such as the change in angle when a person nods his or her head.

pixel   A single-resolution cell in an image produced by a vision sensor. A typical total picture might be made up of 512×512 pixels.

presentation layer   Layer 6 of the ISO/OSI interface specification, which allows data translation between different data formats. Currently not standardized in MAP.

prismatic joint   Provides linear or translational motion.

programmable controller   A solid-state control system similar to a computer but specifically designed for executing instructions that control a robot or related machines.

protocol   A standard format for messages transmitted from one computer system to another, allowing information communication between different types of computers. Protocols are necessary at each data communication layer.

repeatability   The degree of accuracy to which a system will return to the same point under repetitious conditions. Repeatability accuracy is usually better than its absolute accuracy.

resolution   In mechanics, the smallest positioning increment that can be achieved by a motor. In vision, the smallest independent element of the picture.

resolvability The ability to solve the inverse kinematic problem of locating a robot's end-effectors.

revolute joint Provides rotation about its joint axis (twist motion).

robustness A figure of merit for an equation (or system). The more the system is able to minimize sensitivity to small errors in the model it is based upon, the more robust the solution that is provided.

roll One of the three angles used to specify end-effector orientation (along with pitch and yaw). Roll provides twisting motion, such as the change in angle as a person uses his or her wrist to turn a key in a lock.

RS-232C One of the most common standards used for serial transmission of data between a computer and its peripherals, or between different computers. Developed by the Electronic Industries Association (EIA).

sensor A device to convert one form of energy, such as light or sound, into an electrical signal.

session layer Layer 5 in the ISO/OSI interface specification, which provides the system-dependent aspects of communicating between specific devices.

slew rate The fastest method of robot movement in which intermediate positions or velocity accuracy is not controlled.

static torque The amount of torque available to a motor at zero velocity. Usually the maximum amount of torque the motor can produce.

stiffness The ability of an arm or joint to resist motion from external sources.

tactile sensor A sensor usually placed on an end-effector to provide information about touch, force, temperature, or slip.

teach pendant A hand-held keyboard with limited display capability connected to the robot or its controller through a long cable. It is used to control and enter motion commands into a robot. For safety, teach pendants usually limit robot speeds.

telechir A remote-controlled robot. An operator, sitting in one room, uses television and other sensor information to determine what action the robot should take and then directly provides this information back to the robot, often through a joystick. Most often used in hazardous environments.

template matching A method used by computer systems to perform pattern recognition. The computer matches a stored set of visual data about the object (the template) to the information derived from the image, after making necessary size and rotational adjustments. A similar approach is used in speech recognition.

thresholding Setting a brightness level in a picture such that any pixel of equal or greater brightness is digitized as a 1, and any pixel of lower intensity is digitized as a 0. It is used to create binary images, as a first step in pattern recognition.

tool center point (TCP) An imaginary point that lies along the last wrist axis at a user-specified distance from the wrist, such as at the edge of a welding gun or the midpoint of gripper jaws.

tool coordinate system (TCS) The tool locations expressed in coordinates relative to a frame attached to the tool itself.

torque One force associated with rotational motion. A tangential force acting at some distance from the center of rotation.

torque-to-inertia ratio A figure of merit for motors. The higher the ratio, the faster the motor can be accelerated.

trajectory A curve in space through which the TCP moves.

trajectory planning Algorithms provided to move the robot arm through space toward a desired end point (goal) while ensuring that neither the arm nor the tool collide with objects in the way.

transducer A device that converts electrical energy into another form, such as sound or vice versa.

transport layer Layer 4 in the ISO/OSI interface specification. End-to-end control of the data flow between network points once the data path is in operation.

TROLL (*Technion robotics laboratory language*) A high-level language used for robot assembly applications, developed at Technion, Israel Institute of Technology.

via point A location along the robot path that the robot is commanded to pass through without actually stopping. Via points are used to increase path following accuracy.

work cell An integrated operation within a factory in which robots, machines, and material transport systems work together in a production group and perform a set of manufacturing operations. Usually has one cell controller (computer) controlling all machines in the work cell.

work envelope The maximum volume, in three-dimensional space, that a robot can move through. Important for determining limits on the applications the robot can perform and for determining an operator-free area for safety purposes.

world coordinate system (WCS) A fixed coordinate system, often referred to the base of a robot, that does not change as the manipulator moves about.

wrist The connection between the robot arm and the end-effector. It usually contains two or three joints to allow proper orientation of the end-effector.

yaw One of the three angles used to specify end-effector orientation (along with pitch and roll). It is side-to-side motion, such as the change in angle when a person shakes his or her head no.

INDEX